UNDER CONTRACT

UNDER CONTRACT

THE INVISIBLE WORKERS
OF AMERICA'S GLOBAL WAR

NOAH COBURN

STANFORD UNIVERSITY PRESS
STANFORD, CALIFORNIA

Stanford University Press
Stanford, California

Printed in the United States of America on acid-free, archival-quality paper

Library of Congress Cataloging-in-Publication Data

Names: Coburn, Noah, author.
Title: Under contract : the invisible workers of America's global war / Noah
    Coburn.
Description: Stanford, California : Stanford University Press, 2018. | Includes
    bibliographical references and index.
Identifiers: LCCN 2018003015 (print) | LCCN 2018005313 (ebook) | ISBN
    9781503607163 | ISBN 9781503605367 (cloth : alk. paper)
Subjects: LCSH: Afghan War, 2001—Participation, Foreign. | Foreign workers—
    Afghanistan. | Contractors—Afghanistan.
Classification: LCC DS371.4125 (ebook) | LCC DS371.4125 .C64 2018 (print) |
    DDC 363.28/980973—dc23
LC record available at https://lccn.loc.gov/2018003015

Cover design: Christian Fuenfhausen

For Ann Struthers Coburn and Michael Cutler Coburn. I am grateful for their enlightened parenting decisions and consistent love and support.

CONTENTS

UNDER CONTRACT

PROLOGUE: NO SMALL WAR

REMAINDERS

On a mild April day in 2015 with the wildflowers on the hills in full bloom, I stood by the side of the road and watched half a dozen men, armed with crowbars and blowtorches, swarm a shipping container. With quick efficiency, they pried off the tops and sides. As if pulling apart a cardboard box to be recycled, they dismantled the container into neat piles of metal siding. They loaded these pieces into the back of an aged truck, which grumbled to life and, a few minutes later, trundled east toward Pakistan.

Just half a mile from the front gates of Bagram Airbase, these piles of scrap metal marked the end of President Barack Obama's surge that the U.S. government had hoped would turn the tide in Afghanistan. Living in Kabul and visiting the area around the base regularly, I had watched with local Afghans as U.S. troop levels had shot up to almost 100,000 in 2011 as buildings sprang up rapidly on and around the base.[1] With the troops came diplomats, development projects, construction contracts, and tens of thousands of workers. The troops were leaving, and soon, it seemed, not even these shipping containers would remain.[2]

Alongside the road, just below the watchtowers of the base, there were buildings and small compounds heaped with plywood and used office furniture, but also more bizarre sights, such as dozens of porta-potties, stacked up on their sides like blocks, and a small shop bursting with used printer cartridges. Overflow was spilling out the front door. Workers and traders crowded the street haggling over various goods. Anything on the base that could be salvaged had been snatched up and was now being carefully taken apart and repurposed. I walked up the road toward the base, watching shoppers and laborers sort through the debris that had come off the base, the remainders of the American invasion of Afghanistan and the subsequent fifteen years of fighting.

While the war in Afghanistan was not over—there had been a Taliban raid in the area that morning—the period of vast American spending was drawing to a close. These funds had made many in certain industries wealthy. Supplies intended for the base flooded the local market. Aid rushed in, generating hastily thrown together development projects. Certain businesses boomed while others struggled.[3] Now the base was being dismantled, and the scrap was being shipped off first to Pakistan and then beyond. Tens of thousands of soldiers, diplomats, aid workers, mercenaries, businessmen, and contractors were leaving the country, some with terrible injuries, others having made their fortunes. Those who had come for the conflict and had been reshaped by it were all moving on, just like me, creating an ever-expanding, largely invisible blast zone of the war.

All of us were marked by the violence and wealth that war creates.

I first arrived in Afghanistan in 2005 as a graduate student, eager to study local politics in an area north of Kabul, assured by the media and politicians that the war triggered by the U.S.-led invasion was over, or at least would be shortly. They were wrong. Soon there were reports of the Taliban retaking districts in the south of the country. And I, thinking my time in Afghanistan would be brief, perhaps eighteen months on the ground to complete my dissertation, was wrong too. Twelve years later, watching the dismantling of the war, I was still there, my work and life now intertwined with the future of the country.

The timing of my arrival had been lucky. During the period of relative peace, I had time to learn Dari, the variety of Persian spoken in

Afghanistan, and get to know Afghans as they rebuilt from the destruction of the Taliban period. In the year and a half I spent researching a group of potters who lived about an hour west of Bagram in the town of Istalif, the war had spread gradually north, eventually encircling the district where I was working. Sitting in the bazaar, talking with local merchants, watching as prices rose and fell and marriages were arranged, the war touched our lives in unpredictable ways. There was little open conflict in the area, but the shifting politics and economics of the war created opportunities, particularly for merchants and contractors who managed to make deals with those on the base while those outside the elite tended to suffer.[4]

After finishing my graduate studies, with so much attention being paid to the country, I decided to stay in Afghanistan and see what I could contribute. I lived in Kabul, working for a local research organization, the Afghanistan Research and Evaluation Unit, as well as the United States Institute of Peace. Jobs were plentiful, and the funding allowed me to do the type of on-the-ground research on politics, conflict resolution, and elections in particular that interested me. As a white American man who spoke Dari and had lived with Afghans, I was in a privileged position. I could talk to international journalists and meet with diplomats at various embassies, but also visit small Afghan villages and see what some of the results of the ongoing conflict were—the crater from an insurgent rocket in my friend's backyard; the storefronts that were damaged by an American convoy driving by too fast. I saw the war from more sides than most other people did.

Living on the edge of a war zone for five years, I had my share of close calls: a building had been blown up by a suicide bomber directly across the street from where I was standing. I had been in a high-speed car chase when my car in Kandahar had been targeted by a Taliban spotter. Friends and colleagues had been kidnapped or killed. But most of the time, life in Kabul had been more mundane. I had lived in a house with three roommates. We had a dog and two tortoises that hibernated in the winter and munched on our vegetable garden in the summer. We planned dinners and did the things that other young professionals did. The war was a constant in our lives and, at the same time, shockingly easy to ignore.

I worked with a team of Afghan researchers studying aspects of the international presence—for example, how internationally sponsored elections were changing politics on a local level. As an anthropologist interested in how the conflict was affecting others, I conducted interviews with diplomats, soldiers, development workers, and ordinary Afghans. I traveled to massive international bases, government offices, and the homes of Afghan friends. I tried to view the war from as many angles as possible. I became interested in the inadvertent effects of the war: businesses that had failed because aid money had disrupted the market, people who had been promoted since no one else was willing to go to Afghanistan, opportunities won and lost. These changes were often most visible in places like Bagram.

In the bazaar at Bagram, the contrast between those who had benefited from the war and those who had not was stark. Local businessmen and government officials had made money from the building of the base and providing it with supplies. Others had made millions setting up private security firms.[5] The newly rebuilt compound of a local parliamentarian and the brand-new mosque he added next door contrasted the dusty, dilapidated structures on either side of the street, bullet holes visible in houses built decades prior. The Congressional Research Service calculated that through December 2015, $686 billion had been spent on America's Operation Enduring Freedom in Afghanistan, and that had created immense wealth for a select group of people.[6]

Those working in the bazaar and other ordinary Afghans benefited much less. Those who were not in the tiny ruling elite tended to live in houses that still lacked running water, and electricity was available only for those in the largest towns in the area. According to the World Bank, in 2006–2007, 36 percent of Afghans were poor and did not have the buying power to satisfy basic material needs. Five years of international funding later, in 2011–2012, the numbers had changed little: 36 percent of the population was still poor and the bottom 20 percent had experienced a decline of 2 percent in the amount of money they had to spend on basic necessities, while the top 20 percent had seen a 9 percent increase.[7]

Afghans were not the only ones shaped by the war: there were also soldiers and their families, diplomats, and aid workers. Back at home, the fallen had been memorialized, movies had been made, and memoirs had been written.

Inside Bagram, however, was another group that had not received much attention in the international media. These were the so-called third-country nationals—workers from places like Sri Lanka, Bangladesh, the Philippines, Turkey, Bosnia, and Nepal, who were ever present but largely invisible during the war. They did most of the work on bases like Bagram, everything from manning the guard towers and cleaning the latrines to more technical jobs like engineering and accounting.

I had met some of them working at places like Bagram and the U.S. embassy and wondered how they experienced the war. They were different from soldiers. None of them spoke of patriotism or other ideological motivations but were drawn to jobs that were not available in their home countries. Many paid brokers large sums, families falling deeply into debt, to secure their positions. Some had come legally; many had not. Most of them also tended to stay much longer than the majority of soldiers. Many of these contractors prospered working for the American war; they saved and sent money home. Others struggled. The human trafficking networks that supplied many of the workers allowed for easy exploitation. Some workers were scammed, while others were detained, kidnapped, or thrown into jail.

These migrant workers had fascinating stories to tell, but few had gotten a chance to do so. One of these migrant workers was a Nepali man named Teer Magar.[8]

SMASHING TELEVISIONS

Teer Magar's case had not been covered in the international media, but as I began to seek out contractors who had also been in Afghanistan, first in Nepal and then elsewhere, his name came up repeatedly in interviews. In these accounts, Teer was simultaneously something of a legend and a cautionary tale. Depending on your point of view, his detention allegedly was for spying, but others argued that he was simply a scapegoat for those higher up at his company. His eventual escape was reassuring. It proved that workers could extract themselves from the worst possible situations when going to conflict zones like Afghanistan. His story also served as a warning for how easily it could all go wrong, particularly when a company abandons its workers.

Teer passed through Bagram Airbase the same spring in 2015 that I watched the breakdown of the bazaar outside. In his mid-thirties, like me, he came to Afghanistan in part to look for work and also because he felt it would be a good adventure—something to tell his children later in life, he said, when I finally arranged to meet with him at a café near his home in Bandipur, in the center of Nepal. Like me, he first arrived in Kabul and worked on several international compounds. His experience of the war, however, could not have been more different than my own.

During his almost three years in Afghanistan, Teer spent most of that time in prison. He didn't speak Dari or Pashto, and often went for long stretches without talking to anyone. The few words of Pashto he was able to pick up allowed only limited communication with the guards and fellow inmates. His only visitor was a French representative of the Red Cross, who occasionally managed to send some letters home to his wife in Nepal.

I spoke with Teer after he had returned to Nepal and was working on his brother's farm in the mountains. Getting to his village in the hills of the southern Himalayas was not easy due to an ongoing fuel crisis, but eventually I found a seat on one of the buses heading west out of Kathmandu. I rode west for four hours to a small junction where the driver indicated to me that the road to Bandipur headed up into the mountains. After a taxi ride, I arrived at a café, tucked into a scenic corner between the hills.

Teer occasionally paused to take a sip from his Coke. He would then set his bottle down and continue his story, slowly and thoughtfully, like a man used to waiting. As I scribbled down Teer's dark tale in this beautiful mountain setting, the river churning below us, it felt like the long journey for both of us was a reminder of just how far-flung and unpredictable the consequences of the war in Afghanistan were.

For the most part, Teer said, the Taliban prisoners he was mixed in with had left him alone. It did not seem to bother the other prisoners much that he had worked as a contractor for an American-funded project. At one point early in his detention, a large, bearded Talib came to him and demanded that he convert to Islam, he said. Teer tried to explain that he respected all religions. He wasn't sure if the Talib understood him, but after a while, he was left alone again.

The prison was actually nice by Afghan standards, Teer continued. When I first heard of his case from some other Nepalis, I imagined him sitting in the dark jail I visited in western Afghanistan around the same time that Teer was detained. I had visited it as part of my work for the United States Institute of Peace in 2010, and it was one of the more disturbing scenes from my years in Afghanistan. There, in a cramped room filled with thin mattresses on the dirt floor, insurgents were mixed in with the mentally ill. Flies buzzed in the dim light around a stack of dirty tin bowls.

Teer's prison, however, had been newly built by the British specifically to hold insurgents caught on the battlefield. Clean and sterile, it was one of the thousands of structures built for the Afghan government by the international community to help it battle insurgents and terrorists. After getting to know the other prisoners, however, Teer decided that few were terrorists and most were simply local people, inadvertently dragged into the conflict, perhaps found with guns in their homes when the Americans went out on raids. In this sense, Teer fit in with the other prisoners, who felt confused and unjustly detained during the war.

Teer explained, step by step, how he had ended up in this distant prison. The local office of the construction firm he had been working for had been accused by a rival firm of spying for the Pakistani government. When the Afghan secret police raided the office, the Afghans in the office had managed to make it look like Teer was the one guilty of stealing plans for the bases they were building. Since he had no translator during his trial, it was difficult for him to plead his case or even understand the charges, and before he had any real sense of what was happening, he was sentenced to eighteen years for espionage.

Thinking back, Teer said, the most frustrating aspect of having members of the Taliban in the prison was that some of the more conservative prisoners argued that television was against Islam. For supposedly religious reasons, they prevented other prisoners from watching. Some prisoners protested and cited Koranic verses they claimed proved that television was permissible. Nevertheless, this religious dispute continued. Then, during a riot in 2013 when the prisoners briefly took control of the building, the televisions were all smashed, ending the debate over whether they could watch television, Teer said.

As Teer talked, the afternoon moved to early evening. The story of how he ended up in an Afghan prison, and then how he made it back to this peaceful town, was complicated, filled with betrayals and more than one kind stranger. Separated from his family and unable to communicate with them for much of the time was painful, and he paused several times during our conversation to collect himself.

Over time, it became clear to me that Teer's tale was not entirely remarkable. During the fifteen months I systematically sought out and interviewed non-Western contractors, I talked to men who had been arrested, kidnapped, trafficked, and forced to work essentially as indentured laborers. I also interviewed contractors who had built new houses, started businesses, and enrolled their children in better schools than they would have been able to afford before coming to Afghanistan. For these men and women, the outsourcing of war and the use of international contractors was neither clearly good nor clearly bad. The lack of transparency around the international migration of these workers and the vastly different experiences that they had show how personalized the experience of war is and how important it is to understand the repercussions of how wars are being fought on an individual level.

The accounts from these contractors were different from past interviews I had gathered about the conflict from local Afghans, U.S. soldiers, development workers, and diplomats.[9] They told the story of a conflict that was similar to the one I had witnessed as a graduate student and later as a researcher, yet it had a different logic and set of rules. For Teer and other Nepali contractors, the bombs and rocket attacks were dangerous, but they were even more afraid of deportation, losing their jobs, or getting tricked by a broker or their employer and abandoned without a visa in a foreign land. And this was all too common an occurrence. The Nepali government had no embassy or other representation in Afghanistan, and it apparently was unaware of Teer's plight until a Nepali reporter working for a German news organization in Kabul happened to overhear two other Nepali guards talking about Teer in a grocery store two years after his arrest.

But that is getting ahead of our story.

WAR FROM THE GREEN MOUNTAIN STATE

In 2012, after living and working in Afghanistan for five of the past seven years, I moved to a small cottage in the shadow of the Green Mountains of Vermont, 6,500 miles away from the bustle of Bagram. I was teaching anthropology at Bennington College, and it felt strange to be teaching and talking about war and conflict in such an idyllic setting. It disturbed me that students and colleagues on campus, but also my friends and family, did not feel more affected by the war in Afghanistan.

Certainly Afghanistan is far off, and the media's attention span is brief, but this was the longest war in American history. Over 2,300 American soldiers had died in Afghanistan by the end of 2015, and another 20,000 had been wounded.[10] Between 2009 and 2015, 21,323

Afghan civilians were killed.[11] Some estimated the combined costs of the Afghan and Iraq conflicts at $4.8 trillion, which would come out to more than $25,000 for every household in America.[12] More long-term estimates suggest that when pending costs, such as care for veterans and cumulated interest are calculated in, the wars will add $7.9 trillion to the national debt.[13]

And in 2012, the war wasn't over.

In Vermont, the violence may have felt distant to some, but this was often only an illusion. Hundreds of thousands of veterans had returned from the wars in both Iraq and Afghanistan. Portrayals of these conflicts in recent years, in movies and memoirs, have become popular, ranging from adventure movies like *Lone Survivor* to short story collections like Phil Klay's *Redeployment*. But most of these accounts were so dramatic they resembled the fantasy of an action film more than anything else. It was certainly not something most of my students found easy to relate to. For the most part, that America had been at war for fifteen years, ten thousand miles away, did not seem to concern them as much as I thought it should.

This did not mean that everyone was cut off from the conflict. The college's only student veteran came to my office one day, asking to be allowed into an upper-level seminar that I was teaching that looked at violence from a variety of angles—anthropological, political, and literary—despite missing the registration deadline. Late for a meeting, I was walking out of my office, and Ben Simpson trailed me across campus, pleading his case. Ben had served as a medic in Iraq and was now at Bennington on the GI bill, passionate about researching the U.S. government's failings in its recent wars. I let him into the class, and eventually he became my advisee. Later, he helped me do research, mostly because of his persistence. As we debated the effects of America's recent wars in class and during office hours, it became clear that we were both struggling with the distance between these wars and the American consciousness.

In class, the other students did not always know what to do with

Ben's harsh realism. He had little time for sentimentality or academic jargon. He was, however, a determined researcher, willing to dig through piles of government documents, trying to find statistics that would support some of his ideas about a war that he had become so disillusioned with. As a result, he made an ideal companion, helping me dig up research reports, budgeting numbers, and difficult-to-access statistics from the Department of Defense to help make sense of what I was hearing while conducting interviews. As Ben and I worked together, going over our notes at the one coffee shop just off campus, we began to map out the edges of the war and the way in which it spread out across the globe.[14]

It's not entirely surprising that the American public wasn't more outraged at some of the ongoing conflicts. Vast amounts of money and careful political calculations effectively cut most Americans off from this costly war and its effects. These ranged from carefully orchestrated public relations campaigns, with most reports coming from embedded journalists, dependent on U.S. officials for access,[15] to the fact that U.S. troops in Afghanistan reached their maximum level in March 2011 at 99,800, which seems reminiscent of charging 99 cents instead of a dollar to convince U.S. consumers that the war is a good deal.[16]

Contracting work to people like Teer also enabled the expansion of the war by relying on international workers to do jobs that had previously been done by American soldiers in earlier conflicts. And with no draft, as there had been in the world wars and in Vietnam, average citizens, and particularly college students, were certainly less interested than they would be if there was conscription. To the American public, those who went to join the war effort in the military, like Ben did, or as civilians, supposedly went voluntarily. (This was an assumption I came to question increasingly as I met and interviewed contractors who were often motivated by poverty or family debt.) The war didn't create shortages in the United States or demand rationing as either of the world wars had, and the financial impact on many American civilians was buried in tax forms.[17]

Beyond this, American deaths had been minimized through careful techniques. This included extensive air campaigns and drone strikes that put U.S. soldiers at little risk. It also involved outsourcing thousands of jobs to private contracting companies, many of which were not American owned, whose deaths did not generate the same outcry among voters as when a young American soldier was killed.

The American military, in particular, insisted that the war in Afghanistan was an insurgency, a so-called small war, and the international media portrayed the war as largely a battle between local Taliban groups and the American military. Other groups rarely appeared in articles or news story.

This, however, was far from the truth.

Particularly following the 2009 surge, with 150,000 international troops in the country and even more civilian contractors spending more than $2 billion of U.S. government funds a week, it was difficult to call the conflict "small."

A SMALL WAR?

This question of the scope of the war in Afghanistan is not a simple one. Just trying to keep track of the various individuals, groups, and even countries involved in the war is not easy.

On one side of the conflict, the ISAF alliance (International Security Assistance Force) included forty-two countries, some of these traditional North Atlantic Treaty Organization (NATO) allies of America, like the United Kingdom, Canada, and Germany. Others were newer members of NATO, eager to solidify their credentials as allies, like Poland, Slovakia, and Slovenia. Other countries in the region simply provided the coalition with logistical support, like Tajikistan, Kyrgyzstan, and, farther afield, Thailand. Some countries intentionally maintained a more marginal presence; China contributed only to land mine clearance, and Japan provided refueling support for the coalition in the Indian Ocean until a change in government in 2007.

On the other side of the battlefield, the Taliban was hardly just an Afghan or even Afghan and Pakistani phenomenon. While most insurgents were disgruntled local people who felt marginalized by a government that failed to provide significant services and an international military presence that had failed to stabilize the country, the insurgency itself was both deeply local and deeply international.[18] Detainees who were caught in Afghanistan and ended up in Guantanamo came from twenty-six countries, ranging from Algeria, Australia, and Azerbaijan to Tajikistan, Tunisia, and Turkey. Many of the Taliban fighters and their families lived in Pakistan, some leaders were educated in Egypt, and much of the group's funding came from Saudi and other Gulf donors.[19] While few seemed to travel back and forth between Iraq, Syria, and Afghanistan, insurgents did pass along techniques and inspiration on the Internet and sometimes claimed kinship with jihadi movements elsewhere.

Others were more indirectly involved in the conflict. Some countries were more interested economically in the outcome of the conflict—like India, the United Arab Emirates, and Iran, which sent more businessmen than troops. China snatched up mining rights, while Iranian businessmen were involved in Kabul real estate.

All told, more countries were involved in the war in Afghanistan than had been involved in World War II. Over 3 million non-Afghans went to Afghanistan to participate in the war in one way or another over the fifteen years following the U.S.-led invasion.

Small war indeed.

While researching for a previous book, I spent a good deal of time around Bagram, interviewing people in the villages surrounding the base, as well as those working inside the base. Although many of those in charge in the base were Americans, everywhere I looked there were workers from other countries. Inside the base there were U.S. soldiers, but also hundreds, if not thousands, of Nepalis, Indians, Filipinos, Bangladeshis. and other workers from poorer countries working inside. It was hard to call it an "American base."

These contractors cleaned the base, cooked food and guarded convoys; they were, in fact, involved in almost every aspect of the war. Some were working directly for the U.S. military, but most worked for private firms or for businesses benefiting from the war economy. These groups mingled together, forming an international civilian presence that filled many historically military roles that the United States and other countries now outsourced. These workers were not on the base because their countries had a political commitment to defeating the Taliban. Their presence was a result of the tendency of the United States and others to outsource their work to the cheapest laborers they could find who moved across the world's increasingly porous borders.[20]

Even as an academic who spent a lot of time thinking about the war in Afghanistan, I had little sense of what the war had been like for these individual contractors or how it might have shaped the countries that they came from.

Digging deeper, I found that several political scientists looked at the numbers behind the contracting phenomenon and the role of what many refer to as third-country nationals, or simply TCNs, as those in military and diplomatic circles referred to them.[21] Sarah Stillman's 2011 *New Yorker* article, "The Invisible Army: For Foreign Workers on U.S. Bases in Iraq and Afghanistan, War Can Be Hell," provided an exposé on the fate of a handful of these contractors in Iraq and Afghanistan. Al Jazeera also produced a short documentary about these workers in 2015. These media accounts highlighted the extreme situations that some contractors faced, ranging from near-slavery conditions to more routine forms of exploitation, such as long hours of work with no breaks.

For the most part, however, contractors' experience of the conflict and how it shaped their lives remained a largely untold story. The political scientists who wrote about it tended to focus on their impact on U.S. military policy and budgeting, but few had spoken to contractors themselves, particularly those from other countries. While the media focused, rightly, on the most egregious cases of exploitation, less is known about more typical experiences of contractors, some of whom returned to places like Afghanistan year after year. As an anthropologist who focuses on how politics are lived and how individuals experience war from a more holistic perspective, I was surprised that more had not been done to try to understand the experiences of this group.

Just trying to get a sense of the scope of the contracting phenomenon is difficult. In March 2012, when the United States had 88,000 soldiers deployed in Afghanistan, the Department of Defense had 117,227 contractors, only a third of whom were American.[22] This does not include tens of thousands of others who were on contracts for the Department of State, the United States Agency for International Development (USAID), the Drug Enforcement Administration, or any other of a dozen different American agencies working in the country.[23] These other agencies rarely track the contractors working for them and, even when they did count, their statistics were often filled with counting errors, sometimes off by as much as 50 percent.[24] These contractor estimates do not include civilians on contracts from the United Nations, the World Bank, the British government, the Canadian government, or any of the other large donor countries in Afghanistan.[25] Rarely are these figures publicly available, and from my inquiries to officials, often it seems that the numbers are never even counted.

Another group of civilians who often intermingled with these contractors provided services for the international workers in Afghanistan during the war, whether it was to set up the booming cell phone industry, which was driven by international dollars, or the influx of sex workers providing services for soldiers, contractors, and others.[26] As Ben and I continued to dig through mountains of government reports, pulling out numbers, we were constantly finding new groups or subgroups of contractors who were not counted in most instances.

For the most part, these contractors are an afterthought. After a June 20, 2016, attack on an embassy convoy in Kabul, the Canadian embassy received criticism after it sent out a tweet declaring that the embassy staff were "all safe," even though fifteen Nepali and Indian guards working for the embassy on private security contracts had been killed in the attack.

The U.S. Department of Labor keeps figures for contractors claiming workplace injury compensation, and its numbers suggest that by the end of 2015, 3,712 contractors had been killed in Afghanistan and Iraq, but these reports, they admit, "do not constitute the complete or official casualty statistics of civilian contractor injuries and deaths."[27] In testimony before Congress, John Hutton of the U.S. Government Accountability Office conceded that "contractors are generally not updating the status of their personnel to indicate whether any of their employees were killed, wound or missing."[28] These statistics rely on contracting companies themselves to report these numbers. When I spoke with Nepalis who had been injured in Afghanistan, I found that more often than not, their cases did not seem to be included in these numbers. Figures from other agencies contradict these reports, and other agencies, like the Department of Defense, released numbers only of American contractors killed or other particular slices of data. I submitted a Freedom of Information Act request to try to find more information, but received nothing. Ironically, USAID eventually hired a contracting company to keep an accurate count of the contractors working on their various projects.

One of the trends that the numbers do indicate is that, far from a fading phenomenon, the drawdown of U.S. troops led contractors to take on an even more important role in the American presence: a year after my visit to Bagram in the second quarter of 2016, there were 8,730 American soldiers in Afghanistan and 28,626 Department of Defense contractors, making contractors 76 percent of the Department of Defense presence.[29] This is the highest contractor-to-military-personnel ratio of any conflict in which the U.S. military has taken part.[30]

While the media and much of the policy world seem to be moving on from Afghanistan, understanding the lessons of the war can say much about how we might think about wars in the future. A handful of academics and journalists have produced thoughtful critiques of the war in Afghanistan, but the critical accounts that have gained the most publicity by far have been produced by the Special Inspector General for Afghanistan Reconstruction (SIGAR).[31] These highly detailed audits describe hundreds of millions of dollars in waste, highlighting examples ranging from a $36 million, 64,000 square foot command center that the military did not want and never used, to $260 million spent on health facilities, many of which lacked electricity and running water and were not even where the U.S. government thought they had been built.[32]

These reports tend to be dry reading, and I found that anyone who is not an accountant will have difficulty translating all those numbers into a real sense of the actual human costs. In 2013, the Department of Defense had obligated $310 billion to contracting companies. This was more than all other government agencies combined gave to contractors. Out of a total unclassified defense budget of $613.9 billon, this meant that over 50 percent of the defense budget was going to contracting companies of all types and sizes.[33] In other terms, 9 percent of *the entire* U.S. government budget was going straight to defense contracting companies.[34] One company that I came to know well had $11.2 billion in U.S. contracts just for feeding troops.[35]

The numbers are so big they are numbing. Morally, I found it worrying that the American press tended to pick up stories on projects that wasted millions of taxpayer dollars but were less inclined to be interested in how many foreign contractors had been killed, injured, or exploited working on these projects.

As an anthropologist, I was interested in what this meant on an individual level. Typically anthropologists spend time in one place, doing participant-observation research by attempting to become part of the community and writing ethnographies, holistic accounts of the local culture, explaining phenomena from the ground up. In recent years, however, anthropologists have become more versatile; they still use similar tools, but now look at more transnational phenomena, like migration and the international nature of war.[36] I became increasingly convinced that understanding contractors needed a study relying on systematic, ethnographic techniques in multiple locales. It was not enough to visit where the contractors worked; it would mean visiting the homes they returned to and the pathways they took as migrants. It also meant visiting contractors in various countries to see what held contractors together, but also what made them distinct from each other. Such an anthropology of contracting would provide a means of trying to understand how the contractors themselves experienced war.

The experiences of some of these workers that I began to collect felt more tangible to me than the numbers in SIGAR reports. These individuals snuck across borders, lived for years in tents on U.S. outposts, and forged documents to bring home a few hundred dollars to their families living in villages, some with conditions little better than those in Afghanistan. Others had lived in far more comfortable housing, with regular vacations, saving more in a year than they could in a decade at home.

For me, there was an academic interest: I wanted to better understand how wars were being fought and to see if this aligned with how most people tended to talk about them. Personally, I was also interested in the impact of war on others. Particularly for non-Afghans like me who had gone to Afghanistan during the war, what were their experiences of it? How had it shaped their lives? How did they think about it? Finally, from an ethical standpoint, I realized that those who were funding the war, primarily U.S. and European taxpayers, had little sense of the costs being borne by many of the workers who supplied the labor in support of the conflict. What were the global consequences of this increased outsourcing?

Standing outside Bagram as the sun began to go down, I watched truck after truck rumble heavily off into the growing dark to the east, carrying the rubble of war across the border to Pakistan and beyond. I was literally seeing the impact of the war spread out slowly from the epicenter at Bagram, spilling over the rim of Afghanistan to the rest of the world. What were the ripples I could not see? What of those international contractors who provided so much of the labor of war, propping up the conflict on their backs day in and day out, hoping this was a chance to change their fortunes? How had the conflict shaped their lives? And what can their stories tell us about the future of war, migration, globalization, and inequality?

THE RIPPLES OF WAR

Instead of following the trucks of scrap metal flowing off the bases, I followed the people, particularly the individual contractors from other Asian countries. This book largely tracks that journey. My first stop, in summer 2015, was in Nepal, where I spent the next six months, before spending three weeks in the Republic of Georgia, three weeks in Turkey, and then three months in India in spring 2016. In each place, I systematically conducted in-depth interviews with contactors who had spent time in Afghanistan. I compared the experiences of contractors for various companies and studied how gender, ethnicity, class, and education all shaped the contracting experience. I also interviewed the brokers, agents, and government officials who had not been to Afghanistan but were part of the massive migration industry that sent workers to conflict zones. As I conducted interviews, I did periods of participant-observation research, spending time in the towns and villages that workers returned to, sitting in the government offices where travel documents were issued, and chatting in the bars where migrant workers waited for their visas to arrive.

Similar to the winding paths of contractors' own experiences of the war, my research was not a simple straight line, and after I was in India and later in summer 2016, I went back to Nepal to do another month of more interviews. I ended up meeting with Nepalis I first met in Nepal later in India and Afghanistan. On the way back to the States, I spent a week in Sri Lanka searching for contractors and also spent time on Skype with contractors who had been in Iraq, the Democratic Republic of the Congo, South Sudan, Lebanon, the Balkans, and almost every other major zone of conflict from the past two decades. Later, in summer 2016, I spent two weeks on both coasts of the United States interviewing Afghan contractors who had immigrated to the United States and an additional three weeks in the United Kingdom interviewing primarily Nepali soldiers and contractors. Closing the circle, I ended by returning to Afghanistan again in late 2016 and visited some of the contactors I had first met in Nepal.

Over the course of the fifteen months, I interviewed more than 250 individuals who had worked in Afghanistan as contractors or were related to the process of migration to war zones in some way, shape, or form. Most of them were men, as was typical in the hypermasculine world of the war industry. Contracting is a male-dominated domain, but that did not mean that there were no women present, and I interviewed a smaller number of female contractors as well. As I spoke to each new contractor, all of these pieces came to paint a vivid picture of a far-reaching conflict.

This book focuses on some of the individuals I feel best embody the wider lessons from these contractors and the outsourced military industries they worked for. Some of the stories are extreme, like Teer's, while others are more mundane. Together they give a sense of how small, out-of-the-way places have been affected by the war in Afghanistan and what the long-term repercussions of a contracted war may be.

When America invaded Afghanistan in 2001 and then Iraq in 2003, it set in motion events that would send Teer to prison for nearly three years and change the lives of thousands of others, soldiers and civilians alike. While the practice of contracting during wars is certainly not new and much of it is built on historical patterns of recruitment that started under the British Empire (discussed further below), the scale of contracting in Afghanistan and Iraq was unprecedented during modern times. Contracting in conflict zones has become intertwined with labor migration flows from the global south to the north. This means that brokers, agents, and officials who control these streams at various points along the migrant's journey are increasingly important. These flows, along with international partnering, have meant that experiences of the war are uneven—they vary from place to place; they are influenced by your race, gender, and class. Contractors in Afghanistan were shaped by factors such as the economic conditions of their home country and their country's historical relationship with the United States. At the same time, contracting means that citizenship is sometimes now less important than the company that employs you. In many cases, states cannot do as much to protect the rights of their citizens as the companies contracting them can: the fate of workers is very much in the hands of their managers. As Teer found out, once their contracts are ended, workers can find themselves completely adrift in a strange and dangerous land.

More broadly, a study of these practices also suggests some ways in which the economics of contracting reinforce global inequalities, giving opportunities to agents, government officials, and others, usually in the ruling class, with special insight into the practices of contracting, trade, war, and migration. In contracting, war, migration, citizenship, racism, and inequality are all knotted together. They circle back and reinforce each other, sometimes in unpredictable ways.

As the wars in Afghanistan and Iraq have scaled down, new practices are emerging: contractors have been demobilized and sent home to wait, some with serious physical and psychological damage, many anxious to return to conflict zones. This has created a far-flung reserve army, ready at a moment's notice, waiting for a call from America or Britain or Russia or Yemen or ExxonMobil or whoever else is willing to give them their next paycheck. What war will employ and deploy them next?

Contractors' stories, and the story of the American intervention, are not easy to tell. Contractors inhabit a shadowy world where their employers carefully guard details of what they did and where they worked. The outsourced nature of contracting, subcontracting, and sub-subcontracting makes it difficult to track these workers while simultaneously obfuscating responsibility: Whose job is it to protect workers?

Cutthroat competition between security firms means that companies are constantly being sued, breaking apart, rising again, and forming new businesses with innocuous names, working with manpower companies located everywhere from Mumbai to Dubai and Doha to London. Workers moved between them, afterthoughts to most policymakers and their bosses.

Trying to understand the global repercussions of the war couldn't be done just by focusing on Bagram and in Afghanistan. It also couldn't be done in a library, relying on budget numbers and troop levels. I sent my passport off for additional pages and began packing my bags, hoping to uncover how far the ripples of the war extended and how the war reshaped those who had been in Afghanistan, their families and their countries, with the hope of bringing some of these stories in the shadows out into the light. It is my hope that by telling some of the stories of these previously ignored laborers, we can better grasp the deeply personal and truly global consequences of our wars.

CHAPTER 1

MERCENARIES, CONTRACTORS, AND OTHER HIRED GUNS

MATTA TIRTHA

In July 2015, I stood on the hillside of a small town at the edge of the Kathmandu Valley and took in the rapidly expanding metropolis of Nepal's capital city, Kathmandu, from about 10 miles away. I was there to interview several Nepali security contractors who had guarded, among other things, the U.S. embassy in Kabul, but before the contractors arrived, I paused in a clearing to take in the view.

The name of the town, Matta Tirtha, translates to "the path of the mother." The town is known in mostly Hindu Nepal for a shrine that sits above it in a shady courtyard, the hills of the valley shooting up above. That day, the shrine was primarily occupied by six young boys who were jumping in and out of the stone pool at the center of the small religious compound to cool off. A few other devotees strolled by, mostly enjoying the view and fresh air. On Nepali Mother's Day in the month of Baisakh, the town swells with mourners who come to pray for mothers who died in the past year. Visitors are both Hindus and Buddhists. Nepal's 80–20 split between the two tended to create syncretic melding more often than it created serious divides. It was common to come upon shrines with elements from both religions. This was a far cry from the Islamic Republic of Afghanistan, where it is assumed that essentially all Afghans are Muslims.[1]

Having spent the past ten years working in and traveling back and forth to Afghanistan, my eye was drawn to elements like the shrine, which demonstrated the contrasts between the two cultures. In the Matta Thirta, for example, a group of young men sat at a teahouse under a pipal tree sharing tall bottles of beer, while several young women walked by in skirts, and the two groups eyed each other in a way you would never see in Afghanistan, where casual dating is unheard of. Seeing women without a head covering was, of course, different–but for some reason, I was most shocked by the occasional woman driving a motorbike. I was conditioned by my time in Afghanistan, where there were virtually no women who drove.

Despite these differences, there were numerous reminders that while Afghanistan and Nepal are separated by some of the highest mountains in the world, they are not all that far from each other geographically or sociologically. In reality, the Nepalis who guarded American bases had far less in common with the Americans they were protecting than the Afghans they were supposed to be protecting the Americans from.

The politics and the cultures of both societies are shaped by the steep mountains they live below and the fertile valleys that run out of them. In Matta Tirtha, as in Afghanistan, extended families live together, carefully cultivating fertile plots of land. Patriarchal structures mean that parents are respected and families remain close. These practices, in turn, shape how people in these communities see each other and the world they live in. In both Afghanistan and Nepal, most will tell you that those in the mountains are heartier souls, while people on the plains are softer, yet generally better at business and politicking.

While most of the divides in Nepal are along caste lines and most in Afghanistan are ethnic and tribal, both of these splits create low levels of resentment that occasionally boil up, particularly against the ruling elite who control most of their countries' main resources. In both places, there are often land conflicts between neighbors and even more resentment aimed at large landowners who don't live in the area. As a result, most rural communities attempt to regulate their own affairs and avoid the meddling of government officials. Just as important, both countries are extremely poor, burdened by indirect colonial legacies, ineffective and corrupt politicians, and neighboring countries (Pakistan in the case of Afghanistan and India in the case of Nepal) that bully them, taking advantage of their isolation.

Despite these similarities and their close geographic proximity, formal ties between the countries are virtually nonexistent. There is no Nepali embassy in Kabul and no Afghan embassy in Kathmandu. The closest Nepali embassy to Afghanistan is in Islamabad, which serves as diplomatic representation for Nepalis in Pakistan, Afghanistan, Turkey, Iraq, and Iran—not an ambassadorship that I can imagine many coveting. But this lack of diplomatic links belies a long history of Nepalis traveling to Afghanistan, particularly during times of war.

Historically, the kingdom of Nepal expanded westward in the early nineteenth century, and, for a time, the western border of the country was only a couple of hundred miles from the eastern frontier of Afghanistan. Later, when the British invaded Afghanistan in the middle of the nineteenth century, they were supported by tens of thousands of "Indian" troops, many of whom were actually Nepali.[2] On one level, it was tempting to jump to the conclusion that with the U.S.-led invasion, a new, more subtle American empire had taken over for the British Empire in South Asia, relying on Nepali private security contractors instead of Nepali soldiers. However, while the parallels are important, so are the differences and the ways in which the American experience in Afghanistan was unique.[3]

TWO COUNTRIES

Links between Nepal and Afghanistan are not simply historical.

As I stood looking out over the Kathmandu Valley, my eyes were drawn to one of the newest and largest houses in town. Contrasting sharply with its neighbors, it was painted a rosy pink with bright red trim and had reflective neon green glass windows and several large balconies. The clashing colors were slightly disturbing against the lush mountain backdrop. The home looked similar to what my friends in Kabul would have called a narco-palace: monstrosities usually built by Afghan warlords, presumably using money from drug trafficking (though often these figures had profited more from lucrative contracts with the American military) in Kabul's premier neighborhoods, using an obnoxious mix of architectural elements from Pakistan, Dubai, and Bollywood.

The similarities between the house and those Kabuli palaces were not coincidental. Yogendra, the owner of the house in Matta Tirtha, had grown up in a small building that still stood next door, now dwarfed by the new home. He made his fortune driving in armed convoys through Kabul's congested streets. Passing beneath such palaces in armored SUVs for six years, he decided he wanted a house just like those palaces and carefully saved his money. Once back in Matta Tirtha, he told me, he used half of everything he saved to build this new house. The rest he was saving for a rainy day.

He was not alone, he explained, as we sat in his enormous and slightly dim reception room, far larger than any I had seen in other Nepali homes. As Yogendra, well built and stern looking, went on with the story of his time in Afghanistan, he became more animated and smiled at the memories. Seven of his friends, all from Matta Tirtha, had gone to Afghanistan and worked for the same company, DynCorp, one of the key U.S. private security contracting companies there. DynCorp was best known in Afghanistan initially for providing President Hamid Karzai's security detail, but eventually it became responsible for tasks ranging from training the Afghan police to opium eradication.[4]

It was not by chance that so many from the town had worked for DynCorp. After one man was successful at securing a lucrative job in Afghanistan, it was common for him to recommend some of his friends and neighbors to his American or European boss. Getting from Nepal to Afghanistan legally was no easy task, and it was much easier with help from someone in managing the hurdles. Many of those I interviewed over the year not only knew each other, but took similar routes to Afghanistan, both physically and in terms of their life stories.

Yogendra first spent eighteen years as an officer in the Nepali Army. There he served with Resham, Himesh, Bhek, and Surya, all neighbors of his. The army paid frustratingly low wages. Enlisted soldiers in 2016 were paid less than $120 a month, and it was even worse when he was a recruit thirty years earlier.

Hoping to find a way to earn more, he tried to join the Nepali peacekeeping soldiers who were sent to Lebanon in support of the U.N. mission there. Nepalis assigned to international missions received better compensation, but competition for these positions was fierce. As the Nepali saying goes, you had "to keep people in your hand" through connections and bribes to get a posting. Yogendra did not have connections. He was also concerned that the army might get involved domestically in the growing Maoist insurgency, which had destabilized large parts of the Nepali countryside in the late 1990s and early 2000s. So he took early retirement after hearing that several of his neighbors had gotten positions at one of Kathmandu's exclusive, foreigner-only casinos as security guards.

The pay at the casino was better, a couple of hundred dollars a month, but there, while on smoke breaks with the other security guards, he began to hear stories of former Nepali soldiers who made thousands of dollars a month working in Iraq and Afghanistan. Twelve Nepalis had been killed in Iraq a few years before, including one who was beheaded. But while they sat around gossiping, they rarely discussed such incidents. How dangerous could it be inside a base?

When Surya, one of his neighbors from Matta Tirtha who worked as a contractor in Afghanistan, came home on leave, pockets filled with cash, Yogendra decided to give it a chance. So he went into Kathmandu and visited the broker's office that Surya had used, and in early 2008, he found himself headed to Afghanistan.

STATIC SECURITY

Contracting is not a new phenomenon in the American military. During the American Revolution, German mercenaries fought for both sides, the eighteenth-century version of private security contractors.[5] Contracting in its current form, however, increased substantially during the Vietnam War and grew rapidly in the most recent decades.[6] The advent of the all-volunteer army in the United States created a military labor shortage, and contracting simultaneously helped address this problem while lessening public opposition to America's wars abroad. In the first Gulf War, one out of every hundred individuals deployed was a contractor, but by the time America invaded Iraq again in 2003, roughly half of those deployed were contractors.[7]

This made the American military presence much less American: 80 percent of those contractors were not American citizens.[8] In 2010, the United States had 175,000 soldiers deployed to war zones, mostly in Iraq and Afghanistan, and, simultaneously, 207,000 contractors funded by the Department of Defense, like Yogendra, in these theaters.[9]

Many of the contractors were laborers who were not generally thought of as mercenaries, since they mostly built bases and worked in dining halls.[10] Others did resemble mercenaries in the more conventional sense. Despite this, many in the military and the private security industry insist there is a clear divide between private security contractors and other contractors, but from the view of those supplying the labor, Nepalis and others like Yogendra, who were willing to take whatever work abroad they could find, this divide meant little.[11] And though Yogendra certainly did not consider himself a mercenary, the concept was blurry.[12] Yogendra did not go on patrols to attack an enemy, but the convoys he drove in were still called "missions," and he often went out expecting to be attacked, draw fire and shoot back. Some private security companies even set up quick reaction forces, which closely resembled their military counterparts.[13] Those whom I interviewed who were most active in ongoing operations tended to be involved in work such as opium eradication, but most were in support positions, guarding walls, vehicles, or "clients," generally high-profile diplomats or aid workers in need of close protection.[14]

The contractors I interviewed who acted the most like soldiers were generally those who worked on supply convoys.[15] The United States typically did not use American military personnel to defend supply lines to many of their small forward bases scattered across the country or, in some cases, even the roads to their larger bases. Contractors were hired to deliver the necessary supplies to many of these bases. Such convoys were under frequent attack by the Taliban, and the organization of the convoys included complex preparations for attacks. These convoys often received little to no air support or other resources from the international military.

Many of the troops I had interviewed previously tended to assume contracting is not as dangerous as soldiering, a questionable stance, I found. From the very beginning of the conflict, international contractors were deliberately targeted by insurgents, and as early as 2004, for example, eleven Chinese laborers working on a road-building project in Kunduz were killed.[16] These numbers kept increasing. By early 2016, 2,300 U.S. soldiers had been killed in Afghanistan, and the Department of Labor had reports of 1,644 contractors killed in the country.[17] This gap is probably even narrower since these numbers include only cases reported to the Department of Labor by contracting companies, which also meant compensation had been paid to the families of the dead contractor.[18] Reading statistics and casualty lists, I found cases in Nepal where contractors who were killed or injured were not included in these counts, suggesting a potentially much higher actual casualty rate. Considering the intense media coverage of American soldier deaths, it seems likely that the number of U.S. military personnel killed is accurate. This is not to take anything away from the danger the troops faced, but as I spoke with Yogendra and others, I increasingly felt that the idea that contractors were different from soldiers because they were not in the same amount of danger was wrong. As I gathered more interviews, it also became apparent that for some of these contractors, the Taliban threat was only one of many dangers they faced.

When Yogendra first arrived in Afghanistan, he was sent to Herat, in the west of the country, where he was employed doing "static security." This meant he was essentially a heavily armed security guard, usually stationed at a compound gate or tower or just inside the various compounds he was assigned to guard—the most common security job for Nepalis in Afghanistan. Inside Yogendra's compound was a small military base, but other Nepalis guarded compounds that varied from housing small office buildings for nongovernmental organizations (NGOs) to the enormous international military bases.

I visited Herat twice during this period at a time that it was considered one of the safest cities in the country. I had been able to walk around the town freely, and there were only a few districts outside the city that were insecure. Yogendra was not threatened by the Taliban or other insurgent groups, but his job was still dangerous. Somewhat surprisingly, the danger for him came from two feuding brothers who owned a large amount of land near the base where he worked. Yogendra was never told why the brothers were fighting, but they were clearly deeply involved in some dispute, and their men regularly fired small rockets and rocket-propelled grenades (RPGs) at each other's land. When some of these landed within Yogendra's camps, everyone had to run for cover. This was particularly annoying, he said, since they were living in tents and susceptible to the shrapnel and debris that these explosions kicked up. Both of the brothers were supportive of Karzai's government, so there was nothing Yogendra and his colleagues could do to curtail the regular rocket attacks.

Guarding the compound itself was less risky for Yogendra. DynCorp, like most other security firms, also employed Afghan guards. These guards usually worked as the outermost ring of security, making it far more common for the initial casualties to be Afghan when these compounds were attacked. Nepalis tended to guard posts along a second ring of security, inside the first wall. Often it was American and

European contractors who worked at a third, interior station, where they were even less susceptible to danger.

Contractors I interviewed described that this coincided with the strict and predictable hierarchy that security firms had, which largely aligned with the race and nationality of their employees. Americans, Europeans, and white South Africans were in the positions of authority and earning the largest paychecks.[19] They were followed by Serbians, Romanians, and other Eastern Europeans. At the bottom of the chain were Nepalis, though they were usually higher than certain groups of sub-Saharan Africans, particularly from East Africa. Last were the Afghans.[20]

Within most international compounds, this hierarchy was rigidly enforced, with Americans and Europeans running the camps, groups like Eastern Europeans and Turks in more technical positions, Nepalis running security, other Asians doing most of the skilled or semiskilled labor, and Afghans generally having the worst positions, guarding the most dangerous gates with the oldest weapons and least training.[21] This created tensions not between the Nepalis and the Americans and Europeans who tended to run the operation as much as between the Nepalis and the eastern Europeans who were just above them. Similarly, Afghan contractors tended to resent most the Nepali contractors just above them in the hierarchy, whom they saw as taking work they could have been doing.

Yogendra said that for him and the other guys from Matta Tirtha, a turning point was when DynCorp allowed them to start applying for jobs on personal security details (PSDs) — teams that provided protection and traveled around with clients, instead of simply doing static security. With this policy change, experienced Nepali contractors could apply for such jobs, working alongside Romanians, Serbians, and occasionally Americans. It also meant almost a tripling in their salaries. With more money coming in and working on a mobile team driving clients to various meetings around Kabul, Yogendra began planning the house he would build when he returned to Matta Tirtha. Like troop levels, contracting jobs peaked around 2011, and in the following years, fewer and fewer jobs were available. Downsizing became the norm, and eventually, in 2014, Yogendra's contract was not renewed and he returned home for good.

Many of his friends and colleagues followed. As I interviewed more of these men, I realized how personalized experiences of Afghanistan were. Yet there were clear commonalities. Many were like Yogendra: they had earned good money, were not too disturbed by the risk or various forms of discrimination they faced, and were generally satisfied with the time they spent there. Some were almost radically pro-American, having spent so much time with retired U.S. Marines. Others had much more negative experiences, ranging from trafficking and imprisonment to simply getting ripped off by the brokers who helped get them to Afghanistan. I would eventually meet more men like Teer Magar who had been kidnapped and detained for years, but also those who had set up successful businesses, taking advantage of opportunities they never would have had in Nepal. Some cursed Americans for their mistreatment, while others befriended American contractors whom they stayed in contact with on Facebook.

As Yogendra's children ran through the spacious living room, it was clear that the experience was also affecting those back home directly and indirectly. Back in Nepal, largely unemployed but financially comfortable, several of the contractors from Matta Tirtha played badminton each morning. With some capital at their disposal, they had joined together to form a land speculation business, while another served as the government representative from the area. None had managed to become really wealthy, but all were comfortable and relatively unconcerned about money in the short term.

KATHMANDU

I arrived in Nepal to begin my research in earnest in mid-July 2015.[22] The earthquakes of May and April had left people on edge. The 7.3

and 7.8 magnitude earthquakes had killed almost 10,000, damaged 500,000 homes, and affected the ability of 30 percent of the population to produce the food they needed.[23]

While most of the rubble in Kathmandu had been cleaned up, there were still refugee camps in parks and other green spaces scattered around the city, making the place feel more chaotic than it had been on my previous visit the summer before. But in a country that in the past two decades had seen a ten-year Maoist insurgency, the 2001 massacre of almost the entire royal family that led to the eventual overthrow of the monarchy, followed by nine years of wrangling over a new government by two constitutional assemblies (which at the time of my arrival had still not passed a constitution), and where electricity was cut eight hours a day in the capital, people were accustomed to continuing life with challenges of almost every shape and size.

Since I planned to use Kathmandu as a base for tracking down contractors, I found a small studio apartment away from the tourist district and began to take Nepali language lessons from Geeta, a patient and meticulous Nepali woman, in the hopes of at least following along when I needed to have interviews translated.[24]

No one kept actual records of the number of Nepalis in Afghanistan, but my research eventually suggested that probably around 50,000 had worked there since the invasion.[25] Still, I was not initially sure how to find them. I met initially with Deepak Thapa and Bandita Sijapati who ran the Social Science Baha, one of Nepal's best research centers, to discuss research strategies. They had done extensive work on migration and labor out of Nepal, but admitted that no one they knew had looked at the contractors who went to Afghanistan or other conflict zones specifically. Trying to find these contractors presented certain challenges. There was no simple directory or association for those who had been abroad. I was to find that the companies were unwilling to provide any information, and in many cases, the laborers had gone abroad illegally or semilegally. I also didn't know whether they would have any real interest in speaking to me. The contractors were spread out in villages and towns in the mountainous country, and finding them did not seem like it would be easy.

The Social Science Baha, however, was filled with young scholars working on interesting projects, and Deepak and Bandita kindly granted me a carrel in their well-stocked library that served as my base. They also introduced me to Dawa Sherpa, who had been working on several projects on migration and was particularly interested in the Gurkhas—the storied Nepali soldiers of the British and Indian armies. Gurkhas have an extensive mythology emphasizing their bravery and loyalty in battle, and the British Army still recruits a small number from Nepal today. Many retired Gurkhas were in the first wave of contractors who went to both Afghanistan and Iraq.

Dawa confided early on that he, like many other Nepali boys, had thought about joining the Gurkhas when he was younger. He found, however, that he liked meeting new people and going to new places and ended up gravitating toward social science courses and getting a BA in social work. He was working at the Social Science Baha as a researcher, a job that suited him. His friendly, empathetic, and inquisitive nature helped him in his earlier work interviewing those who had been most affected by the earthquake. Turning to study contractors appealed to him since he was interested in the conditions that led so many Nepali youth to look for work abroad. So we started brainstorming ideas for how to get in touch with Gurkhas and other contractors, calling the local pension center and searching for the manpower centers that claimed to send Nepalis to Afghanistan and other conflict zones.

The research progressed slowly at first. Most of the Nepalis I met in Afghanistan were, unsurprisingly, still in Afghanistan or had moved on to jobs in Iraq, one of the Persian Gulf countries, or elsewhere. Some of those who worked for American companies took advantage of an easier visa application process and moved to the United States. Others emigrated to Europe. The formal institutions we contacted were not generally interested in talking to us: the government departments claimed they had bureaucratic restrictions, and private companies were concerned about getting in legal trouble or compromising their client lists.

Still, we kept bumping into signs of the many Nepalis who had been in Afghanistan. Most Nepalis seemed to know someone who had been there or some other conflict zone, even if it was a distant relative or a friend of a friend. On my first week in Nepal, a taxi driver told me that his cousin had been in Afghanistan, but he was now in Iraq. He wasn't sure when he would be back, but the driver assured me that the cousin had felt much safer in Afghanistan where "at least they knew who the bad guys were."

Later the same day, as I sat having a beer at a café in the center of the city, I noticed that my waiter was wearing a U.S. Army cap. I asked him if he had worked for the U.S. Army and he said, "No, but I wish." He had a friend who worked in Kuwait, and his brother worked for the U.S. embassy in Kabul. He asked me if I had been there, and when I said yes, he asked if I knew of any jobs there. Like many other young Nepalis, the pull of well-paying positions abroad was strong, and he had thought about trying to secure one of the lucrative positions in Afghanistan, but all the contracts seemed to be controlled by shady middlemen. He said that he would like to go to work in a place like Afghanistan, but would not go unless he could go "directly," not using the brokers who charged enormous fees and, he had heard, at times got their clients stranded or arrested in distant countries.

I asked him how he felt about the danger of working in a country at war, and he said that he had heard from some of those who had returned that if he worked in a big compound, there was little risk. I told him that rockets can still come, and he responded that in that case, "If I die, everyone dies." He paused for a minute and said, "Where there is money, there is risk," and then went off to serve some other customers.

THE RISK

The risk itself varied.

The nature of the conflict changed significantly over the course of the first fifteen years of the war. When American operatives began helping the so-called Northern Alliance in October 2001 following the September 11 attacks to take back the country from the Taliban, it was largely an internal affair, with funds and weapons coming from the Americans. Most of the international extremists associated with al-Qaeda, such as Osama bin Laden, quickly fled the country.

With many of the Taliban strategically retreating to Pakistan, the United States made little effort to establish much of a presence outside major cities, losing an important opportunity to help establish an effective Afghan government.[26] With the 2003 invasion of Iraq, American troops and officials turned their attention elsewhere, and Taliban leaders began to regroup.[27] More important, the Americans backed Hamid Karzai, a Pashtun, as transitional president, but also allowed Northern Alliance leaders, most of whom were leaders of ethnic minority groups, to maintain many other key government posts. These figures, many of whom who had been recently commanding corrupt and brutal militias, quickly consolidated their positions, usurping both government and international resources.[28]

Karzai, in the lead-up to the first postinvasion elections, opposed creating a system based on political parties, which might have weakened the pro-government warlords. At the same time, the United States largely refused to support his attempts to rein in these warlords during the early years of his administration.[29] This contributed to the creation of a government that was a corrupt web of patronage. Positions were handed out in exchange for cash, and warlords continued to rob and steal, now in the name of the Afghan state.[30] Despite the international development funds that came with the invasion, life for most Afghans improved only marginally. By 2015, the World Bank reported that among poor Afghans, only 63 percent had access to electricity, 40 percent had access to safe drinking water, and 2.8 percent had access to sanitation.[31]

This consolidation of power rewarded the political elite and provided few services for ordinary Afghans. In the following years, local resentment of the government and its international allies grew, leading to an insurgency, particularly in the Pashtun south and east of the country, that started taking back previously government-held territory. By 2006, this was a full-fledged war that threatened to overthrow Karzai's internationally backed government.

Until this point, the number of contractors in Afghanistan was not particularly remarkable. Many of the CIA teams initially coordinating the overthrow of the Taliban were on contracts, but these were mostly American and their numbers were limited. The majority of Nepali contractors I spoke with who had been in the country during this period were based in Kabul, since there was little international presence outside the capital. During these years, it was more common for Nepalis to go to Iraq, where America was concentrating its efforts, to work as private security contractors in the Green Zone. The election of Barack Obama in 2008, who campaigned in part on the fact that Afghanistan was the forgotten "good war," would change this.

In 2009, following a lengthy review of the war strategy, Obama announced a "surge"—essentially a massive influx of American troops and money. [32] This move was part of the U.S. government's embrace of a counterinsurgency approach to the war. The reasoning was based on the assumption that the resurgence of the Taliban was primarily being fueled by young men with "local grievances." Unemployment and a corrupt, unresponsive government were the driving forces of the conflict, proponents like General David Petraeus and General Stan McChrystal argued, and the best way to counter this was by "winning the hearts and minds" of local populations. This included building roads and schools and a pu pu platter of other development projects, small and large.

Most of this would be done by private contractors, not by the U.S. military.

The increase in spending on contracts was astounding (Figure 1). Companies that were already working for the United States in Afghanistan saw rapid budget growth: Supreme Group had $650 million in American contracts in 2008, mostly for feeding American troops. [33] By

2010, this was $2.1 billion. DynCorp had $400 million in Department of Defense contracts in 2008, which was also up to $2.1 billion in 2010, not including another $1 billion from the Department of State, the Department of Agriculture, and other U.S. government agencies.[34] Soon, recruiting firms in countries ranging from Nepal to the Philippines and Kenya to Bangladesh were recruiting for cooks, cleaners, construction workers, and guards. A new era in the war was underway.

The hasty upswing in troops and money was disruptive. Suddenly there was money for almost every project imaginable, which also meant the opportunity for increased waste and corruption. Marines and contractors now traipsing through rural villages previously ignored by the U.S. military meant more casualties and increasing resentment from the local population.[35] Three years later, in 2012, Obama began to withdraw the troops according to the time schedule he had laid out at the beginning of the surge. Troop levels declined rapidly; contractor numbers decreased too, but they did so more gradually.

In some ways, a history of the contractors who went along with the soldiers during the surge has many parallels with the soldiers themselves. These contractors lived in some of the same compounds, were involved in many of the same fights, and faced some of the same challenges as soldiers. But there were some important differences.

Contractors often did not serve fixed tours. They might take leaves but were not rotated home. Soldiers serve set tours that could be extended, and many who served between 2001 and 2015 found themselves recalled, sometimes multiple times. With each tour normally being around a year, this meant the members of the U.S. military I interviewed had typically been in Afghanistan for two or three years total on separate tours.

Contractors had no restrictions on how long they could serve, and many Nepali contractors served five, six, or seven years there. I ended up interviewing multiple contractors who had been there ten years; in one case, a Nepali contractor was closing in on his fourteenth year involved in the conflict. Ironically, this often gave them a sense of the

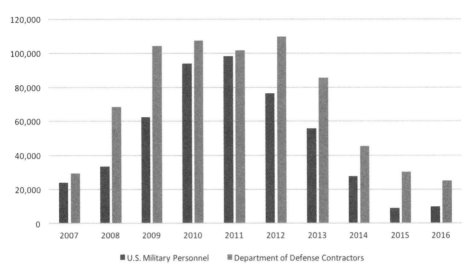

Figure 1. Military Personnel and Contractors. Source: Peters, Schwartz, and Kapp 2017.

long-term ebbs and flows in the war that was far more nuanced than the U.S. soldiers they were supposed to be supporting.[36]

The contractors were also there for profit. This meant that as the surge was winding down, many of the contractors I interviewed expressed regret that the war might soon be over. It was troubling that there were over 100,000 contractors being paid by the U.S. Department of Defense who were actually in favor of the war continuing. The war in Afghanistan has been the longest in American history. To what extent did the fact that the majority of those working for the United States wanted, consciously or subconsciously, for the war to continue, contribute to this? As my year in Nepal went on, this question returned to me repeatedly.

For the contractors themselves, what happened once they returned home? What were they returning to? How was the money they earned reshaping their lives? But first, perhaps, why were so many of the security contractors in Afghanistan Nepali?

CHAPTER 2

NEPALIS AT WAR

A LIFE IN THE INDIAN ARMY

Sitting in his neat living room in a more modern neighborhood east of Matta Tirtha, Bodh Bikhram apologized that his wife was not there as well to greet us; she was off visiting one of their sons in the United Kingdom, so Bodh brought Dawa and me glasses of fruit juice from the kitchen. Once settled into his seat, he asked us what we would like to know about the Gurkhas.

For those without some knowledge of South Asian history, it is not immediately clear why so many of the lower-level security contractors in Afghanistan were Nepali. Yet at the gates of almost every international compound in Kabul, you would find a team manning the checkpoints, and odds were that when you walked past guard towers, a Nepali was looking down at you.

Nepal's low gross domestic product meant that wages were low. But they were low in countries like the Philippines, Bangladesh, and Sri Lanka too, and while I found Filipino, Bangladeshi, and Sri Lankan cooks, cleaners, and truck drivers, there were almost no security contractors from these countries.

What made Nepalis different was their history of serving in foreign armies, particularly the British Imperial Army and, later, the Indian Army and the Singapore Police Force. This gave rise to their reputation as so-called Gurkha warriors, who were said in the media and numerous British memoirs to be brutal in battle and fiercely loyal to those they fought for. This colonial practice began in 1816, but the history continued to shape both the ways that international companies hired Nepalis and that Nepalis viewed the process of working abroad, particularly in military roles. In order to find out more about how Gurkha recruitment shaped the tendency of contracting companies to recruit Nepalis, I approached a series of active and retired Nepali soldiers. The first was also Dawa's "uncle," a Nepali who had spent the majority of his life as a Gurkha officer in the Indian Army.

Bodh Bikram was not actually Dawa's uncle, but a neighbor and family friend. In Nepal, such figures are often referred to as members of the family, which can make conversations about friends and family confusing. Dawa, who had only a biological brother, was always telling me about going out with his sisters who weren't, as he would put it, "my real sisters." Later, when we talked to security guards, they would refer to each other as brothers and cousins; sometimes they were related to each other biologically but most of the time they were not.

We arrived at Dawa's uncle's tidy new home in one of Kathmandu's southern suburbs around midday, and the visit was initially more of a social call than it was research. His tales of the Gurkha regiments and their feats, however, were clearly linked to the continued presence of Nepalis in wars abroad.

Bodh was a retired lieutenant colonel. He was well built, with a vigorous manner, despite the fact that he had suffered a stroke a year before. With his wife out of town and no taxis available the night the stroke occurred, he told us he had driven himself to the hospital. He was almost completely recovered and had been advocating for more stroke awareness in Nepal ever since. He hadn't been to Afghanistan, but he was a great talker and was eager to chat about some of his experiences, which were typical of a Nepali who had served in a foreign military.

Like many of those I encountered, being in the military was not simply a profession for Bodh; it was a family affair. On the wall were large professionally done photos of his extended family, which he pointed to as he talked. Bodh's father had also been in the Indian Army, as had his wife's father, and his sons followed in his footsteps. His eldest son joined the Indian Army, and the youngest joined the British and had served two tours in Afghanistan. Even his daughter's husband was a military man. Bodh had served in the Indian Army for thirty-three years in posts all over the country. His children had grown up there, attending schools for the children of Indian soldiers, returning to Nepal mainly to visit family over vacation.

The term *Gurkha* comes from the town of Gorkha, high in the hills of east-central Nepal, which was the original home of Prithvi Narayan Shah.[1] In the second half of the seventeenth century, Prithvi Shah became the first king of a more or less unified Nepal after conquering the various city-states in Nepal's valleys, most notably the cities of Kathmandu, Patan, and Bhaktapur, all located in the Kathmandu Valley.

For Americans, the idea of having foreign nationals in the military might seem odd.[2] In fact, the tradition of these Nepalis in armies abroad stretched back to a series of battles between Nepal and Britain's East India Trade Company that has now been so mythologized by those on both sides that it's hard to sort fact from fiction.[3] Most stories from my interviews and the extensive British military literature on Gurkhas tell the story that in 1816, the East India Tea Company fought the Nepali king, Girvan Yuddha Bikrama Shah (the great-grandson of Prithvi Shah) to a standstill.[4] As a result, in the Treaty of Sugauli, which ended the war, the British gained permission to recruit troops from Nepal since they were said to have been so overwhelmed by the bravery and fierceness that the troops demonstrated in defeat.[5]

Nepali troops became an instrumental part of the British forces in India, and the Nepalis I interviewed were not the first soldiers from the Himalayas to fight in Afghanistan, Bodh explained. They notably helped the British forces invade Afghanistan during the Anglo-Afghan wars in the nineteenth century and were active on the northwest frontier of the empire fighting against Pashtun tribes. In a parallel with some of the ways in which the United States was to rely on Nepali security contractors two centuries later, one of the chief duties of the Gurkhas was to protect caravans and convoys as they passed through hostile terrain.[6]

Of even more historical significance, in 1857, Nepali Gurkhas in the British Imperial Army sided with the British officers and played a key role in putting down the uprising of Indian troops, later known as the Indian Mutiny or the First War of Independence, India's first substantial bid to throw off colonial rule. The king of Nepal, the ruler of the only remaining Hindu kingdom, sided with the British over their Indian cousins, an affront still remembered in India today.

The sacrifices Bodh described that Nepal made for the British were not imagined: Gurkha recruitment increased dramatically during World Wars I and II, when Britain recruited 200,000 and 250,000 Gurkhas, respectively, with some 10,000 Nepalis killed and 23,000 wounded in World War II.[7] Following Indian independence, the shape of the Gurkha regiments shifted as Gurkha regiments split. Four of the ten Gurkha regiments remaining after World War II stayed part of the British Army and the other six joined the newly established Indian Army.[8]

Bodh described how in the early years, these recruits were usually young men from high villages in the hill. Tales of Nepalis going off soldiering, bringing home wages and goods from the outside world, were stories that everyone in Nepal is familiar with. They built on the British idealization of the geography and attempted to use the mountains to help explain the valor of Nepali soldiers. A British officer I interviewed explained the thinking common among many in his memoir: "A bracing climate, abundant milk, regular exercise, a frugal diet and self-reliancy [sic], to say nothing of learning how to cope with monotony and a simple routine, all help to foster the characteristics which Gurkha soldiers are so well known."[9]

For Bodh and contractors I interviewed, this sort of geographic determinism makes little sense in today's world and had little to do with Bodh's actual history (especially since he was raised in India), yet the

recruitment practices based on this strange colonial logic continue un-
deterred. Following his father from Indian town to Indian town as he
was transferred, Bodh attended military schools and was extremely well
read. His upbringing made his eventually joining the Indian Army less
a bold step for him than a logical choice that his father and grandfa-
thers had also made. He was simply continuing the family business.

Within Nepal in the twentieth century, retired Gurkhas occasionally
played influential roles in Nepal politics.[10] More generally, they were
considered individuals who were likely to amass wealth abroad and
then brought back strange goods and ideas, usually retiring at a fairly
young age to a life of leisure.[11] Such soldiering was a way to quickly
gain status in communities, and it has long been a desirable career path.
Having made their money, many returned not to their home villages
but to cities such as Kathmandu, Pokhara, and Dharan—centers of
Gurkha recruiting.[12] In these cities, the sons of former Gurkhas were
often branded as lazy and inclined to party, relying on their father's
wealth.

During the Maoist rebellion in the late 1990s and early 2000s, Maoist leaders pointed to the continued practice of sending Nepalis into
the British military as evidence of its colonized state, fitting nicely into
their Marxist ideology about the exploitation of third world workers
and made the ending of the practice one of their central demands.[13]
After coming to power, however, no serious attempt was made to end
the practice. While numbers had declined, around 200 were recruited
into the British Army in recent years, with little indication that the practice would cease anytime soon.[14] Even more Nepalis were still being
recruited into the Indian Army, and a few for the Singapore police, another living relic of the British Empire that originally recruited Nepalis
to police the small state.

All of these recruitment streams eventually led into the contracting
pipeline for America's wars.

SELLING GURKHAS

Reading through the histories and memoirs of Gurkhas made this recruiting seem like a distant relic of the past, but Bodh's explanation of the practice suggested just how important it remained to Nepal and to the war in Afghanistan.

Both abroad and at home, the image of Gurkhas had become a marketing tool, and almost everyone I lived and worked with in Kabul seemed to assume that Gurkha protection was a mark of excellence. Security companies realized this and built on it; for example, IDG, a British security firm, used the large crossed kukri knives of the Gurkha brigade in its logo and advertised its ability to supply "genuine Gurkhas."[15] Another contractor's website suggested, "Today, Gurkhas are marked by their graciousness, loyalty and great courage. As gentle in daily life as they are fearless and tenacious in battle, they are a dignified people and ideal soldiers and security personnel."[16] Such branding was effective, and in Afghanistan and Iraq, security companies would charge nongovernmental organizations and other international groups higher rates for the provision of Nepali "Gurkhas," whether they had served in the British or Indian armies or not.[17]

The connection between Nepalis, the British Army, and Afghanistan also remained a real one even if myth and marketing had been layered on top of it. Bodh described how his son went to Afghanistan as a member of the British forces on multiple tours. His son, however, had returned to Britain, where he lived with his wife. Bodh, however, did have other contacts who had been to Afghanistan, and a week later we drove west out of Kathmandu toward the edge of the valley, before turning south up into the foothills where the houses of Matta Tirtha lined the hillside.

On the way, Bodh noted there were seven natural springs in the eastern part of the Kathmandu Valley coming out of carved stone spigots near shrines. There had been a rumor that since the earthquake, the water, which had previously been a trickle, was now rushing out. It was a sign from the gods, some said. Bodh thought this dubious and wanted to see for himself while introducing me to some of the men he knew who had worked in Afghanistan.

We parked near the shrine and walked around to see the springs. The water was indeed coming out fairly briskly, but with nothing to compare it with, it was difficult to conclude that the pace was miraculous. Bodh was even more skeptical than I was, convinced that if there had been any change, there must be a more geological explanation. Maybe the earthquake had simply shifted some of the rocks below.

Wandering up one of the small streets to the west of the shrine, we went to visit a man from whom Bodh had purchased land several years before. With inflation in Nepal rising and falling rapidly in recent decades, Bodh, like others who had earned money abroad, put his savings into the land, which was more likely to hold its value. While he was in the Indian Army, he had rented his land out. Since then, he had sold the plot as property prices around Kathmandu soared, but he remained on friendly terms with the families in the area, some of whom had also served in the Indian and British armies. These personal connections linked together many retired soldiers and, later, contractors.

When we arrived at the man's house, he embraced Bodh warmly, sending his wife off to gather some of the recent harvest from their garden to send back with us to Kathmandu. Unfortunately, he said, his brother who had worked in Afghanistan was out of town. I left my card and returned to Kathmandu, unsure if anything would come of it.

A few days later, I received a call from a number that I did not recognize. It was the man's brother. He invited Dawa and me to Matta Tirtha, and a week later we drove back to the village on Dawa's motorbike. This time, I was introduced to Yogendra and, a few minutes later, to another seven men, all of whom had worked in Afghanistan for between five and eight years. This conversation led to another and another and another, and Dawa and I returned to conduct interviews and simply visit our new friends in Matta Tirtha. The men in Matta Tirtha introduced us to fifteen other contractors in Kathmandu, Pokhara, and smaller towns across Nepal, and these contractors introduced us to still others through Facebook and LinkedIn. We found other contacts from other companies and occasionally serendipitously made new contacts, but ultimately, I discovered, it was this type of introduction that drove most of my research: one guy knew another guy, who knew another guy. As the weeks went by, our contact list grew, and so did the number and variety of contractors we interviewed.

As I met certain contractors from a specific company, they would introduce me to others in that company. If one had left to work elsewhere, that would be a lead to contacts in that other company as well. Over the next five months or so, Dawa and I interviewed approximately 120 Nepalis who had been in either Afghanistan or Iraq or were connected to the process of migrating to conflict zones in some way. Even after I left Nepal to follow up other cases in India, Georgia, the United Kingdom, and Turkey, I continued to build contacts by email and through Skype and phone conversations. I returned three more times over the course of the next year, each time meeting more and more Nepalis who had been part of the conflict while also following up with contacts the contractors had in India, the United Kingdom, and elsewhere. The contractors were all part of a wide social and professional web stretching across the globe, and our research slowly began tracing the various strands.

As I spoke to more and more people, their stories began to overlap and intertwine, building, expanding, and sometimes contradicting each other. And the more this happened, the more a picture began to emerge of how the Gurkhas had transformed into private contractors and what the war was like for a Nepali contractor in Afghanistan. The more conversations I had with contractors about Afghanistan, the more it became clear that these experiences abroad had reshaped life and politics back in Nepal.

CONSTITUTION DRAFTING

In Nepal, a country that had no formal involvement in the war in Afghanistan, the lines between the war and the ongoing political morass in the country were subtle. As I conducted more interviews, it became more apparent that the U.S. government's priorities in the region were directly related to Nepal's shifting political situation and that ongoing political upheaval certainly shaped the lives of the Nepali contractors who chose to venture to Afghanistan, helping to drive the migration of workers to Afghanistan, making them more willing to take risky positions in dangerous parts of the world.

In 2015, Nepal was still recovering from a ten-year Maoist insurgency that ravaged the countryside.[18] The Maoist rebels often sought quarter in villages, demanding supplies from local farmers, who were doubly punished when police came to arrest, and sometimes torture, them for aiding the enemy. Over 13,000 were killed, and 1,300 were still missing.[19]

The timing of the conflict meant that the war was largely ignored in most of the rest of the world. If the Maoist uprising had come a few years before, the war would have drawn international attention as a communist uprising with all the Cold War implications. Starting a few years later, the government probably could have benefited from labeling the insurgents terrorists and linking them to George W. Bush's "axis of evil."[20] As it was, the international media paid little attention to the upheaval in the mountainous country, which has a population similar to Afghanistan's but less geostrategic importance. Of the larger powers, only India seemed interested, though both the Maoists and the Nepali government remained wary of India's attentions.

And so, the war continued until 2006 when, worn down, both sides agreed to negotiate. The peace deal that ended the conflict called for an assembly, which was elected in 2008 to write a new constitution. The assembly immediately voted to end the monarchy by a 560 to 4 vote, but it struggled when it came to the actual drafting of the constitution.[21] The war was over, but no one was sure what was next.

A similar constitutional drafting process had taken place in Afghanistan in 2003, following the Bonn Accords. This brought various Afghan constituents together in Kabul to draft a new constitution. The three-week process was closely observed by international advisors from the U.N. Assistance Mission in Afghanistan, and the presence of tens of thousands of NATO troops in Afghanistan already helped drive the process.[22]

In contrast, the U.N. Mission in Nepal primarily oversaw the disarmament process of the Maoists. While it was involved in election monitoring, constitution writing was not in its mandate.[23] The constitutional assembly had little external pressure to complete its charter and failed to finalize a document, leaving the country in legal limbo.[24] By the time I arrived in 2015, the stalemate was wearing on people.

The Maoists had received a good amount of popular support in the initial election of 2008, soundly beating the older parties, which were seen as bogged down by nepotism and corruption. However, most of the public believed that Maoist leaders had used the last ten years to enrich their own family and friends more than to promote their agenda of social justice. Few could be spoken of as real populist leaders anymore, and the government was seen as primarily corrupt and inept by the majority of people I interviewed. Contractors like Yogendra and soldiers like Bodh were rarely from the elite families that ran much of the country, and they tended to speak with frustration about these politicians.

The political morass was contributing to Nepal's economic stagnation. The country remained incredibly poor, even while gross domestic product (GDP) had increased significantly in other countries in the region, such as India. Even Sri Lanka had seen significant economic growth following the end of its own civil war. The failure of any real industrial investment, declining agriculture due to low global prices, and concerns about instability meant that Nepalis increasingly looked abroad for employment. By fiscal year 2013–2014, approximately 30 percent of Nepal's GDP came from remittances sent home by over 500,000 Nepalis, a figure that does not include the thousands of Nepalis working in India, where no labor permits are needed and fewer records kept.[25] Coming from America, a country struggling with its own relationship with immigration, I found it striking to see how simply taken for granted migration was in this mountainous, rural country.

Nepalis looking for work abroad were most likely to be construction workers or maids in Malaysia or various Gulf countries, but a sizable number did end up seeking work in conflict zones, and the Maoist conflict encouraged the outflow of soldiers-turned-security-contractors to places like Iraq and Afghanistan. During various interviews, former Nepali Army soldiers told me they had left the army to search for contracting jobs because they did not want to be asked to fight against other Nepalis, or they had close friends or family members who were Maoists. On the other side of the conflict, following the peace treaty in 2006, some of the Maoist insurgents refused to reintegrate and simply left the country, looking for work elsewhere. Some of these essentially went from being freedom fighters to fighters working for a paycheck in other conflicts. The communist insurgency had been capitalized.

Countries in the Gulf benefited from the outflow of laborers working for minimal wages, and India saw the Nepali government's failings as helpful in exerting their influence across South Asia.[26] This all contrasted sharply with my experience in Afghanistan, where the international community had been deeply involved in elections and, later, the implementation of various aspects of the new government. [27] Afghan government officials were paired with international trainers to help "develop capacity." I watched these Afghan officials essentially ask for international permission to write their own policies. In contrast, few outsiders seemed very interested in whether the political elite in Nepal were setting up an equitable political and economic system that would benefit its people.

By 2015, the initial optimism from the peace deal of 2006 had faded as politicians bickered, the constitutional drafting process stalled, and the economy continued to stagnate. All of this was felt more acutely following the devastating earthquakes that spring. Talking about this period, Dawa recalled how dismayed he and many others had been by the slow response of the government. In comparison, many Nepalis

praised several youth groups who had sprung into action using a combination of social media and jumping into rubble removal and relief
work.[28] Once back in Kathmandu, Dawa and some of his friends from
a club they were part of started organizing small-scale relief work. This
swell of youth activism was embarrassing to many of the government
leaders who appeared increasingly incompetent. Their legitimacy erod-
ing fast, leaders agreed to a compromise that lifted the stalemate from
the constitutional drafting process.

I was in Kathmandu the evening the constitution was finally signed,
and the streets were busier than normal, and at one intersection, a
group of young people were painting a Nepali flag on the pavement,
but for the most part, the celebration was subdued. After decades of
agonizingly slow development and continued poverty, few were optimistic the new constitution would change things quickly.

BANDHA

The mild optimism of those in Kathmandu was not echoed across the
country, and in the southern plains of the country, known as the Terai,
the agreement, which seemed to favor those living in the central hills of
the country, was not met with enthusiasm. This discontent would soon
spread to the rest of the country.

The Terai, the fertile swath of flatland that hugs the border with
India in the country's south, is one of the most economically promising regions in the country, boasting both agriculture and new industrial
production. Until the early twentieth century, the area had been almost
uninhabitable, when the draining of the swampy terrain and the eradication of rampant malaria had improved the area in economic terms.
The Terai had also increasingly benefited from its close proximity to
India, making it more industrialized than other parts of the country.

Despite this economic growth, local inhabitants faced serious discrimination. The Terai is largely inhabited by several groups known
collectively as the Madhesis, who share many cultural similarities with

the Indian groups living across the border from the Terai. Despite making up almost a third of Nepal's population, Madhesis have been historically underrepresented in the government and military and faced social and political discrimination.[29]

This tension has only increased as Pahadis, or Nepalis from hill areas like Kathmandu, who have traditionally controlled the country politically and economically, have moved south to take advantage of the Terai's economic growth. While many Madhesis cross the border to work in India, Yogendra, like most of the other Nepali security contractors I interviewed, was from one of these hill groups that tended to dominate the ranks of contractors abroad, particularly those in well-paying jobs. This was part of the way in which place of origin, as well as socioeconomic status and caste, shaped the different ways that workers migrated.

These groups saw the new constitution as continuing the dominance of the plains by those from the mountains.[30] The Madhesis' cross-border ties with Indian communities also meant that the Indian government, which had seemed fairly uninterested in the Nepali constitutional drafting process, was now concerned that a key ally in Nepal might be about to lose political influence. It was rumored in Kathmandu that Indian diplomats were, either directly or indirectly, supporting the growing political protests in the south of the country.

In the days following ratification, several Madhesi groups protested in Kathmandu. The afternoon of one of the protests, I met with a local journalist at a rooftop café, and while we sat drinking tea, about a thousand or so protesters marched up the street and sat down, blocking traffic. Police in riot gear lined the road below us, but there was no real confrontation between the two. After an hour or so, the crowd began to disperse in a slow drizzle. The only people seriously upset seemed to be some of the drivers who had been caught in the jam. Several other marches were organized around the city, but they didn't have much of an effect. So over the next few days, Madhesi political parties turned to an older approach: the *bandha*.

Bandhas are essentially city-wide strikes, though in Nepali, the word can simply mean "closed" too. They were a particularly popular political tactic during the Maoist conflict, when a political party would declare a city shut down for the day. Young men from the party would enforce the strike early in the morning by stoning any cars or other vehicles that were out in the street. Such an approach could be very effective: stores and offices would close, and students would stay home from school. These protests helped the movement to bring down the king a decade earlier.

The first Madhesi bandha when I was in Nepal was actually called before the constitution had been passed, and on the day that I was moving into the apartment I planned to rent for the next six months. Going out to get a taxi was eerie. The streets were empty with almost no cars in sight. The guard at the front of the hotel I had been staying in suggested that maybe I should ride the hotel van, which had "Tourist" written on the front. Protesters usually allowed essential vehicles, such as ambulances, press vans, and, apparently, tourist vehicles, to pass through. So we loaded my two backpacks into the back of the van and headed out into the deserted streets.

In the weeks following the passing of the constitution, five more bandhas were held. It was never clear exactly when there would be one or who had called it, but generally rumors would circulate the night before, and by the next day, I would have to cancel my interview if it required driving or was at a restaurant. If it was in someone's home that was within walking distance, I would walk there. This meant that businesses closed, students stayed home from school, and the Nepali economy suffered further. It was particularly tough on poorer groups since they were more likely to depend on daily wages that weren't paid on bandha days.

Most of the contractors I spoke with seemed to greet questions about the bandha and Nepali national politics with frustration. Unlike in America, where military veterans are often seen as ideal politicians, the situation was very different for Nepalis who had fought in Afghanistan or elsewhere. Certainly some of this was that as contractors, they were not fighting for their country as much as they were fighting for profit, but more had to do with the ways in which the networks of Nepal's wealthy ruling elite were largely closed to those from what was perceived as rougher backgrounds. Fighting abroad did not establish their honor as much as it confirmed their status as outside the Nepali political system. Although Yogendra and Bodh were both well off by Nepali standards, they would never qualify as one of the elite.

Still recovering from his stroke, Bodh walked with his wife each morning at 5:00 a.m. and was hoping to be able to get back to the tennis courts soon. He, like others who had been to countries like Britain, knew what a more efficient government looked like. But former contractors had also seen places like Afghanistan and Iraq and realized things could be much worse. Still, the political stalemate cast a pall over most of our conversations and continued to fuel the desire of young people to leave the country.

Particularly for those who had been injured or more negatively affected by working in Afghanistan, however, the ineffectiveness of the government particularly added to their hopelessness and the sense that there was little they could do about the conditions they found themselves in. As I hiked around town to interviews, I found more and more cases of those who had been injured or harmed in some way while working abroad, and in their accounts, the government's ineffectiveness took on more meaning.

CHAPTER 3

ONE BLAST, MANY LIVES

CAMP PINNACLE

On the morning of July 2, 2013, around 4:30 a.m., about the time when most Afghans prepare to do their first set of prayers, a large truck filled with marble and explosives rammed into the gate of a private contracting compound east of the center of Kabul, known as Camp Pinnacle.[1] Sitting on his couch in a small city in eastern Nepal, going through some photos from his time in Afghanistan, Gopal told me that the explosion tore through the first-row housing units. Out of the dust from the pulverized marble, young Taliban fighters emerged firing AK-47s. Dazed DynCorp contractors, the primary occupants of the camp, poured out of their damaged units, some wearing only the underwear they had been sleeping in.

During the initial explosion, Gopal, a short, powerfully built man with a crew cut, was asleep in his cube toward the back of the compound. After the initial blast, he was most struck by the overwhelming amount of dust in the air. He had been involved in other attacks before, he said, but the pulverized marble in the air made it difficult to get a sense of what was going on at all in the choking haze, making this attack seem worse. As he continued his story, he described how he stumbled out into the hall and found an American contractor lying outside his door. The American appeared to have a partially severed arm.

Hoisting him up, the two made their way down to a bunker where there was a first aid kit. A few other contractors were already gathered there, though with his ears ringing and the dust swirling even more thickly as it began to settle, it was difficult to recognize anyone.

The layout of the compound added to the confusion. The compounds that tended to house many of the contractors were generally made up of tight rows of identical, cheaply built temporary housing linked by paths of thick gravel. I had visited friends who worked for the United Nations and other international groups and lived on similar compounds in Kabul. One of the benefits of such visits was that the dining halls often served hamburgers and other American comfort foods that were not easy to find off military bases, but they could also be confusing places, where it was easy to get lost. The buildings were monotonous, and the walls that towered over the compound made it impossible to see out and difficult to really get your bearings inside. The smoke and shooting made this worse, Gopal said.

After spending several minutes applying a tourniquet to stop the bleeding from the American's wound, Gopal pulled out his pistol and cautiously stuck his head outside the bunker. The explosion had torn apart the flimsy housing units. He stumbled on a gravel walkway. Mixed in with the stones were now all sorts of debris. The sides of a couple of the housing units were ripped cleanly off, and he could see directly into their bedrooms. In some of the less damaged living quarters, doors had been pushed in, wedged into their frames and trapping the occupants in their rooms. I had been across from a compound that had had a similar attack in 2010 in Paktya Province, and I remembered how disorienting the minutes following the initial explosion were. It was terrifying not knowing whether the attack was over or ongoing.

Shooting continued sporadically, but by this point, it appeared that the attackers had been killed, though it was not clear exactly who fired the shots that killed them. Before the explosion, several of Gopal's friends who also worked for DynCorp and lived in the compound were preparing to escort a convoy to the airport for an early-morning flight.

They were lined up with their gear on, ready to leave, which meant that they were able to respond to the attack quickly and helped ensure that the shooters did not do even more damage.

Occasional shots heard over the course of the next several hours seemed to be coming primarily from Afghan police, who were not certain if the attackers had actually been killed and the continued firing later lent credence to the rumor that some of the shooters had escaped. As a result of the attack, the main road leading east out of Kabul was shut down for much of the morning, clogged with police vehicles and hazy from the smoking ruins of the truck. Contractors were evacuated from the scene.

Over the course of the attack, five international contractors, primarily from India, were killed, along with several Afghans; four of Gopal's friends were badly wounded.[2] The initial reports said that Nepalis had been killed, though it turned out that these were actually Indian citizens of Nepali ethnicity. Nepali families back at home heard about the attack almost instantly through Facebook and frantically tried to find out more information. Other contractors I interviewed said that confusing reports caused concern for several days because many of the wounded were unable to Skype or email home.

Despite the damage, the attack was only lightly covered in the international press. Most in the media saw the attack as a small part of a larger spike in violence across the country. The Taliban had just announced it would be opening an office in Qatar, and some factions in the group who opposed this move were trying to disrupt the effort toward political reconciliation with an increase in attacks. Even for those Taliban in support of the office, turning toward diplomacy could have been perceived as a sign of weakness. So, over those weeks in July 2013, there was an upturn in all sorts of insurgent activity across the country.

With so much other news coming out about the conflict at that point, coverage of this medium-sized attack in the international press was mostly a few brief stories, none longer than five or six paragraphs and none interviewing anyone involved. The limited coverage probably had to do with the fact that the men killed were contractors, not soldiers, and none of the contractors were American or European. In Kabul, Afghan reporting was more extensive, including interviews from police who had responded to the attack and local onlookers, but it still was only the second story that night on the news at TV One, following a story about an attack in the city of Jalalabad.[3]

The attack on the compound had much longer-lasting consequences for those who were there. It also demonstrates some of the dangers that contractors faced and the difficulty they had getting the follow-up medical treatment they needed.

After interviewing Gopal and eventually a dozen other contractors who were in the compound during the attack, I started trying to put the pieces of the story together, but this was tricky. The news reports sometimes contradicted each other and did not entirely line up with the accounts of survivors. Some of these discrepancies were minor, such as disagreement over whether the truck was an eighteen- or twenty-four-wheeler, and since it was pretty much obliterated, there seemed no way to solve that question. Some of the disagreements over what became known among the survivors as the Pinnacle attack were bigger, however, and many reports said that there were three Taliban attackers, all of whom were killed, while Gopal assured me there were four and that one had escaped during the shooting.[4]

After the shooting ended, most of those who were critically wounded were taken by helicopter to the Craig Joint Theater Hospital at Bagram Airbase, thirty miles to the north. Those with lesser injuries were moved to a nearby compound named Baron, which primarily provided housing for Chemonics, another large contractor. (This was also one of the few instances I ever heard of contracting companies cooperating without apparent financial incentive.)

Arriving at the Baron compound, Gopal noticed that his feet were bleeding badly. He must have cut them on glass on the way to the bunker since he had not been wearing shoes, he said. At Baron, a medic used long forceps to remove the glass and bandaged his feet. By that afternoon, he was back at work, providing protection for a convoy taking some Americans on a United States Agency for International Development contract to a meeting on the other side of Kabul.

Several days later, some of the contractors were allowed to return to the site of the blast. Where the explosion had detonated, there was now a twenty-foot crater. One of the contractors used a video camera to document the damage. Sitting in his cozy apartment back in Nepal, Gopal showed me the shaky footage on his laptop, pointing out who had lived in various buildings. Some units looked fairly undamaged, while others were almost completely leveled. On a piece of paper, he sketched out the layout of the camp and the paths he thought the shooters had taken.

Over the course of the afternoon, Gopal insisted on not just feeding Dawa and me three meals as we spoke, but he also said that he wanted to take us on a tour of his town. As we set out, Gopal pulled on a hat from The Duck and Cover, the U.S. embassy bar in Kabul. Wandering around, looking at a herd of deer and several scruffy coyotes at the nearby zoo, he continued to reflect back on the attack and described picking through the rumble and confirming that most of his possessions had been destroyed. He would have to eventually request a new passport and other documentation, which were express-mailed to him in Kabul. The contractor had spare supplies, so he was quickly issued new equipment and a clean uniform and went straight back to guarding convoys.

Others, he said, were not so lucky.

CHECK OUT ANYTIME YOU LIKE

It was not easy gathering information on attacks on contractors, I found. The first few times I heard about the attack, I was sure that Gopal and others were saying that the compound was "Camp Pinochle," and I was not able to find any record of the attacks. I was a little surprised a camp had been named after a card game until I realized, after interviewing a few of those there, that the camp was actually called "Pinnacle." It was also not a military camp as much as it was a heavily guarded compound that housed the North Gate Hotel run by the Pinnacle Group, an Afghan firm. To complicate things, I later discovered from a subsequent lawsuit in Texas that the Pinnacle Group was technically owned by a company named C3PO.[5] Such confusing layers of contracting and subcontracting were common.

Pinnacle was on the busy Jalalabad Road, which led east out of Kabul. The relative openness of this area attracted contracting companies looking to build new compounds that weren't concerned with being located closer to the center of the city. This created a confusing gauntlet of similar-looking bases and compounds on the road, which had consequences for the contractors inside. Particularly for those who did static security along the walls and towers and at the entrance gate, there was little information about the evolving political or security situation outside. I spoke with several contractors who had at various times been confronted by angry crowds of Afghans outside their compound, yet they had no clue what the Afghans were protesting. A quick chat with some of the local Afghan staff could have helped, though in my experience, these sources seemed to be rarely consulted, and in some of the compounds where I interviewed people, the rule was that all local staff had to leave the compound by a set time, such as 5:00 p.m. In comparison, security contractors in supervisory positions had more information, but often a distorted view of what was "normal" and "safe" in Afghanistan.

On more than one occasion, contractors who worked in the area around Jalalabad Road told me that it one of the safer parts of the city, when it was in fact the place where most attacks were taking place. For the security managers, the large number of guns and having their "clients" in walled compounds made them far easier to keep track of, which suggested safety, while living in an ordinary house with Afghan neighbors, as my housemates and I had, seemed insane. Yet the anonymous walled compounds attracted attacks. In cases like this, a strong security presence actually seemed to create insecurity, while things remained quiet on my small residential side-street. Nevertheless, in the distorted world of security contracting, this heavily fortified road was deemed "safe."

Pinnacle billed itself as "a full service and live support solution company." It was "able to mobilize quickly to fit your needs. Offering rapid deployment camps, permanent encampments and solutions for any catering scenarios . . . even in the harshest of environments." According to their advertisement on YouTube, it provided "independent power, water, sewer, and, of course, security" and used images of "the North-gate secure hotel," which was inside Camp Pinnacle, as evidence of its "critical knowledge of just what it takes to satisfy the demands of personnel operating in all environments for private and military contracts" all at "relatively low cost . . . so that your staff can concentrate on what is vital to the operation of your company."[6]

The advertisement showed high-walled compounds, with sleek interior shots of modern meeting rooms, a computer center, a billiard room, and a dining hall. Playing into the racial stereotypes that were common among those involved in the conflict, although the company was Afghan owned, almost everyone in the video advertisement was white except a few Afghans and other South Asians who did not speak but were clearly taking orders or listening and nodding to Americans and Europeans who were giving instructions.

The Pinnacle compound catered to American and European contractors as well as Nepali security guards, which meant that the accommodations there were much nicer than many of the other contracting compounds where only Nepalis stayed. Particularly for those in the south of the country, compounds were hot and dusty, and the ones in newly established compounds relied on tents and porta-potties. Others were essentially in plywood shacks that housed rows of bunk beds pushed against each other. As one of the key primary contractors for the U.S. government, DynCorp had a lot more money to spend than smaller contractors, so Gopal and his friends ended up in these nicer quarters.

As is typical in security contracting, Pinnacle was an independent contractor and was not actually owned by DynCorp, even though Dyn-Corp was both the primary resident and supplied the security for the compound. After the attack, with the front wall collapsed and many of the housing units in disarray, DynCorp moved its personnel to a new compound. Pinnacle responded by suing DynCorp for breach of contract (it had a $12 million lease, Pinnacle argued). DynCorp responded by pointing to the obvious security issues with the current arrangement, which Pinnacle claimed were DynCorp's fault, since one of its guards had let the truck through the initial security checkpoint.

In response to DynCorp complaints, Pinnacle upgraded its security, but DynCorp refused to return to the camp. Pinnacle sued it for compensation for the upgrades it had already made. Furthermore, Pinnacle argued, an "independent security assessment revealed that [the Baron Hotel, where DynCorp had moved its staff] was not any safer than North Gate."[7] The court documents did not mention how this "independent" assessment was made and in the competitive world of contracting, it would have been difficult to find a group that had no interest in the conflict between C3PO and DynCorp, since most other companies in town were hoping to partner with or were in direct competition with these groups. Notably absent from these American court documents outlining the battle between a multimillion-dollar and a multibillion-dollar corporation over who was responsible for paying for an upgraded facility that was not used was any account of anyone who had been there during the attack.

While these companies battled back and forth about who owed what, most of those injured in the attack had a much more difficult time getting compensation for their injuries.

AFTER THE BLAST

A few weeks after interviewing Gopal, I climbed an incline alongside an idyllic mountain lake. Almost at the top, Bhim took my hand and held it against his hip. Through his pants, with each step, I felt the ball of his femur popping out of the socket, pushing up on the skin of his waist. It was unclear to me how he was walking at all since he seemed to be missing most of his hip bone. I felt queasy, though Bhim only gave me a slight nod. It almost seemed to be verging on a smile, as if to say, "See, I told you it was bad," and we continued on to the top of the hill, with Bhim limping only slightly.

Bhim looked older and was much thinner than Gopal. After his initial injury in the Pinnacle blast, a metal plate had been inserted into his pelvis, and he was dismissed by DynCorp. At first, his recovery had been smooth, but then the joint started to hurt him again. A Nepali doctor told him that there was some risk the metal in his hip might cause cancer or some other serious condition. Concerned that his insurance might expire, Bhim decided to have the plate removed immediately. Since then, his pain had increased, and his limp had worsened. He now wore a brace on the lower half of his leg, and as we walked by the lake, he rolled up his pant leg to demonstrate how he could push his toes downward but could no longer pull them back up.

On the lakeside promenade, he pointed to the far shore where he grew up in a small village, several hours' walk from the road that ended here by the lake. Talking about those days, Bhim's eyes looked sunken and tired. Around us, Nepali families, mostly from the city, picnicked raucously. Unlike the larger Fewa Lake near Pokhara, this smaller lake was not well known enough to attract international tourists. Music blasted from speakers, and a group of young Nepalis clapped and laughed as some of the mothers and grandmothers in the group got up to dance. Bhim, however, was somber, and although he greeted some other men on the road who were sitting at the local café, we did not stop and chat. He asked if Dawa and I wanted to go out in a boat to see the lake better, but his tone suggested that he would not enjoy such an excursion. Back at his house, sitting on its roof, as he finished telling us the story of his time in Afghanistan, he said, "Now my own life is over," and as he looked out toward the lake, he told us, "I only have hope for my son."

Bhim had been in the Nepali Army for eighteen years before leaving to take a contracting job in Afghanistan. His résumé was particularly appealing to DynCorp because he had experience abroad, serving in the contingent of Nepali troops assigned as U.N. peacekeepers in Lebanon during the 1990s. In Afghanistan, he worked for another security company before a friend recommended that he move to DynCorp, where the pay was better. He worked for most of the next six years primarily as a static guard at the U.N. compound in Gardez, in the southeast of the country. I had visited the office that he guarded in 2009 and 2010, and we discussed security there, since this was a much more dangerous assignment than working in Kabul. Bhim was at Camp Pinnacle the day of the attack only because he was spending the night there while traveling through Kabul to go on leave the next morning.

At the time of the explosion, he said, he was asleep. Luckily, he fell off his bed in such a way that the bed prevented the collapsing ceiling from crushing the upper half of his body. But his legs were pinned under the rubble. He lost consciousness and told me that the next few hours and days were confusing as he slipped in and out of a hazy reality.

Bhim thought the doctors who initially evaluated him apparently did not feel that he was injured enough to evacuate him to Bagram with those who had serious injuries. Instead he was taken, like Gopal, to the Baron clinic, where he was originally checked. Here, he said, someone found that his internal injuries were severe enough to warrant being transferred to the military base, but Bhim doesn't remember who decided this. He was in pain, but the lack of urgency of those treating him was reassuring; perhaps it was something minor.

After being transported to Bagram where the other wounded contractors were sent, they performed some other checks but no surgery. It was not clear to him, he said, what was wrong with his legs or what was going to happen next. All he knew was that he was having trouble moving his legs, but they kept him on enough medication that the pain was bearable.

Instead of being operated on, he was flown to New Delhi, where he remained for three days. They were headed back to Kathmandu, but, he was told in Delhi, the weather was too cloudy to fly. He spent the three days sleeping at a clinic in India and the doctors still didn't perform any procedure on him there. He was given more medicine for the pain, which helped some, but it was increasingly apparent to him that something was seriously wrong with his leg. Back in Kathmandu, there was another series of checks, and according to the hospital records he showed me, the Nepali doctors finally operated on his pelvis twelve days after the initial blast, inserting a metal plate.

Bhim had few documents about the injury or his treatment, which we flipped through sitting on his roof (in contrast with these, there were plenty of other documents from the DynCorp bureaucracy that he had collected before the attack; his binder was thick with training certificates). He did have a report from the hospital in Kathmandu, however, and I noticed that it was dated a full eight days after the date on his DynCorp document stating his weapon had been turned in as part of his discharge. The same document noted that he would receive the balance of his wages in his next paycheck, but that he was no longer considered an active employee and would receive no more payment for his work. He had essentially been fired eight days before receiving any real treatment for his injuries from the blast.

Bhim called for his son, a quiet, studious-looking boy, to bring up drinks. While telling his story, he continued looking through his folder of documents. While Bhim's English was not great, he understood some of the documents clearly, such as the papers from DynCorp declaring him no longer medically fit for employment. Others were more puzzling, and after we chatted about them, he let me send some copies to a friend of mine who is a doctor. Unfortunately, at this point, the doctor said, it would be seriously difficult for Bhim to put together a case for neglect or mistreatment on the part of DynCorp or any of the clinics that had treated him. On occasion, a doctor might wait for the swelling to go down before operating, my friend told me. Still, she said, the long wait for Bhim's treatment did seem odd.

For Bhim, the delay and eventual flight to Kathmandu for the operation was difficult to interpret as anything other than a cost-saving measure. It's common for rich Nepalis to fly in the opposite direction and use hospitals in Delhi, which are better equipped than those in Kathmandu. Why hadn't Bhim been treated in Delhi or even at Bagram? He wasn't sure, and the delay in receiving treatment continued to bother him, particularly given that DynCorp had not delayed his dismissal. He phrased his answers carefully and clearly didn't want to accuse his former employers of anything, but he was suspicious about how he had been treated.

The insurance payments were similarly confusing to both Bhim and me. The initial insurance company that was supposed to be paying Bhim's continuing medical expenses had been bought by a different company, and while he had papers for the first, the second had not provided him with an insurance card or other documentation. A representative had called, and said that the only way he could get his medical treatment paid for was by traveling to Kathmandu and going to the hospital accompanied by a representative of the company, who also worked as a travel agent, according to his business card, which Bhim did have. The agent would then pay the bills out of his pocket. Bhim found this agent vexing but was also worried if he was considered "healed," the payments would stop.

Throughout our conversation, Bhim never seemed angry as much as he seemed sad and baffled by his lack of luck. In this small and beautiful lakeside town, it felt that there were few options left for him and no one to help him or even to really talk about the experience much with. No one in Nepal spoke of posttraumatic stress disorder, and even if they had, Bhim was not going to find anyone to treat him or have a means for paying for that treatment. For contractors, when the contract is over, it is over. This was particularly true in Nepal, where there wasn't even the idea of some type of legal recourse.

At the end of our visit, Bhim walked us down the road to the bus stop and waited with us. Dawa and I usually debriefed animatedly after interviews comparing impressions and recording key details, but this time we just sat silently. There seemed to be nothing to say. After the bus started moving, Bhim stood by the side of the road, waving as we disappeared into the distance.

AN AMERICAN SAVED ME

Jivan's story was different. He wasn't a security contractor, but a supervisor of some of the mechanics who worked at DynCorp, when the blast at Camp Pinnacle occurred. Hearing about him through one of his friends at DynCorp, I assumed his story would probably be similar to Bhim's. It was not.

The first strange thing about my interview of Jivan was that after contacting him on the phone, he refused to give Dawa and me his address. Instead, he said, he would send a private driver to pick us up, which he did midmorning the next day. Once in the car and on our way, he called again to make sure that we were comfortable and that the car had shown up on time. When we arrived at his neat and modest house on the outskirts of town, Jivan took us into his living room, which was dominated by family photos. In several, his eleven-year-old son featured prominently. The boy was studying at an elite private school just over the border in Darjeeling, India, Jivan told us proudly.

As we sat, his wife brought out plates of food for Dawa and me. Despite the early hour, she offered sausage, fried chicken, potatoes, and salad. Jivan and his wife both sat with us, pleasantly refusing to eat, a Nepali display of supreme hospitality that we nevertheless found slightly uncomfortable.

As we ate, Jivan asked if I would like some beer to go with the meal. Since it was still midmorning, I declined. He looked so disappointed, however, and when he asked again, I said I would take a beer if he were joining me as well. This seemed to make him much happier, so we sat sipping our beer while he told us that he had been a schoolteacher before taking the job at DynCorp, pausing every few minutes as his wife brought out more dishes and piled additional pieces of chicken onto our plates.

Jivan said that, like Bhim, he was in his room at the time of the attack and had no memory of the explosion. He only remembered lying in the rubble, calling out for help. Out of smoke emerged an American, he recalled, making it sound almost like an action movie. The man lifted him up, pulled him onto his back, and carried him away from the shooting. Considering that Jivan was over six feet tall and broad, this was an impressive feat.

As he was being carried, Jivan said he could not see any of the attackers but could hear the firing continuing around then. His face was badly swollen and it hurt, but initially the pain was not severe. Once the two were safely with the paramedics on the side of the compound away from the attack, the American left him. This man, Jivan repeated on several occasions during the recounting of his story, was the only reason that he survived, and he was disappointed that he had never been able to thank him. He didn't know his name, he sighed.

After the shooting died down, the paramedics flew Jivan immediately to Bagram. He was clearly given some priority or status that Bhim did not qualify for, and at Bagram, the doctors began treatment on the part of his face that had been injured in the blast. By this point, the pain was worse, and he realized he could not see out of his left eye. He said he was surprised when the doctors not only treated him well, but as soon as he was lucid, a doctor brought him a laptop so that he could Skype with his family at home and let them know that he was all right. Once connected, he assured them he was okay despite the rumors that had been going around about Nepalis killed in the blast.

His wife, who was sitting next to him, jumped in and said how pleased she had been when she saw the Skype call on the computer. Still, the conversation was traumatic, she said. The Internet connection was not good, and at first she didn't even recognize her husband since his face was so swollen. She was left worrying about the type of condition he would come home in despite his reassurances.

From Bagram, DynCorp sent him back to Nepal. He had lost his left eye, but the scarring on his face was not particularly noticeable. He also walked with the slightest limp and had some hearing damage as well. To assist him, his wife sat next to him, often subtly rephrasing and repeating questions we asked in case he didn't hear them correctly.

Because of his injuries, a DynCorp representative contacted him, offering him either continual payments over the next years or a lump-sum payout of $100,000 as compensation for his injuries. He had decided to take all the money at once. He still lived in the same house he had lived in before going to Afghanistan but had made a few renovations. He invested most of the money in starting a dry goods company; registry books clogged his desk, and a supply truck was parked in the driveway.

Later in our conversation, thinking back to some of the problems that Bhim had, I asked Jivan how he felt he had been treated by Dyn-Corp. He looked at me, surprised at the question, and said, spreading his arms wide: "DynCorp made me rich."

As his story wound down, Dawa and I prepared to leave, but Jivan was reluctant to let us go. He kept repeating how indebted he felt to America. He took us around to the back of his house and showed us his well-organized garden, filled with squash and tomatoes. We sat in the garden for a while and watched some of his chickens peck aimlessly while Jivan tried to convince us to stay for more beer. He wanted to make sure that we had enjoyed our visit. With other appointments scheduled, we finally made our excuses, saying goodbye to Jivan and his wife, and got back into the car that Jivan had arranged for us.

As we drove away, Dawa commented that Jivan seemed honored to be visited by an American. His hospitality had been almost overwhelming. I could not help but think that perhaps he did much of it because he hadn't had the chance to thank the American who had pulled him out of the rubble. Still, Dawa, concluded, $100,000 seemed like a steep price for sacrificing an eye.

I got another call about an hour later. It was Jivan checking to make sure that we had reached our next destination safely and asking if there was anything he could do for us.

CHAPTER 4

COSTS AND COMPENSATION

OTHER INJURIES

As I continued interviewing those who had been in the camp during the blast, it was striking how unique each experience seemed to be, so it was difficult to draw many conclusions. Some contractors were bitter, but others, like Jivan, seemed almost cheerful about the experience. While some accounts were more traumatic, most saw the attack as simply part of the job. Most were satisfied with DynCorp's compensation, but others were not. Increasingly, I tried to get a sense of the calculus that the workers used to judge these experiences. How did they feel about their treatment? What compensation was available to them? Did they feel it was just?

I also became increasingly convinced that anthropological techniques, relying on interviews and exploring these issues from the ground up, provided more nuance and insights than more data-driven statistical studies that were prevalent in most studies of contracting. A finergrained approach was need to understand the ambiguity for many of these contractors, who tried to balance the financial benefits of contracting with the risks of injury and exploitation.

Bipul, for example, told a story that seemed to fall between Jivan and Bhim's accounts. As we sat in a dark café near Kathmandu's main bus station, he described how he had been particularly unlucky. He spent eight years in the country working for DynCorp, based almost exclusively at the DynCorp compound in Herat, where Yogendra also worked. That camp had been closed, and he arrived at Camp Pinnacle only two days before the blast.

After the initial blast, he was also pinned under rubble. The shooting he heard was so intense that he assumed that at least twenty Taliban had stormed the compound. As he lay there waiting for help to arrive, he lost feeling in his legs and became convinced that he was going to lose one, if not both, of them. In the rubble, he said, he could hear his cell phone ringing, but despite his groping, he couldn't find it. (Later, Bipul found out that several of his fellow Nepali contractors who had retreated to the bunkers had been trying to contact him. Unable to reach him, his friends assumed that he was dead and his name was reported in the Nepali media as one of those killed in the attack.)[1]

After what Bipul said felt like hours later, several other Nepali contractors found him. Fear of unexploded bombs had slowed clearance of the rubble. The building he was in had only partially collapsed around him, and he was eventually dragged out through the only small opening. Several medics treated him and put him into the back of a Humvee. He was then flown to Bagram Airbase, he said, where the medics gave him some painkillers that made things blurry. He was unconscious for much of the coming days. He spent five days at Bagram before he was deemed fit enough to fly. His return flight flew through Delhi, but did not stop there as Bhim had. It was only once he was back in Kathmandu that he realized that during the period when he was unconscious, he had been operated on and now had a metal rod fused to his right femur.

Bipul spent an additional eighteen months in the hospital in Kathmandu recovering and learning to walk again, with the insurer paying all of the bills while he was in the hospital. Unlike Jivan, he chose initially to take monthly payments from the insurance company, afraid that he would spend it faster if he took it all at once. But after the April 2015 earthquake destroyed his family's home, he decided to take the rest as a lump payment to pay for rebuilding his home. Like Bhim, he was also afraid that at some point, the company might simply stop paying him and there would be nothing he could do about it.

The metal rod in his leg still gave him some problems, he said, straightening the leg a few times as he spoke as if to demonstrate the tightness. Despite this, he jogged every morning and was as fit as he'd ever been, he said. Toward the end of the interview, he asked me if I had any American contacts at DynCorp. At first, I thought that he was looking for more disability payments or perhaps considering legal action, but it soon became clear that what he really wanted was to get his job back. He demonstrated how he was ready to return to Afghanistan by jumping up and taking a few quick steps before returning to the table. He also wondered aloud to us whether removing the steel from his leg might not help his chances of getting rehired. When he emailed his old boss, he got no response, he said with an unhappy expression.

He had become convinced while trapped beneath the rubble that he was going to die, which meant that he was grateful to be alive at all, he explained. Since he was receiving a pension from the Nepali Army, he was not interested in taking a low-paying job in Nepal. He still had some of the insurance funds left, so it didn't seem to make sense to him to work unless he could find something higher paying abroad. For the time being, he decided, he would take his morning jogs, try to strengthen his leg, and see if any better opportunities came along.

COMPENSATION

After listening to Jivan, Bhim, Bipul, and others who received various forms of compensation from DynCorp and other contracting companies after sustaining injuries, I was confused. The various accounts and the sums they mentioned seemed random. Some received large sums, others nothing at all, and I couldn't tell how much of this was a systematic problem and how much was a question of Nepali contractors not being aware of their rights.

An activist I interviewed in Kathmandu mentioned an American lawyer who became involved in some lawsuits involving Nepali contractors after an incident in which twelve Nepali contractors were killed in Iraq in 2004. The contractors had been employed by Daoud & Partners, a Jordanian company that worked for KBR (Kellogg, Brown and Root, KBR, at that point was a Halliburton subsidiary). According to some reports, the twelve Nepalis had not even known that they were going to work in Iraq.[2]

The killing of these contractors sparked protests and several riots in Kathmandu in September 2004, during which protesters ransacked the offices of several manpower firms that sent workers abroad, as well as damaging a local Nepali mosque and businesses associated with Nepali workers in the Gulf, such as Gulf-based airlines.[3] These events were remarkable because they were the only real evidence of serious anticontracting protests that I found in Nepal during my research.

I eventually tracked down Matt Handley, the lawyer for the families of the contractors who were killed, and we emailed back and forth while I was in Nepal. Eight months later, back in the States, I finally had the opportunity to visit his office and have a longer conversation.

Matt himself was not a disinterested party. He had served in the Peace Corps in Nepal and had developed deep ties to the country. The thing that he said frustrated most was the ways in which Nepal's poverty put people in positions to be exploited, whether it meant contracting in war zones for low wages or having a kidney harvested and sold on the black market.

For migrant contractors, the upside to contracting on U.S. government dollars, as opposed to receiving funds from other countries, he explained, was that workers were then protected under certain American laws, particularly the Defense Base Act. The act was established in the wake of World War II, meant primarily to provide protection for U.S. civilians working on Marshall Plan projects in Germany. Workers who were injured while working on these projects were then legally entitled to medical treatment and compensation. In the case of death on the job, immediate family members are entitled to compensation based on a complex formula that includes the salary of the contractor and the number of years of work that person would have been expected to do, including potential salary increases. For the twelve killed in Iraq, this meant primary beneficiaries collected some $223 a month for the rest of their lives, or around $150,000.[4] In other cases these figures could be much, much higher. Matt represented an Iraqi family recently that received more than $1 million through a settlement he arranged.

The wording of the act is fairly expansive, Matt explained, and did not state that the worker needed to be an American citizen. This meant that contractors, Nepalis and others in Afghanistan and Iraq who were working on projects funded by U.S. government dollars, were protected by the act, even if they weren't the primary contractors, but were on sub- or sub-subcontracts, all the way down.[5]

The problem was that the act left to the contracting companies the obligation of reporting and filing claims under the Defense Base Act. The costs of claims were expected to be covered by insurance, but the companies still seemed concerned by the potential exposure that the press attached to these figures could create.[6] So there was an incentive for companies not to report casualties and send the wounded home with cash payments in the hope that their cases would not be reported.[7] If Matt or one of the handful of other lawyers working on such cases heard of them, they could represent the wounded or the family of the deceased. But more often than not, once these workers were back in Nepal, with little documentation of their injuries, they were not able to reach out to an American lawyer to assist them or, more likely, did not even know they had the right to such compensation. None of the contractors I interviewed, for instance, ever mentioned the act, and it was not clear how the families of those killed would learn about their rights either.

Matt put me in touch with a British reporter who worked on cases of companies avoiding the obligations of the Defense Base Act. We spoke several times over the next couple of months, trying to come up with better ways for tracking cases of injury among contractors. The issue she kept running into while researching the story was that it was so easy to avoid detection: companies would just send injured workers home and there would be no public record of the incident. The only reliable approach we could come up with was to find a specific attack reported in the media, like the attack on Camp Pinnacle, and then to try to track down those involved. This is labor intensive, however, and for attacks not reported in the media, it would be rather easy to avoid detection entirely.

Still, the Defense Base Act was by far the simplest way to for workers to protect themselves legally, Matt explained. In cases of negligence or potential human trafficking, things were even more difficult. In one case, Matt had been able to clearly show that the worker had been trafficked to Iraq, but since the worker was not American, had not been trafficked in the United States, and had not been trafficked by an American company (it was a subcontracting company of an American company), the judge threw the case out. That the trafficking was paid for with U.S. government funds was not enough to give the judge jurisdiction over the case or to do anything to rein in practices that were being paid for by U.S. taxpayers and facilitated by U.S. companies.

In the case of Jivan and others I interviewed who had received some compensation, it was impossible to tell whether they had received the proper compensation through the Defense Base Act. They received no paperwork, just cash. When I emailed DynCorp to ask about the case, I received no reply. The only representative in Nepal I could find was the travel agent in charge of Bhim's payments, but he refused to speak to me and I didn't want to push the issue since it could have caused problems for Bhim's continued coverage.

Looking at the figures and the details of the cases, Matt suspected that they had not been formally registered in the Defense Base Act— the amounts of payment seemed too low compared to what he had seen in other cases—but it was impossible to tell. In the case of some other contractors working for smaller firms that I interviewed, no payments

were made at all. Most cases, however, were ambiguous, and this became a theme in the contractors' stories. The workers themselves back
in Nepal were not in a position to reach out to helpful resources like
Matt, and in the meantime, everyone else involved seemed to have a
real desire to forget that attacks like the Pinnacle ever happened.

STORIES LEFT UNTOLD

There are several reasons the story of the Nepalis in the Pinnacle blast
was not better told.

First, private contractors already had a bad name. In recent years,
DynCorp and other large contractors sustained a large amount of criticism in the international press. Much of the criticism of private security
contractors during the height of the war in Iraq was aimed at Blackwater, which gained notoriety for their involvement in numerous incidents
including the killing of civilians and inappropriate use of force.[8] This
culminated in 2007 when a team of Blackwater contractors, protecting
a U.S. embassy convoy in Baghdad, opened fire in Nissour Square, killing seventeen Iraqi civilians.[9] This incident led to a series of congressional hearings and eventually legislation providing more congressional
oversight of contractors. While these incidents were embarrassing for a
few firms, little changed. Companies were still largely expected to self-
regulate, and Blackwater rebranded itself, changing its name to Xe and
later Academi.[10] Other contractors simply moved from a company that
had fallen out a favor with the U.S. government to whatever company
was replacing it.

The uproar seemed to teach contracting firms that it was better to
avoid any sort of press in order to survive, and the involvement of contractors in various incidents continued largely unremarked. Few noted,
for example, that many of those involved in the torture of prisoners at
Abu Ghraib were not actually U.S. soldiers but contractors working for
CACI and Titan, firms based respectively in Arlington, Virginia, and
San Diego.[11] Both companies largely avoided public condemnation in

the aftermath, and even as the court cases dragged on against the two companies, Titan secured a $170 million contract to do information technology for the North American Aerospace Defense Command.[12] This was even more remarkable considering the deep involvement of contractors in almost all aspects of the scandal from translation to allegedly committing acts of torture. Later government assessments of the scandal revealed systematic faults beyond the actions of a few rogue individuals, such as neither the U.S. military nor CACI "require[d] contract interrogators to be trained in military integration procedures, policy and doctrine."[13]

For the contracting companies, giving the injured Nepalis cash and handing them over to a secondary insurance company was a good way to have the problem dealt with quietly. Keeping the wounded contractors in Afghanistan meant they would be able to remind others of the incident and might end up talking with a foreign journalist. Back in their villages in Nepal, this was much less likely to happen. In fact, only a couple of the Nepali contractors I interviewed had ever spoken with a journalist about their time in Afghanistan. Once back home, many reported that their American and European colleagues they had previously emailed with stopped responding to their emails or defriended them on Facebook.

Just like that, they were forgotten.

Another reason the story of the Pinnacle attack was not widely known was that none of the governments involved seemed to have any real concern about these contractors. The Nepali government provided them with no support in Afghanistan, the American government was little concerned with the fate of non-American contractors, and the Afghan government provided visas only as long as the right palms were greased.

For the American military, the attack occurred at a critical junction during the surge. The Obama administration was waging a media campaign in both America and Afghanistan, trying to convince American voters to continue to support the war efforts and to convince Afghans that it was better for them to back the Karzai government than the Taliban. Casualty counts were being carefully kept of American soldiers in the media and were a major factor in turning the American public against the war in Iraq. The presence of all these international contractors helped since there were now more targets for potential Taliban bombers, targets that were cheaper in terms of both their salaries and the potential for political blowback.

At the same time, the Nepali government had a reason not to press the issue diplomatically and turned a blind eye to the undocumented migration to Afghanistan. The remittances that these contractors were sending back home filled an economic demand in a country that was struggling. The $4 billion sent home by Nepali workers in places like Afghanistan constituted 30 percent of the country's gross domestic product.[14]

The lack of reporting on the Pinnacle attack was not simply a story the media missed. Contracting companies and governments carefully downplayed the danger to contractors, while paying just enough attention to make sure they didn't receive any bad press. Compensations seemed almost deliberately designed to keep the injured quiet.

The silence around tragic incidents like this one had wider repercussions and meant that young Nepalis looking to go abroad themselves had few sources of reliable information. The stories and rumors of those who had traveled abroad took on great importance and tended to glamorize the work. For eager young job seekers, the lure of riches to be earned abroad was simply too great, and many young men continued to anxiously seek out anyone who could help them figure out some of the mysterious ways of connecting with international firms that would pave the path for their own futures abroad.

CHAPTER 5

MANPOWER

GET IN LINE

The departures gates in the Kathmandu airport tend to be filled with tired trekkers and adventuresome tourists making their way home from vacation. In addition to these groups, however, there are tightly bunched groups of Nepali laborers hoping to make it through immigration without being asked for a bribe. Every day, an estimated fifteen hundred Nepalis go through the small airport to work abroad.[1] This wave of labor has made manpower firms that facilitate employment big business. Some offer legitimate job opportunities, while others rely on shady brokers whom the workers must pay questionable fees to in order to help them navigate the murky, often corrupt, immigration system.

In the airport, those who have successfully found positions abroad often wear baseball caps or sometimes shirts or jackets from the company that has recruited them. Those going for the first time are most noticeable for the way in which they grasp tightly to their passports and paperwork and look anxiously at the official who is demanding to know where they are going and why.

Several laborers I interviewed said they had been shaken down and asked for bribes by the Nepali officials and airport workers. Workers with previous experience abroad, however, reported in our interviews that they could usually navigate the lines easily in the Kathmandu airport.

Yaksha, a contractor in his mid-forties from outside Kathmandu, told me he had passed through immigration once a year, and sometimes more, for six years while working in Afghanistan, never with a complete set of the proper paperwork. Simply asserting that all his documents were in order and giving the impression he was not willing to pay a bribe was the best way to make sure the Nepali officer would let him pass through, he said. It was the younger, poorer men who had the most difficulty. The trick was to not look like the easiest mark in line.

Nepalis do not need passports to travel to India, so Yaksha's trick was to put his passport in his checked baggage on his way to Delhi; if the immigration officer asked him, he would just say that he was going to Delhi to visit family members. The workers got in trouble with the authorities, he said, when the immigration officials saw Afghan or Iraqi visas in the passport. Then they knew you were probably making good money, and that was when the requests for bribes and other hassles started.

Despite these headaches, the lines in the Kathmandu airport are not the ones workers really needed to worry about. It is the arrival halls in Doha, Dubai, and other Gulf countries where a worker's documents, including contracts and work permits, were seriously scrutinized. Especially in the early years of the war in Afghanistan, before direct flights to Kabul from Delhi began, almost all Nepali contractors dealt with officials at these immigration counters before continuing on.[2]

While transiting through Dubai on numerous occasions, I had noticed that for Nepali workers, the usual strategy was that there was safety in numbers. Groups of workers for each company tended to get into line together, hoping to pass through as a group. Despite this, I had noted that the Emirati immigration officials never seemed to scrutinize those farther back in the group with any less care, even when they were studying the twenty-fifth version of the same document from the same company from a group of Nepalis who were passing through together.

They would linger over a specific document, though it never seemed clear what they were looking for. The other Nepalis were left to fidget in line in the cavernous hall. Occasionally the workers glanced nervously to the left where there were three ominous office doors. Whenever someone's paperwork was found lacking, they were escorted away for further questioning and potentially faced deportation if everything was not in perfect order. For those standing in line, Yaksha said, it seemed that regardless of how many went into those offices, no one ever seemed to come out.

Those staying in the Gulf were generally met by a broker after immigration. The broker would take them to the work sites where they would work long hours, before returning to dormitories or small apartments where they shared beds in shifts with other workers.[3] For those headed to conflict zones like Afghanistan, the future was less certain.

As each laborer passed through, the steady stream of workers from poor countries, working for low wages in questionable conditions in rich states in the Gulf and elsewhere, strengthened. In the meantime, the Europeans, Americans, and other Westerners moved through the airport immigration line, generally oblivious to the high-stakes game going on around them and the value of the documentation each worker held tightly.

How did the workers get these documents?

For Yaksha, Yogendra, Jivan, and many other Nepalis who went to Afghanistan, the first step before getting to one of these lines was to go to a manpower agency in Kathmandu. These agencies were spread across the city, with a cluster of them not far from the airport and others by the bus station and near the center. These offices were the gateways for the 2 million absentee Nepalis who worked abroad.[4] The agencies would post listing for jobs, mostly in Malaysia or the Gulf countries, and different firms would specialize in construction, hotels and restaurants, or security.

One of the first agencies I spent time doing research in was off a small side street in Kathmandu. Several tall trees shaded the quiet parking lot. Initially I had expected that more people would be there. But, when I arrived, it was just a sleepy secretary who seemed surprised, but not excited, about my visit, along with two or three older men sitting on a bench. They seemed to neither work there nor were applying for one of the current positions, simply passing the day.

Manpower offices were often slightly strange places because, as I found out, very little typically occurred there. They were not places where things happened but where connections were made. Potential workers would stop by to drop off paperwork and then, they hoped, return to pick up their contracts and plane tickets. But because the firms did little more than link laborers with employers in foreign countries, little work actually went on in the offices.

The contractors I interviewed tended to refer to those working at these firms as *dalals*, or brokers, a derogatory term that they never used to describe themselves, and most seemed to consider these brokers a necessary evil. Nepalis used the English word *setting* to describe what these brokers did, and they often said that working abroad was impossible without a dalal to help with the "setting up" of a job. These brokers had contacts with large international companies that used them to recruit locally. Just as important for the workers, the brokers had the know-how to navigate the bureaucracy of crossing the various borders and getting the right permits.

For many workers, particularly those in lower-paying jobs such as construction, there were actually multiple layers of dalals to deal with. First, for many who lived in rural areas, was the village broker, a local agent. These dalals then would be connected to a dalal in Kathmandu, who knew people working at the various firms in Kathmandu. In many cases, the manpower firms would have agents or brokers in Saudi Arabia, Malaysia, or whatever the country was that they primarily worked in, who would find the actual jobs. Often the recruit would pay only one dalal directly, and that dalal would then pay the others. Others paid multiple times, but regardless of the payment plan, each broker got a cut. For many of the poorer workers, this meant taking out loans against their future salaries, each layer of broker driving them deeper into debt.

In the case of Yogendra and others who had friends or relatives who already had jobs abroad—something that was more common with re- tired military contractors who had worked together—it was sometimes possible to jump over certain layers of brokers, but for those with less experience or fewer personal connections, the layers of brokers were the only way forward. Particularly for some of the younger men I in- terviewed, who did not have much education or work experience, these brokers were the gatekeepers and the only way for them to begin the process of seeking work abroad.

The office I first spent time at specialized in security, particularly for hotels and casinos in the Gulf and in Macau, China. It was run by retired Lieutenant Colonel Gurung. When he left the British Army, Colonel Gurung hoped to set up a trekking firm, but he quit this busi- ness when the flow of tourists into the country slowed to a trickle during the Maoist insurgency. Although Kathmandu remained fairly safe, the insecurity scared off many of the foreign trekkers, and Gurung's mili- tary background made security contracting the most appealing option available to him.

The government made attempts to regulate manpower firms, he told me, while shuffling through some of the paperwork he had to file with the Department of Foreign Employment. In much of the twenti- eth century, the government strictly regulated foreigners entering the country and determined exactly which Nepalis could work abroad. But now, in this era of open markets, it was left to private companies like his to manage emigration. Mostly, he said, regulation seemed to lead to haphazard demands for more paperwork. The signs of this bureau- cracy were visible in the piles of paperwork on his desk and on the eight-foot-high government-mandated sign on the wall by the entrance, written in bright red letters, outlining the rights of the Nepali workers coming to the office and exhorting them not to pay any additional fees to the brokers.

The lieutenant colonel asked his secretary to bring him the key to the file cabinet. After opening it, he took out a folder and put it on the table. Leafing through it carefully left to right, was the story of how thousands of Nepalis migrate to the Gulf for work each year with his firm.

In the case of the first file he pulled out, the Nepali recruits were to work at the Beach Rotana resort in Abu Dhabi. The initial documents included what's called a labor demand letter—a document drawn up by the employer the colonel was finding laborers for. On the sheet was a list of positions (for example, room attendant, waiter, cleaner, pool attendant), the number of laborers required for each, and the pay. For most, it was 1,050 dhiram per month (about $285 at the time), with some of the skilled positions slightly higher, but nothing was above $400 a month. After this came a cut-out of the advertisement he was required by law to take out in a national newspaper announcing the position. This was followed by documents from each of the workers who had been sent over, including their registration forms and training certificates from a government-run center.[5]

As the colonel sorted through this large pile of documents, all stamped by various offices, he assured me it was even more difficult to secure permission for jobs in Afghanistan, Iraq, and other conflict zones, which were occasionally said to be "banned" by the government, though it was never clear exactly what was meant by this term.[6]

Instead, the colonel described all sorts of bribes that he was asked to pay by corrupt Nepali officials and the various tricks that agents used to circumvent the bureaucratic barriers the state imposed. He assured me, of course, that all the work his firm did was absolutely legal—an assurance I unsurprisingly received at almost every agency I visited. But having spent time in a few government offices, I found it easy to see how the paperwork could lead to opportunities for corruption. It was also possible to see why workers would feel the need to come to one of these firms as opposed to trying to find work on their own. Who could ever figure out all this by themselves?

The perception that most of these agencies were really only trying to make money off unsuspecting workers who did not know how the system worked meant that some manpower firms were hesitant to speak with me. Others, like the colonel, were eager to defend themselves and the other legitimate firms in the industry. There had been a series of articles in the Nepali press about corrupt firms, and he wanted to set the record straight.

Those firms sending workers into conflict areas such as Afghanistan and Iraq were even more hesitant, given the ambiguous, but also highly lucrative, nature of work there. As a result, most of these were the more established, older firms or brand-new ones that were more reckless and willing to take chances.

EARLY DAYS

Firms sending people to Afghanistan were more difficult to find since they tended to keep a lower profile given their questionable legal status. Even when contractors we interviewed were happy to give the names of the firms they worked with in Afghanistan, they often did not want to share the names of the manpower firms that had gotten them there. These firms were run by their associates and neighbors, and the repercussions of talking about them seemed greater than discussing far-off multinational corporations. Luckily, however, I eventually interviewed Kiran, the head of a manpower company who had worked in Afghanistan until 2010. He was happy to outline the work that he had done.

Kiran had a long background in manpower, and his office with its sleek modern furniture appeared more established than the one run by Colonel Gurung. Set back in a pleasant garden, sitting in his sunny office, Kiran apologized that his office was somewhat disheveled despite the fact it looked neat to me. He had to move his office after the building had been damaged during the first earthquake that spring, which occurred in the middle of a training season for future McDonald's employees in Dubai. Kiran assured me that like a ship's captain, he made sure he was the last one out of the building. Although the building had not sustained serious damage, it had been declared unsafe, hence the move to the new place.

Kiran, wearing a sharp suit, looked more like a businessman than some of the former military men running other manpower firms. In his younger days, he said, he was the American Express representative in Kathmandu in charge of personally verifying cases of theft in the days long before the Internet.

After Kiran left American Express, he began his own company primarily recruiting Nepalis to work for companies in Saudi Arabia. Not long after the U.S. invasion of Afghanistan in 2001, he was approached by a small Canadian firm looking for support staff, mostly office workers, for its engineering and construction work in Afghanistan. The Canadians found him through a friend of his who worked at a trekking firm in Kathmandu. This led to a three-year relationship with the firm, which took a few more Nepalis from Kiran each year. The pay was good, the firm was happy with the work of the Nepalis, and the Nepalis were happy with their jobs, Kiran said.

Each time they sent a new group over, however, the Nepali government seemed to make the paperwork a little more difficult. Kiran had to bring letters from the Canadian embassy in Kathmandu for the first contract, and in order to secure permission for the second, a letter from the Nepali embassy in Canada, verifying the reputation of the firm. By the third time the contract was being renewed, he made eight trips to the Nepali Department of Foreign Employment to bring the various new documents they were demanding.

The other problem he discovered by around 2004 was that everyone knew that work in Afghanistan was lucrative and wanted to get in on the action. That brought more recruits but also more brokers. Brokers were charging laborers to bring them to his firm, he said. He was concerned that they were also lying to the recruits, getting them into debt and not explaining the risks of work in a war zone. Later, when the Canadian firm lost its contract in Afghanistan, Kiran got a couple of contracts from larger companies in Afghanistan. This time, however, the company's representatives themselves came through Nepali agents, and Kiran found that he was being asked for bribes and payments to both those who were supplying the workers and those who were supplying the jobs. It was simply too much.

Around this period, he also secured a large contract supplying the labor for all of the McDonald's restaurants in Dubai. With this large contact guaranteeing him a steady income, he no longer wanted to risk his reputation and financial well-being dealing with the shady brokers who dominated the Afghan contracts. Besides, one of the things he liked about McDonald's was that they treated the Nepalis well. Compared to some of the more exploitative companies in the Gulf, the McDonald's pay and benefits were transparent, he said.

All that was left of his work in Afghanistan were several framed letters of appreciation hanging on his office wall and a photo of one of the U.S. camps that the Canadian firm had helped build. In 2010, Kiran gave up working in Afghanistan for good, right around the time that President Obama was beginning his surge of troops, which coincided with a surge in demand for contractors. The stakes were so high that the contracting market became increasingly risky for legitimate businessmen, Kiran concluded while sipping tea and looking out at his peaceful garden.

RUNNING ON FUMES

As my research continued, the political situation in Nepal worsened. No longer convinced that their protests and bandhas in Kathmandu were making an impact, the Madhesi parties began to refocus their efforts on the south of the country, where they had the most local support and where India could play a more influential role. Their greatest strength was Nepal's extreme geography and the fact that over the past fifty years, almost all of Nepal's development had been done with an eye toward India. That meant that Nepal in general, and the Kathmandu

Valley in particular, was almost entirely dependent on goods brought in from India at checkpoints along the southern border. Using bandhas just in the south of the country to close these checkpoints would result in a slow economic strangulation of the country.

The first evidence of real disruption was a growing line at several of the major gas stations. At first, the drivers of the fifty or so motorbikes lined up at the station were considered foolish. Surely, the crisis would be over in a day or two, so there was no need to wait, many people said. But it was not over, and after a week, those lines of fifty motorbikes had grown to several hundred. At a couple of the main gas stations, the lines were miles long. Dawa and I stopped to fill up his bike at one of the black-market shops where we could buy a bucket of gas to carefully pour by funnel into the bike. Later, as prices increased, watered-down or tainted gas became common.

It was hard to keep up with what the real cause of the shortage was. Rumors were rampant. Some outlets reported that protesters had shut down the highways, while others said that India had closed the border. It was unclear who was more at fault, but India certainly did seem to at least be supporting the protesters.

One of the differences between living in Afghanistan and Nepal was how shockingly out of the international consciousness Nepal is. The international media have little presence in Nepal except after a few major events like the earthquakes of March and April 2015. In Afghanistan, when there was a suicide bombing or some other attack, it was quickly picked up by the media, and it would be fairly easy to figure out at least the latest information on the incident. The outside world, however, seemed to take little notice of the growing crisis in Nepal.

This also meant that it was hard to get any information about the blockade that was not highly influenced by the biases of the Nepali or Indian government. The Indian government insisted that there was no blockade; truck drivers were simply afraid that they would be killed by protesters on the other side of the border. They were only protecting them, they insisted. The Nepali press built on popular sentiment in

Nepal and claimed that this was an "unannounced blockade" by the Indian government.

In the meantime, only a little more than a week into the blockade, the president of the Federation of Nepalese Chambers of Commerce and Industry was concluding that the "blockade" had already incurred more than a billion dollars of economic costs, more than the two earthquakes that spring.[7] The American and other embassies in Kathmandu were largely quiet on the issue.[8] There were a few halfhearted statements about both sides paying more attention to humanitarian issues created by the political statement and U.N. Secretary-General Ban Ki-moon expressed "concern."[9]

Again, the part of me that was used to Afghanistan, where the United States routinely applied pressure to the government on all sorts of issue, was surprised that there was not even any informal pressure on the sides to resolve their differences. With hospitals complaining that they were running out of necessary medical supplies, all of which were imported from India, Nepal was entering a real humanitarian crisis, but since it was political, not an earthquake or some other natural disaster, the international community seemed disinterested.

The effects of the shortage grew more pronounced. Every day, fewer cars were on the street. Public transportation was overtaxed, with young people riding on the roofs of buses and minivans since there was no room inside. Gas to heat homes was increasingly important as the nights grew chillier. It became more obvious why young people in Nepal had little faith that their government would create economic opportunities for them.

Work abroad appeared the only hope.

CHAPTER 6

TWO HUNDRED YEARS OF GURKHAS

COMING OUT OF RETIREMENT

When the wars in Afghanistan and Iraq started in, respectively, 2001 and 2003, there was an immediate increase in demand for experienced security contractors. International private security companies saw Nepali Gurkhas who had served in the British or Indian armies as an easy solution to this need. Most retired in their late thirties or early forties, so were still young and fit enough to be qualified. They were well trained, had clear military experience that could be verified using British or Indian documents, and, most important, were far less expensive than a white soldier from a Western country with similar qualifications.

During these early years, the different types of Gurkhas fared differently as demand increased. British Gurkhas were by far the most sought after, and many of those seeking work quickly signed contracts. As the supply dwindled, firms increasingly began to look for Nepalis who had been in the Indian Army or the Singapore Police Force, finally turning to former members of the Nepali Army or police. Security companies used the term *Gurkha* in an increasingly loose manner in talking to their clients, in the hopes that they could build on some of these orientalist images of the "brave Gurkha warrior," even though more and more of these Nepalis were civilians with no military training.

Firms that were able to market this image while delivering Nepali contractors with more-or-less legitimate paperwork thrived in the early years of the two conflicts. One of the leading firms I visited in Kathmandu was IDG, named after the initials of its founder, Ian Douglas Gordon, who had served with the Gurkhas. IDG had begun by sending over 25 retired Gurkhas to Afghanistan in 2004, and eventually that number grew to over 800. In 2012, its contract with the U.N. Assistance Mission in Afghanistan (UNAMA) was over $10 million.[1] IDG primarily supplied the United Nations and branches of the organization like UNAMA and the U.N. Development Programme (UNDP), but it eventually used its reputation working with the United Nations to win contracts from other international organizations as well.

The United Nations required that their guards have at least twelve years of prior experience, which usually was not a limiting factor, since most Nepalis wanted to stay in the British military at least fifteen years to qualify for full pension before retiring. The United Nations, however, also instituted a ban on hiring contractors over forty-two years old (or forty-seven if they had been officers). That meant there was a rather narrow band of eligible recruits between about thirty-five and forty-two years old.

As the number of willing British Gurkha recruits fell, IDG recruited heavily from retired members of the Singapore Police Force. Pun, who ran the office in Kathmandu, had, like many other IDG employees, spent two decades in Singapore. Retired Singapore police provided a good alternative to former British Gurkhas, Pun explained, because they were well trained, albeit generally with less active military experience, and had been vetted through the same selection process as the British used. Moreover, their contracts with the Singapore government clearly stipulated that upon retirement, they and their families had to return to Nepal.[2] This group of 6,000 to 7,000 retired Singapore police had all lived together in Singapore in a large compound complete with facilities including a school, and when they returned to Nepal, they generally remained close. They held an annual meeting and celebration, which Pun described as a rather raucous affair. He later took me to the

Kathmandu Singapore Police Health Clinic, which had been set up by donated funds by various retirees to serve Singapore police and their families.

This solidarity and close personal connections made the Singapore Police Force an ideal group to recruit from, but even so, the demand for contractors far exceeded the supply, and so even IDG had begun recruiting Nepalis who had retired from the Indian Army. Their contingent in Afghanistan was not as large as some of the bigger contractors, and doing static guard work at U.N. compounds was not generally thought of as one of the more dangerous postings. Despite this, four IDG guards were killed when an angry mob attacked the U.N. office in Mazar-e Sharif and Pun had a letter of condolence from Ban Ki-moon hanging on his office wall. Six months after my visit, another Nepali IDG employee was killed. It could be dangerous work, Pun said grimly.

While we spoke, Pun made a few phone calls to other retired Singapore Police Force officers who had been in Afghanistan, and one joined us for lunch at a local restaurant serving Malaysian food. Most of those in Kathmandu whom he knew, however, were still in Afghanistan. Others tended to live in Pokhara and Dharan, where the British had set up their recruitment camps. And so Dawa and I, armed with phone numbers of contacts, started preparing to head east toward Dharan.

THE ROAD TO DHARAN

Crossing the plains of the Terai toward Dharan during the fuel blockade was eerie. The mountainous geography of Nepal means that the route anywhere is never straight, and we headed west for two hours before turning south for another hour, before finally turning east on the plains of the Terai several hours after the sun had set. We ate some dinner, and Dawa laughed that several of the buses alongside ours had broken windows that had been replaced by plastic. He complained about the state of Nepal where bus companies wouldn't even replace their windows.

Before we left Kathmandu, it was difficult to get a straight answer on the ongoing protests and the related security concerns in the south. Since my funding for this research had come from the Fulbright Foundation, I had to register my trip with the Regional Security Office at the U.S. embassy in Kathmandu, which I emailed to ask about the situation. All I had received in response was a forwarded message that said, "We received a travel request to Dharan and cannot tell the person not to go." While these double negatives were not reassuring, the bus company running buses down to Dharan said that they were still going, but only at night, so we decided to give it a shot.

There was nothing out of the ordinary with the first six hours or so of the trip. We settled into our seats much more comfortably than the two men across from us. They told us they were returning from working in Qatar and had numerous bags of gifts to prove it, shoved under their feet and above and between them.

A couple of hours after dinner, the driver abruptly pulled over by the side of the road behind another bus and turned the engine off. When a couple of passengers went up to try to find out what was going on, all the driver would say was, "We have to wait here." So the two workers, a few other passengers, and I climbed out to have a look around. Another bus pulled up behind us. Then another and another. Ahead of us, the line of buses and trucks stretched as far as the eye could see in the moonlight. After a few minutes, the line behind us also was so long that it too disappeared.

Smoking cigarettes, the passengers debated what was happening. A couple who had passed through recently said there was a road blocked by protesters, but we couldn't see anything going on ahead. The road was just dark and quiet. Others thought we would begin moving once a police escort arrived, but we had to wait for it to take vehicles through from the other direction, and there certainly weren't any cars coming from the other direction; we were standing in the opposite lane next to the bus. The lack of good information about the political situation became more real, and I thought particularly back to some of the contractors I had interviewed who complained about traveling home during the Maoist period when there were similar protests. Travel during this period, they said, was difficult since they never knew exactly how long it would take them to get home or to get back to work.

One passenger talked about being on a bus that had been stoned by protesters who had lined the road. I turned around to look at the line of buses behind us, and it became clear why almost half of them had plastic-covered broken windows: their windows had been broken by protesters. As we prepared to drive through the same stretch again, why would the bus company want to spend money on new windows that would just be smashed again? The two workers home from Qatar seemed annoyed that their homecoming was being further delayed.

Then, in the distance, one of the buses turned on its lights and then another. The passengers threw away their cigarettes, and we climbed on the bus quickly. As the driver's assistant put on a motorcycle helmet, I quietly hoped he was being melodramatic.

The driver shouted at us to pull the shades on our windows closed as he started up the engine. By this point, it was well past midnight, and even the excitement of a potential blockade couldn't keep me from nodding off, before jerking awake again a few times. It was hard to see much other than the back of the bus in front of us (Laxmi Express, Comfort Seats, AC!) and the occasional moon-lit riverbed as we trundled over a bridge. At one point, the siren from the police escort coming from the opposite direction woke me. I could make out three police vehicles racing along, followed by what appeared to be an infinite line of trucks and buses. I dozed off again.

At one point, we stopped at a police checkpoint, where we were glared at for a few minutes before being allowed to move forward. I couldn't see any protesters outside, but the focus of the driver and the silence of the passengers made the trip tense. We later heard that during this period, the police had set up an escort, usually leaving around midnight, with one headed east and the other west. Protesters on occasion had still stoned vehicles passing these escorts, and occasionally the escort was completely canceled. Others said that it was just the police being lazy.

As the sun began to come up, we were dropped off in Itari, a junction town in the middle of the plain in eastern Nepal. The bus was headed south toward the border, and we were headed north, back up into the hills, toward Dharan.

GURKHAS AT REST

As we rode on a local minibus into town, it was clear why Dharan was one of two towns the British had used to serve as their base for recruitment from Nepali mountain tribes. Nestled between two ridges tucked into the foothills of the Himalayans, it is a place where hill people come down to mingle with those on the Indian plains below. At the start of the town, the road climbed slowly, but then turned sharply upward as we arrived at the central depot. The square where the buses left from was packed with minibuses and shared taxis connecting to villages in the hills below and larger buses set to take these passengers south.

Gurkha recruiting had left a stamp on the layout of the town. The former British camp at the center of the town had been turned into large teaching hospital, surrounded by park grounds and a golf course. The British still used one small corner of the grounds for military activities. To the north, the Indian Army also had a large compound that was still in use, largely to pay the pensions of Nepalis in the area retired from the Indian Army. Farther up the road, on the ridge above town, was a pensioners' home run for twenty-six retired British Gurkhas and Gurkha widows. The view from the complex was stunning. To the south, there was nothing but plains and haze stretching into India toward the Bay of Bengal. To the north, rings of hill upon hill spiraled upward toward Everest, just eighty-five miles to the north.

The location of the town also said much about Nepal's divided politics and the ongoing struggle between the groups on the plain that have historically been disenfranchised by Kathmandu and looked south toward India for support, and the groups that live in Nepal's hills and valleys and resent the social and economic intrusions by their bulky southern neighbor. Many of the local residents in Dharan had moved down from the hills, and it was common for Gurkhas in either the British or Indian armies who came originally from hill villages to retire in Dharan.

Its strategic location also meant that it served as a point of departure for many Nepalis working abroad, not just as British soldiers. One report estimated that in 2005, ten years before my visit, the town was receiving $5 million in remittances every month.[3] It seemed likely this number had only climbed in recent years.

While the entire town is fairly well off, particularly compared to the mountain villages above it, there was an odd cluster of streets in the middle of town where the homes were particularly large, with lush gardens. They appeared more luxurious than some of the other neighborhoods, and it turned out that they were organized to a large extent around the militaries that the fathers in each family had served in. One street next to the former British military camp where recruiting took place had the largest houses, some three or four stories. Many had ornate gardens and decorated mailboxes that felt distinctly English. These homes were inhabited almost entirely by retired British Gurkhas. The next street over, farther from the British camp, with only slightly smaller houses, was inhabited by retired Indian Gurkhas.

Many of the retired men sat in their gardens or on their roofs chatting with each other. Their wives walked down the streets, heading to the stores a few blocks up while their children played together. While extended families were the main social group in Nepal, these families had grown close through their husbands' military service and created their own extended military families. As we walked down the street, a friend of Dawa, whose father had served in the Indian Army, pointed to half a dozen houses where there were men she knew who had served in Afghanistan or Iraq. After a few phone calls, we were off to interview several who had worked as contractors abroad.

At the first house we visited, we climbed three flights of stairs to a pleasant roof-top patio, where Jeshwal, who was about sixty years old, was reading the paper. Jeshwal had served in the British Army for twenty years, as had his father, who had served during World War II. Once Jeshwal retired, a broker firm from Kathmandu called him after a friend had passed his number along. They were recruiting for Global Security, which at that point was the security firm that protected the U.S. embassy. His neighbor across the street lived in almost an identical house and had also gone with him to work at Global.

Jeshwal, like many others we interviewed, was straightforward about his reasons for becoming a contractor. It was the salary, he said. Through a quirk in the British Army's changing treatment of Nepalis, those retiring after 1997 received a pension equivalent to what British soldiers got, which, depending on rank and time served, could be thousands of pounds a month, while those retiring before 1997 received only a few hundred pounds a month.[4] For Jeshwal, who had retired just before 1997, this meant a steady but small pension, and the attraction of this supplementary income from going to a conflict zone for a few years was strong, particularly when compared with the few hundred dollars a month one might make doing security work in Nepal.

After being offered a monthly salary of $1,700 to work at the U.S. embassy, Jeshwal flew to Delhi and then on to Afghanistan. It was static security, so fairly dull and the salary could have been better. He rarely left the embassy or the compound nearby where they were housed and had no real interactions with the Americans inside the embassy he was protecting. The Nepali guards mostly stuck to themselves and interacted with their American and European supervisors only when they had to. The conditions there weren't particularly good, but he had been a soldier so was used to barracks. Mostly it was just dull, he said.

When Global lost its contract to ArmorGroup in 2007, Armor-Group asked him to stay, but the officials were planning on adding other Nepalis with less experience to their team and said that they would cut his salary by $600. Instead of taking a cut, he decided to return to Dharan and enjoy his retirement. Rather notably, soon after taking over the contract, ArmorGroup was accused of failing "to meet contract standards." There were calls to replace the firm, but no action was taken until a series of sensational photos of Western security contractors (who were Jeshwal's former superiors) wearing coconuts over their genitals, drinking shots of alcohol (officially banned in the compound) held between the butt cheeks of other contractors, and performing a series of other perverse acts gained attention on the Internet.[5] Since the living quarters were divided by nationality, Jeshwal and his Nepali colleagues hadn't taken part in any of this, but looking back, he didn't seem too surprised it happened; the company was lax and not particularly disciplined, he said.

Despite the public embarrassment created by the incident, the situation did not get much better: It took three years after the initial complaints arose for ArmorGroup to be replaced by another security firm, Aegis Defence Services, which also had a questionable track record. Before winning several major contracts in Iraq, Aegis specialized in antipiracy work, and its CEO, Tim Spicer, previously led Sandline, a private security firm that was accused of breaking U.N. sanctions by selling weapons to Sierra Leone as well as being one of the main players in the so-called Sandline affair, during which its potential contract to end a domestic conflict brought down the government in Papua New Guinea.[6] Since ArmorGoup cut the salaries of the Nepalis at the embassy so drastically, Aegis had trouble finding retired British Gurkhas who would work for the company. Shortly after the company started its contract, it was found to have not filed proper documentation, suggesting that its employees were not trained to the levels claimed.[7] Further investigation suggested the company was overbilling in addition to providing undertrained security guards.[8]

Working for a corrupt, unprofessional private security firm guarding one of the most important compounds in Afghanistan was not what Jeshwal had signed up for. As more non-British Gurkhas were recruited, the gulf between those who had been in the British military (and in some cases the Indian military) and those with little or no military experience, but who were often more tolerant of questionable, at times risky, practices, grew more pronounced.

Later I interviewed Sanjog, who lived just a few houses down from Jeshwal in a similar-looking home. Sanjog also served with Jeshwal in the British Army, but instead of joining him in Afghanistan, he found his way to Iraq shortly after the initial U.S. invasion in 2003. There he started as what was called a platoon commander for a private security firm, leading a team composed mostly of Nepalis who did not have experience in the British Army.

Sanjog complained about these Nepalis who were essentially civilians. He described an ambush in which a convoy protected by a team of four of his men, two former British Gurkhas and two with no military experience, was attacked. When the shooting started, he said, the two with no experience fled, and the other two held off the attackers with great difficulty. It could have been worse, he said; he had heard of other cases where untrained security guards started shooting haphazardly and injured their own men. As the demand for security guards increased in the late 2000s, so did the number of undertrained contractors, making his job increasingly tougher.

After a couple of years, Sanjog was moved into a supervisory position, overseeing several teams, and his salary jumped to $2,500. After five years in Iraq, he was promoted again and reclassified as an expatriot, earning $5,000 a month.[9] Ex-patriot status, which essentially meant European, American, or white South African, was highly coveted among Nepalis, and few were able to make it into the classification. Still, even with the increased salary, Sanjog was quick to point out that most of the Americans and Europeans were making at least $10,000 a month doing the same job he was. These racial divides in terms of pay and benefits annoyed many of those interviewed, particularly those who had served in the British Army.

A few streets over, we interviewed Satish, a contractor whose experience was rather different. Satish's house was in a state of chaotic, partially finished renovation, and we had to climb over several half-completed walls to get into his living room. He had served in the Nepali Army, and probably because he did not have experience in the British or Indian armies, he had to pay a $2,000 fee to an agent to secure his first job in Afghanistan. There he worked as a security guard for a Turkish construction company that was supposed to working on removing rubber from Bagram's runway. The company, however, ran into some issues. One of its contracts had been revoked, or perhaps it was just that a contract they thought they were going to get fell through. Satish was not entirely sure. Before the company could fire him, however, he found work at Supreme, a much larger company also working at Bagram.

Since Satish's English was not very good, Dawa interviewed him in Nepali, but he struggled to make it through some of the answers, since Satish's first language was actually Limbu, a language spoken by a small ethnic group of Tibetan origin living in the hills of eastern Nepal, and Satish kept slipping in Limbu terms and phrases. Since Satish's military experience had been in Nepal instead of India or Britain and since his English was limited, his earning potential was far below Sanjog's or Jeshwal's. Now he was scraping together what he could to finish the renovations of his house and had his sons contributing some of their income as well.

Sanjog didn't regret going, he said, but it was not a rewarding experience. He had to watch out for the Afghan police, he said. They stopped the contractors sometimes when they left their compounds and demanded bribes, threatening to deport them.

After several days of interviews in Dharan, Dawa and I went over the pages of notes we had gathered, covered in salary figures, job titles, and dates served Afghanistan. We had already been in contact with half a dozen contractors in Dharan and would meet ten more over the next week, most of whom lived on these few blocks—a virtual retirement community of former soldiers and contractors. The size of the houses on the street reminded passers-by of the promise of wealth employment in the British and, to a lesser extent, the Indian military. It was no wonder that for young men in town, the prospect of being a British Gurkha was enticing. The shadowy world of private contracts was a mystery to all those outside it, so no one really knew how to get a job or whether an agent claiming he had connections in Afghanistan was legitimate. However, everyone knew that joining the British Army could be a good first step in the process.

Earlier, in the twentieth century, a young man could simply show up at a recruiting station and join. Now there were arduous physical and mental tests that went into narrowing the candidates down. With so much at stake, businessmen tried to take advantage of the recruits' desire to get the slightest edge. In Dharan, private training academies provided help to young men through the recruitment process. There were even more in Pokhara, where the British selection process took place. After spending another week in Dharan and conducting interviews with contractors living in some surrounding towns, we headed back to Kathmandu and then continued west to the other center of British recruitment, Pokhara, to look at how British recruiting shaped those seeking work in conflict zones.

CHAPTER 7

"WHO WILL BE A GURKHA?"

THE STADIUM

The Pokhara stadium complex is just off the highway, leading out of town. With the fuel crisis inflating taxi prices and the buses crammed, it was easier to walk the two miles from the lakefront where most of the hotels are located. As I climbed up the hill from the lakeshore where some of the highest peaks of the Himalayans were visible when the weather was clear, it was easy to forget that I was in one of Nepal's premier tourist towns. Beyond the bars and restaurants near the shore, the rest of Pokhara was bustling with villagers selling crops from their farms and buying clothes, tools, and other supplies. People heading to work and buses headed to Kathmandu competed with farmers' tractors on the narrow road as I walked up to the stadium complex.

I arrived at the gate at 7:00 a.m. Most of the groups of young men had been training there since 5:00 a.m. I felt a little lazy, given my tardy entrance.

The stadium was full of tight packs of recruits running in different directions, doing sit-ups and heaves, and sprinting up the stadium steps. Most groups had a trainer, some only a year or two older than the recruits themselves, barking out commands as the recruits did lunges across the field. One even seemed to be using his whistle lanyard as a whip, hitting one of the boys who had fallen behind.

Each group had twenty to twenty-five recruits, though some were a little bigger, and at one point I counted twenty-one different groups. I had to keep starting my counting over, since while most were near the track, others were running in the soccer field behind the stadium and some were doing circuits along the road, coming in and out of the complex, and I'm sure I missed some of them. Initially I stood on a short but steep hill, overlooking the complex. A couple of groups of older men were also sitting, watching the organized chaos.

Before beginning my research, I had watched the poignant documentary *Who Will Be a Gurkha?* by Nepali filmmaker Kesang Tseten, which tracked a group of recruits through the British selection process. It was easy to see why he had chosen this process: the human drama of these young men physically striving, trying to make the cut to secure a better life was moving, as were the tears of those who failed to make it. With two thousand recruits between the ages of seventeen and twenty-one running laps around me and another ten thousand or more training elsewhere, the scale was overwhelming.

All of the recruits that day at the stadium were enrolled in private training academies and prepping for the British selection process, which was due to start in three weeks. It was possible to apply to the British and Indian selection process independently, but for most, the real goal was the British Army.

The instructors overseeing the recruits were unfailingly stern. One of the more senior ones was a former Indian Army captain whom I had interviewed the day before. He carried a list of names and scores from a recent time trial, red marks next to those who had failed to meet the minimum standard. We exchanged a quick greeting, but he clearly was in character and marched off behind his group of recruits, looking disappointed in all of them.

There was an odd combination of seriousness and jovialness about the young men. Many had fixed stares as they strode along the track or waited for their turn on the pull-up bar with a look of concentration. Several greeted me with a sharp, "Good morning, sir." This was not a place where other Westerners visited. Perhaps some thought I had something to do with the selection process for the British. Others were maybe simply showing off their military demeanor to peers. It was clear at other moments that these were young men at play, some only sixteen and seventeen. One young man in a baseball cap danced among those in his pack, singing to them loudly while running, causing some to dodge out of the way. Another came up behind someone who had fallen behind while running with a pack on the track and attempted to physically push him ahead in encouragement.

By eight, most of the exercises were starting to wind down. The late fall day was hotter than the previous ones had been. Sitting in the sun, I could feel the sweat trickling down my neck even though I hadn't done anything more strenuous than chat with some of the recruits.

The young men streamed out of the complex with their water bottles in hand. Most headed toward the nearby training academies for their morning lessons, some stopping at nearby teahouses for a quick breakfast. The teahouses all served boiled eggs and beans, especially for the young men who were training. The recruits wolfed down their food, leaving dirty plates scattered around the table. I ordered tea at one of the teahouses, and the disgruntled owner refilled the communal water pitcher that the young men had drained in a matter of minutes. Several older men were there as well, drinking their morning tea. Sitting across from me was a man who appeared to be at least seventy with a Gurkha insignia on his jacket. I asked about it, and he said that he was a former Indian Gorkha. His son was on the Singapore Police Force, and his grandson had recently joined the British military. It was the Nepali Gurkha version of a success story: each generation exceeded the rank of the one before it.

Now, even though he didn't live close by, he liked to come to the stadium to take his morning walk and watch the hundreds of young men striving to follow in his footsteps and make his story their own.

THE CENTERS

Over the next few days, Dawa and I visited ten of the recruiting centers, located on the roads surrounding the stadium. Each of the centers had a hostel, some more than one, for the boys who were not from Pokhara. The large number of hostels filled with recruits gave the entire neighborhood almost a college town feel. Sports equipment was hanging from the balconies of many buildings, and young men were running off to class with their books and chatting with the few girls brave enough to venture near. Two older women walked by scowling at the young men's antics.

The office of one of the centers we visited, Gurkha Strength, was up a flight of stairs behind one of the teahouses just off the main Kathmandu highway. A pretty young woman greeted us in the office. (After visiting a few more academies that also had young women working the front desk, despite the fact that it was not particularly common in Nepal to have women working in office settings like this, I started to wonder to what extent this was a conscious strategy to recruit more young men.) Her desk was surrounded by photos of Gurkhas in various fighting poses. There was also a photograph of Prince Harry during one of his deployments to Afghanistan when he had served with several Nepalis, and, rather oddly, another photograph with the caption, "Once We Were Warriors: How Gurkhas Went from Pet Warriors to Victims," which was clearly a reproduction of a magazine cover. When I looked it up later, I found it was an essay on the exploitative nature of continued British recruiting in Nepal. The message was clearly lost on those visiting the office.

This office manager had just taken an application form from a man who was there signing up his younger brother. The younger brother was hanging back, not speaking, while his brother asked questions. The woman took the paperwork and handed the older brother a tracksuit uniform with the company's logo on it, which I had already seen on the backs of boys in town.

After they left, she explained the boys' daily routine and the rigors of the training. Besides a morning and an evening workout that mixed cardiovascular endurance training and more circuit-type training, there were lessons in English, math, and science, and for those applying for the Indian Army as well, something called "General Knowledge." At different centers, the scheduling differed somewhat, but the subject areas remained the same, mostly occurring in the morning or late afternoon hours with physical training both before and after the classes.

Another center we visited was in the hall adjoining an active Buddhist temple, and on the wall at the entrance to the office was a list of the benefactors who had made the building of the hall possible. The temple was renting out the hall to the recruiting firm, since it made a good place to hold classes. Along with the prerequisite shots of "brave Gurkha warriors," the office also had photos of the Dalai Lama and other religious figures.

In the main hall, I watched as a General Knowledge lesson was delivered to about a hundred boys lounging in plastic chairs. Three more recruits pulled up on a noisy motorbike about half an hour late, shuffled in, and sat in the back with their feet up. Outside in a covered section of the courtyard, a slightly smaller class was going on simultaneously, and the makeshift board had geometric equations on it. Dawa and I stopped several recruits on the way out to ask about the training they were doing. This was one of those places where Dawa and I complemented each other nicely. While those working at training centers were more willing to explain their processes to a visiting American scholar, it was clear that the recruits were more willing to chat with Dawa who was jovial and only a few years older than most of them.

After the classes were over, some of the boys headed up the road toward downtown Pokhara, but most walked back toward the streets where the hostels were located to relax or do some laundry. Outside the hostels, running shoes sat by the doors, along with a jumble of woven doko baskets. These cone-shaped baskets, which were most commonly used by villagers in Nepal to carry loads through the mountains or by laborers carrying bricks, had become oddly symbolic of the entire Gurkha recruiting system. Each recruit during the selection process would be asked to run 3 miles up a steep hill located behind the British Army camp while carrying a doko basket filled with 25 kilograms (about 55 pounds) of weight on their backs, with a supporting strap across their foreheads.[1] The British officer describing the race said it was useful since it was a test of "grit and determination." Most of the young men we spoke with were from more urban areas in Nepal and had rarely used such a basket before (or at least not in the traditional sense; I had one in my apartment I used as a laundry hamper). The baskets, however, symbolized the mountain culture that the British assumed the Gurkhas came from, and the recruits at the centers seemed to be learning to fit into that role as best they could.

Few of the recruits actually fit this stereotype: more were from Kathmandu and other urban areas than from hill villages. One said he had just finished his first semester in computing at a prestigious university program in Kathmandu. The lure of the British Army, however, was too great to keep him in class, and he had dropped out.

The recruits invited us into one of the hostels, the interior of which was grimmer than the sunny exteriors. The first we visited was a cramped room with about twenty beds and virtually no space for personal belongings. Some of the hostels were larger. One was a converted storefront and had recruits in large bunk beds, four to a bed, in a large ground floor room that had at least thirty beds in it. There were more rooms upstairs. Gurkha Strength also had a dining hall next to the dormitories, all across from the main office. Unsurprisingly, the young men seemed to prefer lounging outside. One played a guitar, while another group kicked a soccer ball back and forth.

They joked and roughhoused like typical twenty-year-old Nepalis, but they were working for something bigger: a two-hundred-year-old tradition and the opportunity to provide their families with almost unimaginable opportunities, or so they were reminded by the posters and billboards decorating their hostels.

SELLING THE DREAM

The streets around the stadium were covered with signs trying to entice young recruits to join the various centers. These had photos of Gurkhas in action poses: doing jumping jacks, aiming rifles, jumping over trenches. Most offices also had a large billboard photo of their founders, surrounded by recruits wearing garlands around their necks who had supposedly been selected from among their ranks in the British recruitment in previous years—alleged evidence of the program's success.

Joining one of the training programs cost 25,000 Nepali rupees (about $250 at that point), regardless of whether the young man entered the program for a month before recruitment began or a full year. This got him daily training sessions, plus the classes that met during the day. In addition, another 10,000 rupee fee ($100) kicked in for those who made it past the first round of British recruitment. The first round in the selection process was separated from the second round by a few weeks, which meant that those who passed the first round remained at the centers to train even more intensely for the next round. Those who had failed went home.

The centers seemed to make their money on the string of charges that followed the initial enrollment, such as fees for the hostels run by the centers. These were more variable depending on the reputation of the program and the type of room and board requested. These costs tended to run around 9,000 rupees a month ($90), though some cost more. This did not include some of the necessary training equipment and the required textbooks. I polled a series of these young recruits and found that once everything was paid for, most paid somewhere between $600 and $1,000 to compete in the selection process. That was much more than a typical Nepali laborer earned in a year and made the courses a significant burden for recruits from anything other than an upper-class background. Some of those from poorer families whom we interviewed told us they were forced to take out loans, and one young man said his family had mortgaged their two cows to send him to the center.

There were a few less professional training centers, which some recruits used in the effort to save money. I visited one of these, which was just a rented room in a building that was primarily a health clinic. To get there, we had to walk up three flights of dark stairs in the back, past several examining rooms, which were in use, before greeting the rather surprised-looking receptionist.

Most centers were much sleeker. The director of one had written his own textbooks. These were available in the bazaar for $3 each, and I picked up a couple of well-thumbed copies. There were separate editions for the British Army, the Indian Army, and the Singapore Police Force. The books themselves were filled mostly with sample tests, but also essays by retired British Gurkha officers on the history of the Gurkha regiments.

The same center that produced the textbooks was also the only one that appeared to be thinking about ways to diversify its services. In addition to recruitment training, it advertised that the center could assist with "study and work abroad," which included "Body Guard oppertunity in Europe" and "Security Guard (Training/Job) International." While the wording would have benefited from some spellchecking, the huge photo next to it of a James Bond–type muscular European-looking man in a suit speaking into an earpiece was probably convincing enough for many young men.

Despite these assurances of success and promises that hard work would lead to success, most of the recruits we spoke with believed firmly that the system was rigged and that they needed the connections of the recruiting centers, and perhaps additional bribes, in order to succeed. Others mentioned that during their interviews or other parts of the selection process, those inside the camp would ask for money. One young man outside the stadium said he had heard the going price was 1.5 million rupees ($15,000) to buy one of the coveted spots. Yet no one seemed to know who or when such payments were made, and the recruiting centers and other shady characters took advantage of this ambiguity.

One of the scams the British officers described to me later included several seemingly well-connected men who would claim they could get a recruit accepted in exchange for a bribe, usually in the range of $5,000 or so, despite having no actual influence on the process. If the recruit failed, the broker would explain to the family that his contact on the inside of the camp had moved on or was sick on the day the recruit was there and then repay the family the money they had given him. If the young man was successful, the broker would claim credit for it and keep the payment. Stories of such cons and the general confusion and misinformation about how the system worked all impressed on the young men the need to stick close to the centers that appeared to be run by men who allegedly understood how it all worked and were similar to other scams used by brokers who allegedly helped young men find more conventional jobs abroad.

When I asked at each of the recruitment offices about their actual success rate, most told us that 80 to 90 percent of their recruits were accepted. A couple of more realistic ones suggested that perhaps the rate was closer to 50 percent. None of these were close to the truth (and, unsurprisingly, none offered precise numbers of recruits from their centers who had been successful in previous years). The previous year 7,865 had officially applied for the British positions, with only 236 selected, a 3 percent success rate.[2] Despite this, or perhaps because of it, the recruiting centers had become the norm. An internal British assessment suggested that 85 percent of recently successful recruits used training centers before the recruitment process; if anything, this number probably underestimates the actual number since recruits might be hesitant to acknowledge they received help.[3]

When we spoke with recruits in the teahouses and outside the stadiums, no one mentioned the minuscule chances of success. Each hoped that he could beat the odds. As the days of the selection process drew closer, Pokhara filled with even more recruits. These men were from the training centers in Dharan and a few other towns across Nepal, such as Bhutwal to the south. The increase in the numbers still gave the sense that everyone in town was competing for a few coveted positions. Those who were successful would immediately join the British Army, while those who failed often ended up looking for work in other places, perhaps Afghanistan. For the time being however, all the recruits were thinking about was the selection process.

INDUCTION AT DAWN

Dawa and I walked up the hill in the dark. Ahead of us was a group of five young Nepali recruits. The early winter dawn was still almost an hour away, and the floodlights from the front gate of the British military camp caused us to shield our eyes when we explained ourselves to the Nepali officer who stopped us outside. After giving the name of the British officer who was to be our escort that day, we were allowed through.

For the young Nepalis walking up the hill to the camp with us, the process of entering must have been even more jarring. A retired Nepali officer barked out commands, lining up the young men who had walked up the hill behind us. They joined the twenty recruits who had arrived before us and stood rigidly, waiting for the next command. Another officer then came down the line, writing each recruit's number in thick black marker on his arm. Each was given a bib with the same number on it that they were to wear while inside the camp. Most of the recruits carried only a small duffle bag; it was a chilly morning in December, and a few wore winter caps.

The recruits who had worked with training centers were dropped off in groups at the bottom of the hill. Their leaders, knowing they were unwelcome at the camp, were sitting at the teahouse by the taxi stand below, chatting about the odds of various recruits and telling stories about their own attempts.

In order to convince the defense attaché at the British embassy to let us observe the process, I had needed to have the defense attaché at the American embassy vouch to the British embassy that I was a legitimate scholar. This took several rounds of phone calls and a visit to the embassy. Just before the process started, however, the call finally came through: Dawa and I were allowed to observe as long as we stuck to a strict schedule, photographed only in certain areas, and did not speak to any recruits.

That day the camp was taking in a batch of 125 recruits.[4] After passing through the gates, each was given a medical test to make sure that he was healthy enough for the testing, and the recruits' paperwork was scrutinized (forgeries were common). These young men had already beaten the odds: initially almost 6,000 had signed up for the testing, and only around 500 remained. From this group, 118 would be selected from the east of the country and another 118 from the west.[5] During the initial selection round, which had occurred two weeks before our visit, all 6,000 recruits were given medical tests, their paperwork checked to make sure they were between seventeen and a half and twenty-one, and English and math tests were given, along with preliminary fitness tests. Candidates who were too short or too light or did not have wide enough chests were dismissed. The British and Nepalis working at the camp had spent five days processing them all, working twenty-hour days to get the number from 6,000 down to the 500 who remained.

We watched as the remaining young men were processed. They handed in their paperwork again and their bags were searched to make sure they were not bringing in cell phones or other contraband. They were assembled by a group of retired Nepali British Army soldiers who were formally senior recruiting assistants, but most still referred to them as Galliwallas, a name they had acquired decades before when they used to hike through the mountains to distant villages to recruit young men. The balance had shifted in previous decades, and now there was no need to go to the hills and recruit. As we sipped our coffee to stay warm, the Galliwallas lectured the recruits on how to stand at attention and formed them into rows.

Over the course of the next five days, the recruits took part in a series of standardized physical tests, including the 3 mile doko run, which had to be completed in 48 minutes, and a mile and a half run that had to be run in 9:40.[6] Those who met these standards would be then graded on a series of other interviews and tests, and a computer would generate a final ranking of all the remaining candidates. The top 236 would be inducted into the British Army and sent to the United Kingdom for training.

The sun began to rise as the young men marched into the central courtyard of the camp and the light turned the Annapurna range to the north a soft shade of pink. Even for those accustomed to Himalayan views, it was difficult not to be struck by the subtle colors of the snowy mountains looming above. The compound itself was spread over several acres and looked almost like the small New England college campus I teach on. The central grassy quad was surrounded on one side by small, neat houses where the officers lived and on two other sides by long buildings that held classrooms and dorm rooms for the recruits. A large hall along the northwest side of the complex was where, in a week, the recruits would learn their fates, the successful ones exiting through one door leading farther into the camp, the unsuccessful ones leaving through the opposite door out of the camp.[7]

We were escorted around the compound by Mark, a junior officer who organized logistics at the camp. On our way into the quad, we stopped to listen to Lieutenant Colonel Peter Hunt, the senior officer based at the camp, lecture the incoming group in Nepali. All of the British officers in the Royal Gurkha Brigade receive Nepali classes, though the predominant language around the camp remains English. Mark had been in the country for only six months, so his Nepali was not as good as Peter's and that of some of the more senior officers around the camp, yet it still was far better than the Dari or Pashto of any of the American soldiers I had known in Afghanistan.

On the far side of the courtyard, a group of young men were lining up for their mile and a half run: four laps around the center of the camp. The runners danced in the cool morning air, responding to the Nepali officer who asked if they were ready, with a uniformed, "Yes, sir!" On the officer's "Mark!" they toed the line.

With most recruits maintaining stern expressions, trying to perform their roles as soldiers, it was difficult to determine how nervous they were. Two of the sixteen young men went off the line like a shot, sprinting around the first turn. Two minutes later, both looked pained as they came around for the second lap and then faded into jerky jogs by the third lap. They were clearly going to miss the cut of 9:40. Their time in the camp would soon be over.

I watched the runners who paced themselves poorly, trying to figure out whether this had been a case of nerves or poor strategy on their part. Perhaps they felt certain they would not make it and going all out was the only option for them. Certainly those in the training centers had run this distance before, though perhaps the moment had been too much for them.

A few of the smarter runners held back, settling into comfortable paces. A shorter runner in a red shirt had clearly practiced the distance and with each lap his stride seemed longer and smoother. He finished first and was cooling off in a small finishing corral by the time the second runner finished. For him, this was only one more test done, and instead of watching the rest of the runners finish, he seemed to be looking off across the courtyard at the pull-up bar where the next test would be.

CHAPTER 8

THROUGH THE COLONIAL LOOKING GLASS

THE OFFICERS' MESS

At the camp, Dawa and I were served breakfast in a wood-paneled officers' dining room, lined with photos and paintings of well-known Gurkha soldiers—a moment that seemed to highlight some of the differences between the remains of the British Empire and America's spreading outsourced imperialism. An extensive body of literature, written mostly by retired British officers, celebrated the roles of Nepalis in key British victories. This room, like the Gurkha Museum a bit down the road, tended to celebrate their success. For example, one celebrated Kulbir Thapa, who was awarded the Victoria Cross for saving both Nepali and British soldiers in Fauquissart, France, in 1915 after he returned behind German lines despite being injured.[1]

There was less to remind those in the room of the various defeats and tragedies the Nepalis in the British Army had faced. There was no mention of the thousands of Nepali soldiers who were killed or wounded in the middle of some of the fiercest fighting at Gallipoli or the even lesser-known ill-fated Mesopotamian campaign during which the British lost 93,244 soldiers from their Indian-based forces.[2] The history was all delivered with a decidedly British flair, heavy on the nostalgia, telling the past through the bronze trophies sitting on end tables and the oil paintings hanging on the walls. It was harder to find mementos from the past fifty years other than one contemporary photography book of Gurkha portraits.

The breakfast menu, like most of those eating with us, was English— eggs, sausage, and baked beans—and the officers were served by Nepali waiters in uniform on nice plates with silverware. Dawa leaned over and said quietly that it seemed like we were in a fancy hotel more than in an army camp. He was not the first to note the importance of breakfast for Gurkha officers. One account of the life of British Gurkha officers during the period between the two world wars noted that the "legendary" breakfasts were meant to set the tone and fortify the soldiers for the day. After breakfast at "about 11:30 am, the battalion commander held court and he and the adjutant were busy while the other officers wrote reports or studied. By the time the colonel's work was finished, the other officers were usually free for the day and changed clothes to play tennis, squash, football or other games. In the late afternoon the servants scurried to prepare baths and lay out clean mess dress, polished boots and starched white shirts. Mess was more of a ritual than a meal."[3]

Eighty years later, the servants had been replaced by Nepali waiters and the life of the British officer had been professionalized, but the mess was still a ritual and the tennis courts behind the administrative offices appeared to be in good shape. The officers ate and chatted about the results of some of the recruits. It was easy to forget the British Empire had collapsed some time back.

Indeed, the camp, and all of Pokhara, was part of a living tradition. In town two days before, I had tea with retired British officer and local legend J. P. Cross in his pleasant bungalow where he lived with his adopted Nepali son and the rest of his son's family.

A warrior scholar who seemed to blend T. E. Lawrence and Indiana Jones, Lieutenant Colonel Cross had entered the British Army in 1943 and excelled, in part due to his linguistic aptitude, eventually learning nine Asian languages. He first fought against the Japanese during World War II and was a part of the Gurkha forces during "the Malaya Emergency," between 1948 and 1960, a counterinsurgency campaign that would later inform U.S. strategy in Afghanistan. At one point, he and nine Nepali soldiers set up an ambush in a tennis court–sized patch of jungle where they stayed for fifty-three days, unable to move lest they reveal their position. All told, he had spent ten years "under the canopy," the term the military used to describe being stationed in the jungle.[4]

Cross came to Nepal in 1947, and he eventually became the chief recruiter for the British in Pokhara beginning in 1976, where he eventually settled down after his retirement. Since retiring, he had published eighteen books, primarily on Gurkha history and his experience with the Gurkha regiments, and he had a few more books in the pipeline, he assured me. In 2015, at the age of ninety-one, he had no official title, yet he still gave the annual welcoming speech to the new soldiers.

Times had changed, he said. When he ran recruiting, there were no training centers, and the Galliwallas had to trek to remote villages to encourage recruits to come down.[5] When I asked Cross about the private training centers just a few blocks from his house, he shook his head in disgust. They were distorting the true nature of the Gurkhas, he said, and steered the conversation back to the history of the various regiments.

Cross had also given up British citizenship in 2002, one of the prerequisites for securing Nepali citizenship, a fascinating flip-flop of the Nepali recruits who were fighting so hard to get into the Gurkha regiment and would eventually be eligible for British citizenship. Cross's process ran into some political snags, despite the fact that he fulfilled all the key requirements, including speaking fluent Nepali. After being officially stateless for twelve years, existing in a limbo not unlike those Nepalis citizens fighting in the British Army, he was finally granted Nepali citizenship in 2014. In his book-lined study, there were two clocks, one set to Nepali time and the other to British, which seemed indicative of the way in which the Gurkha process was trapped between two different worlds. For those on the British base, this produced an almost frantic need to justify the process at every step.

For example, over breakfast and during the course of the day we spent on camp, almost every British official we spoke with (and many of the Nepalis working on the base too) described the Gurkhas and Nepalis more generally in terms that naturally justified the continued recruitment process. This almost always began by praising the "brave competitors" and describing their "inexhaustible work ethic" and "fierce loyalty." The implication was that the British Army was the lone way to channel this energy correctly. The army was "the only honorable job" for these young men. There were no other good options for Nepalis, I was told.

While these claims had some economic logic, they also created a self-congratulatory system that did not ask what the actual repercussions of the recruitment was and whether there was a better way to do it. The "problem" with Nepali culture, I was told, was that while Nepalis were unfailing brave and loyal, with the ability to "carry on," they were also equally corrupt and likely to take "the easy way."[6] This was why it still made sense for the brigade to have Nepali soldiers and British officers (the latter exclusively white, from the ones we met). These officers were better able "to deal with change" and avoid the corruption that was "endemic in Nepali society."

While much of life in the camp was rooted in the decades-old traditions of the Gurkha regiments, the officers complained over breakfast about some of the more recent changes and challenges to the recruiting process. And throughout our time in the facility, Mark, Hunt, and the other Nepali and British officers we chatted with repeatedly emphasized both the fairness and the logic of the continued recruiting system.

To try and stop some of the rumors of corruption, the British had done work instituting mechanisms to make sure any manipulation of the results would be difficult. Each recruit was given a number, and when the time trial or math test, for example, was scored, the result was input directly into a computer that ranked the recruits by number, with no names attached. To ensure that no one was perceived as favored, the British officers went out of their way to not speak to the recruits.

Instead, the Nepali Galliwallas were the ones who interacted with the recruits the most, giving them their instructions and organizing them before the various events.[7]

To prevent cheating by the recruits, the tests they were given were changed regularly. The cell phone ban inside the camp was meant to ensure that recruits couldn't communicate with training centers about the test and couldn't get outside coaching while in the camp. Yet they were also aware that outside, the perception remained that recruits could somehow buy their way into the army, like foreign investors who were able to purchase British citizenship. But what seemed more worrying to those running the camp was Nepal's embrace of free market capitalism that was seeping into the process.

The training centers were the most obvious example of this, but there were other ways the market was upending traditions. The hill where the doko race was held was in a village not far from the camp. To better prepare the recruits, the training centers would take recruits there once a week to run up it in order to familiarize them with the course. Knowing how the incline changed, where the hill was steepest, and when the end was in sight surely helped the recruits avoid the mistakes of pacing that the two front-runners I had watched that morning had made.

With all this traffic, the village where the hill was located saw an opportunity. By claiming the hill was too congested, the village started organizing and selling time slots to the training centers. Each center was given a morning or afternoon to make sure they did not overlap. At the same time, the training centers also convinced the villagers to stop individuals who might not have the money to pay a center from running up the hill on their own. Just training on the legendary doko hill was now a commodity only certain people could buy.

The officers looked for other reasons to criticize the centers. Several of the British officers complained that the interview training they received taught them only to parrot answers back, and that the interviewers could easily see through this charade. They wanted young Nepalis who were "naturally" the best fit for the process so that they could shape these young recruits themselves, not ones who had been coached by the center. It was almost as if by studying the questions from recent years, the recruits had made themselves too qualified, upsetting the image of the Gurkha hill boy who would make a new life for himself by soldiering for the British Army.

Watching a young man strain on the pull-up bar, eager to follow in the footsteps of his father and grandfather, it was impossible to simply dismiss the system as an obsolete colonial holdout that would soon fade into the past.

WELFARE AND OTHER SCHEMES

The more time I spent at the camp, the more unsurprised I was that most Nepali politicians had essentially ignored it. Even the Maoist politicians who had been so vocal in their criticism before coming to power had done nothing to dismantle or reorganize the system when they led the Nepali government after the 2006 peace agreement. The British presence had clear colonial aspects, but it was also providing opportunities for these young men, and the officers there were earnest in their belief that they doing something that was mutually beneficial for Nepal and Britain.

Undeniably the British presence had an upside. The base did not just house the recruits and the British officers overseeing the process; it was also the home of the Gurkha Welfare Scheme. I visited the center, getting a briefing from the coordinator of the center, Lieutenant Colonel Miles Gibson, and interviewing others who worked with the center. Miles was by training an engineer and served as a member of the Corps of Royal Engineers. As a self-proclaimed "logistics guy," he had previously served in Afghanistan as the British advisor for procurement at the Afghan Ministry of Defense. Three months before, after leaving Afghanistan, he had been assigned to the Gurkha Welfare Scheme for a tour.

The Gurkha Welfare Scheme was tough to label, Miles explained. It was structurally independent of the British camp, but it was given space for its compound inside the base since its primary function was to deliver pensions to elderly Nepali veterans who had retired from the British military or their widows. It also provided grants, funded development projects, and ran medical clinics. The money for pensions came directly from the British government, but the rest was from private donors. It functioned much like many of the international development organizations in the country but was officially registered under by the Nepali Ministry of Defense.[8]

The Gurkha Welfare Scheme ran community and educational programs, building schools in communities that had traditionally given a large number of Gurkhas to the British Army. For programs like its traveling medical and dental camps, anyone was eligible to stop by, but still these camps went only to certain areas. This meant that while much of the work made the scheme look like a development organization, it did not target the neediest, as many relief groups tried to, but worked with families and communities with connections to the British military (who, according to the logic described to me during recruiting, were actually better off than their neighbors since families received remittances from those members of the community in the British military). This created a strange tension, and watching Miles go thought his slide show of pictures of their various medical camps, I wondered what their neighbors thought of these communities receiving this extra attention.

I was well aware that most Nepalis thought that aid was delivered in a far from equitable manner. I asked Miles whether he coordinated the Gurkha Welfare Scheme health camps with any of the other nongovernmental organizations (NGOs) or international nongovernmental organizations (INGOs) running health clinics across the country. For example, what if their camp came through just a few weeks after another group had visited the area? He looked a little surprised, and said no; there was no coordination. When I followed up with other INGOs, they too appeared not to focus on coordinating their efforts. So for most

Nepalis, it seemed, the Gurkha Welfare Scheme grants were just one more piece of aid that was delivered in a somewhat haphazard manner.

The pension delivery aspect of their work was, of course, more directly linked to those who had served in the British military. The process of delivering those pensions was difficult, since many of the pensioners lived in remote villages that were inaccessible by road. Out of a set of offices surrounding a small courtyard in the back of the British compound, Miles organized the work of Gurkha Welfare Scheme's 404 employees who ran twenty-two area welfare centers that they operated out of. Still, many of the retirees had to hike down, or in some cases be carried down, to collect their monthly pay from the welfare centers across the country.

The pensions of those who had served thirteen years or more began at 240 British pounds a month (around $350 at the time) and increased based on rank and time served.[9] The pensions were high enough to be significant, particularly for those living in rural areas, and many entire communities depended on these retirees, Miles said. This meant communities were particularly likely to take care of their retired Gurkhas, since after their deaths, the money would cease (unless the late Gurkha had a widow). At other times, the community would try to keep the news of a death from the British so the money would keep coming in. This, he explained, was one of the reasons it was so important to make the retirees come to the camps themselves to receive their payments.

A DIFFERENT KIND OF EMPIRE

After Miles's presentation, Dawa and I debated the continuing role of the British in Nepal. Their stories seemed to contradict each other, Dawa pointed out. They claimed to be helping those most in need, but also tended to focus on communities already benefiting from having community members who received Gurkha pensions. Miles's office served as the command center for this great network, stretching out across Nepal, that continued to assert Britain's role in the country. At

the same time, as Miles outlined the commitment of the British government to these Nepali citizens who had fought for them more than fifty
years ago, I thought back to Bhim and the others who had been injured
in the Pinnacle attack. While there was certainly something ridiculous
about the photos in Miles's slide show of World War II veterans being
carried on chairs down mountains to receive their pensions or their
widows receiving thin envelopes of cash and shaking hands with British
officials sixty years their junior, this was a scene that no employee of
DynCorp would ever be a part of.

No one was climbing a mountain to give those contractors pensions
or see if their wounds had healed. Quite the opposite. The DynCorp
workers and others had also given themselves bodily to a war effort,
but clearly there was no enduring commitment by the United States to
them beyond some questionable insurance payment. When I followed
up with an official in the defense attaché's office at the U.S. embassy
on the issue of compensation for injured Nepalis, he confessed he had
never thought about the issue, despite the fact he had also served a tour
in Afghanistan.

In recent years, the U.S. military finally begun to take seriously the
long-term effects of the war on U.S. soldiers. Books like David Finkel's
Thank You for Your Service and movies like *The Hurt Locker* have chipped
away at the taboo around speaking about posttraumatic stress disorder
(PTSD) and the psychological impact of war more generally. Other programs, like Wounded Warriors, were contributing to efforts to reintegrate returned American soldiers so that they might be productive parts
of the U.S. economy. Despite this progress, resources for such efforts in
the United States were still limited, and PTSD was only beginning to
be taken seriously in the American military.[10] The national conversation
did not consider the impact of the disorder even for American civilians
working as diplomats or humanitarians during the conflict.[11]

For contractors, especially those from foreign countries, there was
nothing. The sense was that they had volunteered for the work, so they
were somehow more responsible for whatever the outcome of their in-

volvement was. No one was asking whether America had any further commitment to the hundreds of thousands of contractors who had served in the war effort. All this put the work of the British camp into a different light.

While many of the contractors I interviewed did not go out on patrols seeking out the enemy the way American soldiers did, this did not mean that they avoided the same sort of horrific events that soldiers were involved in, whether these occurred directly to them or whether they simply witnessed it. In a typical interview, one Nepali described watching the SUV in the convoy in front of him hitting an improvised explosive device, and getting flipped up twenty feet into the air. After attending to the injured, including one of his friends, he was back at work without missing a shift.

For many of these men, violence and the constant threat of attack was the only world they had known for the past decade (those who went over in their early twenties had spent virtually the entirety of their adult lives in a conflict zone), and yet no one from the U.S. or Nepali government or even one of Nepal's many NGOs had even thought of asking whether these men needed help. A few of the Nepalis I spoke with still hoped to gain some sort of compensation for physical injuries, though even this was limited, and none mentioned seeking support for trauma like PTSD.

As much as I wanted to criticize the British for their odd attempts at keeping up the empire in this mountain bastion, it was also clear that their commitment to these soldiers who had fought for Britain seventy years before far exceeded any sort of responsibility that the American government felt for those who had defended it in Afghanistan in the past decade.

TO BRITAIN AND BEYOND

Beyond the support that the British government provided to retired Gurkhas, the young men at the center that day were also competing for something that their fathers had not: the right to live in Britain.

Over several years in the 2000s, there had been a political uproar in Britain over the status of retired Gurkha soldiers that had been championed in particular by British actress, activist and former Bond girl Joanna Lumley, whose father had been a Gurkha officer. Until that point, Nepali soldiers in the British Army who retired had been discharged in Nepal since that was where they had been recruited from. This meant, given the rather strict visa regulations, that most could not return to the United Kingdom, despite having just spent years working there for the British government. Those who did make it in often remained illegally on expired visas.

The campaign to grant retired Gurkhas settlement rights in the United Kingdom gained intense media coverage and news reports on the protests often featured Lumley, surrounded by retired Gurkhas, occasionally brandishing a kukri, the long curved knife associated with the Gurkhas. Bowing to public pressure, the British government made Nepalis who had served in the British Army for four or more years eligible to settle in Britain.[12] The settlement also allowed most Nepalis to bring their families to Britain eventually as well.[13] This meant that for those struggling through the selection process now, success would also almost certainly mean the right to settle in United Kingdom with his family, and potentially apply for citizenship.

For many young Nepalis, seeking a path to Europe, Australia, or North America made the allure of the British Army even stronger. In previous generations, some Nepalis preferred entering the ranks of the Indian Army instead of the British Army, mostly because the proximity of their postings to their homes allowed for longer and more frequent leaves. Now the Indian Army was a clear second choice for the recruits we interviewed.

British citizenship also made some of the reasoning that the British used to defend the practice of recruiting Nepalis more questionable. For a long time, the British had explained the program as partially an act of largesse, hiring young Nepali boys from mountain villages who would have had few economic opportunities otherwise. The remittances they sent back aided their families and their villages more generally, especially once the recruit retired and returned home. The officers still professed a strong inclination toward "boys from the hills," even though the numbers from the Terai plains had increased in recent years. Those from the hills were "good strong farm boys," I was told, and they also had their virtues. None of the British officers, however, mentioned urban recruits or the better educated as desirable when discussing the recruits in broad strokes, and the continued use of the doko race seemed to confirm the preference for recruits fitting this mountain image.

But what would happen now that these retired Gurkhas were getting citizenship and retiring to Britain instead of returning to their villages? While the change in policy was recent enough that it was difficult to make definitive conclusions, the early results did not bode well for Nepal economically. Some recent retirees still traveled back to Nepal and owned property there; however, I did not meet a single retiree from the British Army who had retired in the past five years who did not reside primarily in the United Kingdom. (Ironically, Cross and two other British Gurkha officers I met had chosen to retire in Nepal.) One British military expert already has concluded that between 2007 and 2009, remittances to Nepali from serving Gurkhas dropped by 97 percent, with that money essentially remaining in Britain instead of returning to Nepal.[14]

Interviewing both potential recruits and retired Gurkhas about the precise qualification for British citizenship and what benefits different groups should receive in the form of pensions, it was easy to forget the striking difference between these soldiers and the contractors who had fought for the United States in Afghanistan. There was no talk of giving these contractors citizenship,[15] and in several cases, I heard of former Nepali contractors struggling to secure visas to visit members of their immediate family in the United States despite having worked for U.S. companies for a decade. At the U.S. embassy in Kathmandu, they were dismissed as security risks: Why else had they been in Afghanistan? seemed to be the implication.

The way the United States established relations with these private contractors in Nepal and elsewhere was fleeting and largely left to private companies, meaning that there is nothing similar to the lasting relationship the British and the Nepalis created. At the camp, the British officers and officials were aware of the ongoing criticisms of the colonial world they continued living in, but perhaps due to the weight of tradition or the inertia of military bureaucracy, not much seemed to be happening to try to adapt the system for the changing times.

As Miles talked about the Gurkha Welfare Scheme, he described how they relied heavily on Nepalis retired from the British Army to help deliver pensions and organize health clinics. These retired soldiers were ideal because many of them were not particularly old, they were still physically fit, and, more important, they understood "the Gurkha mentality." With the Nepali soldiers receiving British citizenship, however, and most retirees now staying in Britain, it was hard to find replacements. He said that in the future, perhaps they would start hiring retirees from the Singapore Police Force.

Other changes seemed more difficult for them to imagine. I asked why the doko race continued when clearly such skills were no longer necessary in the modern army. Why not replace this with testing on computers or some other skills that were useful today? Not only would this bring in smarter recruits, but it meant that the training centers would shift their focus in response and teach skills that might help failed recruits to get other jobs.

The British officers I mentioned this to seemed uneasy. The selection has been this way for years, they said; you couldn't just change it. One was blunter. The recruits who were best at taking orders were generally not the most intelligent. "We wouldn't want the recruits to be too smart," he said.

BACK DOWN THE HILL

At the end of the day with the time trial inside the camp, Dawa and I watched five young men being escorted to the gate by a Galliwalla. It was not clear whether it was the race that had done them in or some other aspect of the selection process, but the duffel bags in their hands made it clear that they were headed home. The British officers at the camp repeatedly commented that they did not like to use the term *fail*. Instead they tended to prefer clunkier terms like "those not selected." Yet the expressions on the young men's faces as they walked out expressed the agony of failure.

The Galliwalla held a clipboard in one hand and collected bibs from each of the young men with the other. The first four were stern, determined, standing upright as they left. The fifth trailed behind, barely holding back tears. The officer told the first, "Better luck next year," and "Study your papers" to the second. He shook each of their hands, while a camp guard solemnly rolled open the heavy main gate and they started the long walk down the hill to the main road.

Later, after we handed in our security badges, the boys were gone. It was a long walk down the drive. At the bottom of the hill, taxi drivers were sitting around, looking up the road; they seemed more somber than they had been when we passed them that morning, perhaps understanding the gravity of the moment for those who came walking down the hill.

Those young men and the 6,000 others who would walk down that hill were headed toward uncertain futures. From our interviews with those who had failed in past years, it became apparent that there was no clear path for these young men; many had left school or jobs to enter the selection process. A few were able to get back into the schools they had left or find jobs in Kathmandu, but most still dreamed of working abroad. This was usually much more difficult than they first imagined.

For example, Rahul had failed in the selection process eight years before. He trained for a year before the process but still came up short. He then spent six or seven months after his attempt at home, not doing much. Through a broker, he eventually got a student visa for Singapore, which he hoped would help him find a job, but the visa meant that he could work only twenty hours a week. Things were so expensive in Singapore that he needed to ask his family to send him money instead of the other way around. A couple of years after that, he paid a broker in Dubai to get him a position doing basic construction labor in Afghanistan for a company that brought him into the country without a visa, leaving him to difficult and dangerous labor, trapped in a compound that was in part guarded by Nepalis who had probably been successful in the recruiting process a few years ahead of him and were now making five to eight times his salary. He had done this for three years, until the contract ended, and was now back in Nepal with little sense of what he was going to do next.

While most of those young men who failed to join the British Gurkhas and were successful in securing work abroad ended up in Malaysia or, preferably, one of the Gulf countries, a good number, like Rahul, ended up in conflict areas like Iraq or Afghanistan, often not doing security work but performing much more mundane tasks. These workers tended to have very different experiences from those in the security companies that recruited from the British and Indian armies. In these cases, young men had often spent up to three years training for the British Army, but that training did little to help prepare them for working abroad.

With the fuel crisis still in full swing, the black market for fuel was thriving. The rest of the economy was stagnating. Who wanted to start a business or invest when even simple things like transportation seemed impossible? The country's limited industrial base was being strangled, making the remittances of workers abroad even more important and increasing Nepal's dependency on the largesse of other countries.

The opportunities for young men for international jobs were almost all in Kathmandu. So on a cool December morning, Dawa and I, like many of the young recruits who had failed to be selected in the 2015 process, climbed on a bus headed back to Kathmandu.

CHAPTER 9

THE LABOR OF WAR

THE BAKER

I came to discover after several months of research that security contractors had a distinctive garb that often made them easy to spot from a distance. They favored the baseball caps and Oakley sunglasses of their Western counterparts and often wore fatigues or track suits. If I was meeting a contractor for the first time at a bus stop or a busy intersection, I made a game of trying to spot them before they spotted me. I was a little surprised at how successful I was. I could usually spot the DynCorp or ArmorGroup guy even at a crowded bus stop. But as my contact list expanded and I began interviewing Nepali contractors in nonsecurity work, I learned there was more than one style among those working in Afghanistan.

When I met Ashim, he was dressed fashionably, and his ripped jeans, earring, and gelled hair were more Nepali urban chic than mercenary tough. His experience was also vastly different: he had gone to Afghanistan at just twenty years old as a laborer instead of in a security position. As he told his story, I realized that he had much in common with those contractors in terms of why he had gone to Afghanistan, but how his time there played out was different in several key ways.

I met Ashim at one of Kathmandu's mostly vacant central malls. He led me up to the food court on the fourth floor, which was a climb since the electricity was off and the escalators were not moving. This, however, did little to dampen the friendly conversations of some of Kathmandu's youth who had gathered there drinking coffee in the dim shop. Ashim spent several minutes shaking hands with people at other tables before we sat down at our own.

As we began talking, instead of discussing security, Ashim wanted to make sure I understood the importance of mixing. In baking, measurements matter, he said, but really, you need to make sure that the dough gets mixed enough, but not for too long. It was a delicate balance. Every day for almost seven years while working at the bakery that supplied most of the bread and other baked goods consumed by the U.S. military in Afghanistan, he was given 80 kilograms (about 175 pounds) of flour, and he spent the rest of the day mixing. After this, he was free to roam most of the compound that housed a variety of other internationals. In his years of work there, he rarely left the compound itself, he said.

Compared to many of the security contractors I interviewed, Ashim was less comfortable talking about his experiences before going to Afghanistan. This was true of many of the laborers I interviewed, who generally came from families with lower socioeconomic standings than those of security contractors. Ashim had been "stressed," he said, before he left Nepal for the first time. This stemmed from family pressure and finances. Ashim's father was a farmer, and Ashim considered himself lucky to graduate from high school, but he had no real hope of going to college. After graduation, he was convinced by a broker that if he paid $1,000 and went to Delhi, the broker could find him a good job in Afghanistan or someplace similar. So, he went, spending a month in Delhi waiting for the job that never came through.

Back in Kathmandu, the same broker convinced him that he had simply been unlucky and if he paid $2,000, this time a job would definitely materialize. Ashim paid again. He still did not get a job, but this time, the broker did get him a visa and a plane ticket to Afghanistan. And so he went, arriving in Kabul in 2007.

His broker arranged for him to be picked up at the Kabul airport, and the car took him to a compound where he stayed with a group of other Nepalis also looking for work. He was lucky, he said; it took him only a couple of weeks to find a job paying $500 a month at the Supreme bakery. Supreme was the contracting company that fed most of the international troops in Afghanistan, though it branched out into a variety of other services and had billions of dollars in U.S. contracts. Ashim knew little about the company and the pay was less than he was hoping for, but after falling into debt and months of waiting, he felt he had no choice but to accept the offer.

His ordeals were not over yet, however. While he was in a taxi on the way to his new job from the place he had been staying, the driver pulled over in a quiet area and demanded his money, passport, and cell phone. The driver wasn't armed, but Ashim had no idea where he was or how to get to the Supreme compound or even back to the compound he had been staying at. Not sure what else he could do, he convinced the man to just take his money and leave him with his passport and cell phone. The driver agreed and took the money. Ashim arrived penniless outside the Supreme compound.

Ashim, like many of the other laborers I interviewed, had lower expectations about salary than those who worked in security. These workers also tended to have much less information about Afghanistan or the potential trouble they could get into, legal or otherwise. Before going to Afghanistan. "I thought Kabul would be like Dubai or Saudi Arabia. I was hoping to see skyscrapers," he said while looking out the fourth-floor widow of the mall at the Kathmandu skyline. He was surprised, he continued, that all the buildings outside the international compounds were made of mud.

When he arrived, he was scared of many of the Afghans he was working with at Supreme. The Afghans could all speak to each other and greatly outnumbered the Nepalis and other internationals who did not understand what they were talking about most of the time. There were a good number of Nepalis in the compound, and they tended to stick together. From this group, he eventually made friends. He even became friendlier with a few Afghan workers as time went by, and he spent his free time in the rec center, which had a gym.

The work itself wasn't that bad once he was used to it. All he did in the kitchen was mix, and as long as he finished in a timely manner each day, his boss tended to leave him alone. On most days, when there were no special orders, he would work quickly, watching the 80 kilogram bag of floor gradually empty.

He was allowed leave twice a year, and while the company covered the cost of the plane tickets back to Nepal, it did not always give him the correct paperwork to make his travel easy. In particular, Supreme did not always give its employees Afghan visas. This meant the Nepali government would not give him a work permit, since it was not clear whether he was working in Afghanistan legally.

As several other contractors explained, Supreme initially stated that its employees did not need Afghan visas since they were working directly on U.S. bases; only those working on other contracts or on other compounds needed visas from Afghanistan. After the Karzai government complained that Supreme had not been paying taxes to Afghanistan, the government started denying visas to its employees and argued that those on bases did not actually have an exemption since they were not direct U.S. employees. Supreme still did not get visas for most of its employees, putting the employees themselves in a legally precarious position, Ashim said.

For Ashim, this legal ambiguity was a problem. Neither he nor any of his colleagues knew what the rules were. This meant he had to fly in and out of a military airport in Afghanistan (where passengers' visas were not checked), and from there to Dubai, where he would catch a flight to Delhi and finally a third one to Kathmandu, where he would have to lie to immigration officials about where he had been. His leave time was reduced by the logistics of this travel, particularly since he was concerned about getting delayed at one of the transit airports without a visa. Usually he tried to build some extra time into his travel schedule, cutting his three weeks of leave down to two.

In 2014, with the troops scaling back, Supreme also cut back on its workers. His contract not renewed, Ashim returned to Kathmandu, but like many others drinking coffee in the mall that day, he was looking to go abroad again. His salary had been too low to have saved much from his work in Afghanistan once he subtracted the money he paid to the broker. Also, as a young man working abroad, many members of his extended family assumed he was making large amounts abroad and expected expensive gifts each time he returned home. Not bringing these would be shameful. But they didn't understand that not everyone working in Afghanistan was making thousands of dollars a month, Ashim complained.

These were common themes in interviews with other laborers: Ashim did not feel directly exploited, but during his time in Afghanistan, he was greatly constrained by debt, legal restrictions, and the machinations of the company he worked for. The lack of information around the migration process made him take risks that in retrospect he was not comfortable with. He was taken advantage of by Nepali brokers and American contractors.

Now searching for work from Kathmandu, he didn't have much of a preference, he said, as to whether it was a conflict zone or not. At this point, he was used to working in a war zone, so he wouldn't mind returning now that he knew what to expect. He got used to life shut up in a compound, he said. The work in Afghanistan had been fine, but here he was, eight years later, still stressed about money.

THE DIFFERENCE A COMPANY MAKES

Before beginning my interviews with contractors, when I tended to think about the experience of a war, I thought first about nationality. What was the experience of an American soldier during World War II as opposed to a German one? What about a Russian soldier? The flag a soldier fought under seemed to determine a great deal of what that person's experience was.

As the number of interviews grew, it became clear that in the more recent wars in Afghanistan and Iraq, this is changing in some strange ways. By the time I gathered a hundred or so interviews in December 2015, it was increasingly apparent from the variety of stories that people told that it was the company they worked for that most shaped their experience. When I did interviews in India later in spring 2016, in most cases the experiences of an Indian working for company A and a Nepali working for company A were more similar than the experiences of a Nepali working for company A and another Nepali working for company B.

To better understand these similarities and differences, I sought out individuals who had worked for certain companies. Over my last months in Nepal, I went out of my way to meet with a series of other Nepalis who worked for DynCorp and Supreme on various security and nonsecurity contracts, ranging from warehouse managers to mechanics. The comparison seemed useful since both companies had contracts for more than $10 billion in Afghanistan, and both employed thousands of Nepalis as private security contractors and as ordinary laborers. DynCorp worked more with the State Department than Supreme did, which had most of its contracts with the Department of Defense. DynCorp was best known for providing private security, while Supreme began in food service, but both companies adapted quickly to the needs of the U.S. military. Both secured contracts for a wide range of activities and ended up working with multiple branches of the U.S. government. Despite these similarities, as I conducted interviews, it was striking how different the experience of Nepalis working for these different companies clearly was.

I went to Nepal knowing a few things about DynCorp—none of it particularly good. When I was working in Afghanistan a few years before, a YouTube video had circulated among some of my colleagues with some Afghan police recruits flailing through jumping jacks, while their DynCorp trainers went ballistic—a clear clash of cultures. Other research I had done previously, or even just a general perusal of the

New York Times, painted a more menacing picture of the company. Contractors working for DynCorp had been accused of a variety of misdeeds in other conflict zones ranging from trafficking Bosnian girls as sex slaves to aiding an Iraqi minister's escape from prison, according to a Department of State cable released by WikiLeaks.[1] More generally, a Department of State–funded study found that while DynCorp and other large contractors "officially profess to having zero tolerance for TIP [trafficking in persons], they do not meaningfully scrutinize the behavior of their subcontractors, on whom they rely for the bulk of their recruiting."[2]

Over the past decade, DynCorp had received numerous U.S. government contracts in Iraq and Afghanistan despite a questionable track record. Its various missteps reported in the international press included everything from overcharging $685,000 for fuel at a Jordanian police academy, to performing $4.2 million of unauthorized work including building a pool and twenty VIP trailers for Iraqi officials. Its invoices and records for a $1.2 billion contract for Iraqi police training were found to be "in total disarray."[3]

Yet early on, as my anthropological training had taught me, I strove to keep an open mind in my interviews. As I interviewed Yogendra and other former DynCorp workers, I was somewhat startled to find that almost all Nepalis who worked for DynCorp were positive about the company. I heard again and again that "DynCorp is the best company." Even some of those who worked for other companies said they had heard it was a good company to work for, and while many Nepalis left other companies to join DynCorp, I could not find, among those I interviewed, any cases of contractors leaving DynCorp to work for other firms.

DynCorp paid well, they said, the accommodations and food were good most of the time, and while there are always some jerks, most liked their bosses. There were a few outliers, most notably Bhim, who had been injured in the Pinnacle attack, but even his criticism seemed to be more aimed at the insurance company in charge of his payments and he professed to hold no ill will toward DynCorp as a company. Jivan, who had also been injured in the attack, was still hoping Dyn-Corp would hire him back, and others who been at the Pinnacle attack were still complimentary about the company.

This was not as simple as just good pay and accommodations. Several DynCorp contractors described working under rather extreme, dangerous conditions but were still generally upbeat about the company. One I interviewed described being dropped with six other Nepalis and three Americans into an open field where DynCorp was planning on setting up a compound. They had to defend the perimeter, getting fired at by AK-47s and other small arms nightly, before the walls were actually built. They mostly slept during the day and spent nights crawling around in trenches, returning fire. The Nepali contractor went on to work for DynCorp for nine and a half years and was convinced that it was "the greatest."

I thought perhaps the contractors were just being positive about the company because they thought it was what I wanted to hear, but this was not true of other companies, where the Nepali workers were much more negative about their experiences. Some of this seemed to be due to the fact that at DynCorp, Nepalis were fairly well paid and fairly treated. But clearly more was going on.

DynCorp had a close relationship with the U.S. military. Often living on the same compounds with American soldiers or at least nearby provided some important benefits. DynCorp employees, relative to other contractors, seemed to have good intelligence and a sense of what the surrounding threats were. To a certain extent, this also meant, to use the phrasing of one soldier I spoke with, that Nepalis working for DynCorp were more likely "to buy into the mission." There was a sense that everyone had a collective aim and that they were working in coordination with other groups in the war effort. More concretely, this close working relationship with the U.S. military meant that if things went wrong, DynCorp could often count on quick assistance from American forces in the area. In contrast, other contractors might be left to fend for themselves or call the Afghan police, who were unlikely to come if the Taliban attacked.

Subtler policies also shaped the experience of these workers. In particular, DynCorp was not as rigid in its segregation of Nepali employees from the other Europeans and Americans that they worked with the way that many other companies did, contractors told me. Early on, Yogendra explained, Nepalis were allowed to work only in static security, guarding bases and compounds, but that changed later when they were allowed to be a part of personal security details (PSDs), which provided close security protection for diplomats and others. This was one of the best parts of the job, Yogendra and several others told me.

For DynCorp, this was not altruistic but a cost-saving measure—the Nepalis on PSDs were paid much more than they were when doing static security jobs, but this was still a mere fraction of what they would have had to pay an American to do the same job. Nevertheless, the Nepalis on these details felt much more a part of an international team, valued for the contributions that they were making. Beyond this, they spoke to me warmly of the friends they now had from America, Australia, and the United Kingdom whom they had been teammates with.

In other security companies, Nepali teams were segregated. Often one Nepali supervisor would oversee a team, the only one who actually interacted with the South African, British, or American who managed them. This also explained why DynCorp employees often spoke much better English than those who working at other companies (though this was often a special type of English—at the end of one interview, a former DynCorp employee, asked, "Can you explain one thing to me: what is a redneck? The Americans kept calling each other rednecks."). Those who worked for companies that segregated contractors had little reason to speak anything other than Nepali, whereas in DynCorp, teams spoke English with each other constantly since they worked together daily.[4]

Perhaps I shouldn't have been so surprised at how subtle social interactions had long-lasting effects on perceptions. We all remember

certain slights and acts of kindness for far longer than might seem logical, and this was the case for many of the experiences of the war that
I collected. One DynCorp contractor told me of one time when he
was in line at the dining hall in Kandahar that primarily served American troops and an American general got in line directly behind him.
Quickly he tried to give up his place and let the general go ahead, but
the general refused, pointing out that the Nepali had been there first.
Seven years later, he was still impressed by the general's sense of fairness. This never would have happened in Nepal or even Britain, the
contractor pointed out, where hierarchy matters more. For the general,
it was a passing moment, but for the contractor, it was why several years
later he was still telling his children and anyone else who would listen to
his stories about how "America is the best."

DynCorp also seemed, at least in eyes of many of the contractors
I interviewed, to worry about conforming to both Afghan and American regulations in Afghanistan, which made the Nepalis working for
them less stressed than some of their counterparts. Several interviewees showed me their cards from the Afghan Ministry of the Interior,
certifying them as trained in and "authorized . . . to possess the following weapons M9 Pistol, M4 Carbine, M240MG, M249SAW, Glock
19, Remington 870 Shotgun, AKM/AK 47 Rifle, MP5 SMG, CZSA58
Rifle, M16A2 Rifle." Others still had U.S.-issued letters of authorization, which were valuable since they allowed the holder access to certain
U.S. bases and to some of the resources on them. They could eat in
the dining halls, shop at the stores set up for American soldiers (usually referred to as PXs), or, even more important, use military flights
out of the country. All DynCorp workers underwent regular training,
and while these sessions did not always sound thorough based on their
descriptions, they were important to the Nepalis I spoke with primarily
because they demonstrated to them that the company had some interest in investing in them and providing them with new skills, which they
valued.

These experiences were different from the interviews I gathered from some of those working for other companies. The contrast was most striking when I began gathering stories from Nepalis like Ashim, who had worked at Supreme.

LIKE A ROCKET IN A FUEL TANK

Supreme, like DynCorp, faced a number of scandals that were publicized in the international press. At least initially, this did little to decrease the funds it received: between 2000 and 2015, Supreme received $11 billion from the U.S. government. Its annual contract for supplying troops in Afghanistan peaked at over $2.8 billion in 2012, then plummeted to just $360,000 in 2015 with the withdrawal of American troops and increasing concerns about its accounting practices.[5] All of this spending took a significant amount of manpower, and Supreme's offices in Dubai recruited workers from all across the globe.

Supreme began by providing food for troops at 4 key bases in Afghanistan in 2005 for $726 million. This was eventually expanded to 68 and later to 150 sites across the country through "a verbal change order rather than a formal contract modification."[6] These alterations led to a U.S. government lawsuit claiming that Supreme had overcharged the government by $757 million, with U.S. Representative John Tierney concluding that the "Pentagon lost control of this contract from the beginning and even today may be unable to recover hundreds of millions of dollars in potential overpayments."[7] Supreme eventually settled the case, paying $434 million in fines to the U.S. government and an additional $101 million in a related whistle-blower lawsuit.[8]

In response to some of this press, Supreme publicly claimed to be working on cleaning up its practices. Supreme's chief ethics and compliance officer, Emma Sharma, speaking at the Twelfth European Business Ethics Forum in 2015, explained: "Organizations around the world are gradually realizing the importance of ethics in their business."[9] This was apparently more gradual for Supreme.

When criticized in the international press or in U.S. courts for overcharging and other crimes of accounting, the complaints almost always referred exclusively to misspent taxpayer dollars and falsified accounting. Few even noted that Supreme, with 295 reported deaths, had by far the greatest number of employees reported killed in Afghanistan and Iraq of any other company directly contracting from the United States.[10] While Supreme was being accused of overcharging the U.S. government, few asked if its questionable ethics might extend to its treatment of the thousands of workers who were filling these contracts on the ground.

Unsurprisingly, the U.S. court cases around overcharging did not come up in any of the conversations I had with Nepali contractors. Issues of financial oversight mattered little to those who were trying to do their jobs. They were more concerned with Supreme's labor practices on a day-to-day basis. This generally meant poor living conditions and the tendency of the company to restrict their movements and keep them working under questionable conditions, with what they described as few opportunities for recourse.

While several Supreme workers told me they worked twelve-hour days, with no days off, it was not the hours or the work itself that seemed to concern most of the Nepalis I interviewed. The most common complaint was about their legal status and the difficulty they had getting visas and other documentation. This was not much of an issue when the contractors were on a Supreme base or working on an international military base. Those arriving on military and contractor flights at bases like Bagram and Kandahar did not need to show visas at all because there were no Afghan immigration officials on these bases. The issue was when contractors traveled between bases, as many did when they rotated between the compounds and camps Supreme had set up or were connected to. If the workers were stopped outside a base and had no visa, they could be arrested or, more likely, shaken down by the police. This made Supreme contractors hesitant to leave the bases they worked on except under extreme circumstances.[11]

The questionable legal status of their workers was actually extremely advantageous for the company. If contractors were unable to leave the compound, they were unlikely to be able to meet up with friends working for other companies and hear about better job opportunities. Even if they did learn of an opening, it would be impossible for them to conduct an interview in person. Making things more difficult for their workers, Supreme also categorized employees according to the security risks they said that they presented. (In reality, there was no evidence that this had anything at all to do with actual security; instead it was a reflection of how much one earned, with those from poorer countries earning the least, receiving the fewest privileges, and facing the most restrictions.) For those at the bottom of the hierarchy, cell phones and computers were banned, so workers couldn't call potential employers or even their families back home. I never heard of this as an issue with any American or European employee, and the Indians I interviewed who worked for Supreme were mostly ranked high enough to be allowed to have these electronics. In contrast, for most of the Nepalis working for Supreme, being caught with a cell phone could result in termination.[12] This, contractors for Supreme explained, made the work atmosphere hostile, and individuals were deeply suspicious of other workers and their bosses, afraid of being punished or even dismissed for possessing contraband like a phone.

While primarily focused on the day-to-day aspects of their jobs and concerns about the security of their compounds, this did not mean that the Nepali contractors I interviewed weren't aware of some of the corruption charges against the company. One contractor, who had a better administrative job managing some of the Nepalis and other workers on one base, said that his boss, an American, was stopped at a security checkpoint and searched while boarding a plane at Kandahar Airfield. At the checkpoint, the guards found that he had $40,000 in cash on him. Among those he worked with, it was rumored that he had been accepting kickbacks from Afghan companies in exchange for subcontracts. Since it wasn't immediately clear where the money came from and there was no direct evidence that he was taking kickbacks, he was allowed to go home and Supreme simply did not renew his contract. This was ridiculous, the Nepali I was interviewing said: a Nepali caught with cell phone or even a SIM card faced the same punishment, dismissal, as an American caught with $40,000 in bribe money.

Yet for the Nepalis like Ashim who worked for Supreme, their biggest complaints seemed to be primarily about poor management, which often led to exploitation. This could infect all aspects of the worker's experience and made it difficult to determine at times whether managers were incompetent or corrupt or whether they were simply a product of the system. One Nepali who had worked previously in Dubai told me he had to pay a broker $1,000 just to find out what time his interview with Supreme was, since the recruiting firm Supreme was using refused to post the schedule.[13] In this case, he suggested, it probably wasn't anyone at Supreme who was getting the payment, but that brokers had convinced the hiring agency to set up their nontransparent hiring practices in such a way that made it possible for their associates to demand bribes.

More generally, I was told, as troops began to withdraw, which meant less oversight, there was something of a Lord of the Flies atmosphere at some of the Supreme camps, particularly at some of the smaller bases where there were no other companies or international soldiers working. I interviewed Bashu, a fuel operator at a small Supreme base in the south of Afghanistan. He said his bosses tried to compensate for lower numbers of employees getting sent from the main office by having the fuel operators work other grueling, physical jobs, like filling sandbags to build another line of walls. This enabled the company to simultaneously save money by not building more secure concrete walls at a base that was likely to close soon.

Bashu's base was attacked regularly, and a South African colleague of his was killed in a suicide attack similar to the Pinnacle attack. The problem was that the base was situated in a volatile province and the Afghan police had no real desire to help when they were attacked. So the workers headed into the bunkers when attacks started, but Bashu said that the workers started to notice it was getting more and more difficult to find space in the bunker. After a while, they realized that this was because the security guys working on the base had started crowding in with everyone else during many of these attacks. At first, the head of the camp chided them for abandoning their posts, but they were allowed to remain after he realized the security unit in charge of protecting the base had cut back on the number of weapons allocated to the base. With only ten guns in the compound, it didn't make much sense for those who had no weapons to stand on the walls weaponless, Bashu explained.[14]

After one of these attacks, they emerged from the bunker to find that a rocket that had been fired into the compound had pierced an empty fuel truck and landed inside but did not detonate. Bashu and his associates were initially relieved, considering that an explosion at the fuel facility would have probably set off a disastrous chain reaction. However, immediately after, Bashu's boss told him to climb into the tanker and see if he could determine whether the rocket could be removed. No way, he responded, if you want someone to go in, why don't you. His boss then threatened to fire him if he didn't climb into the tanker.

Recollecting the event, Bashu described the tension. He tried to remain calm and reminded his boss that he was a fuel operator, not an ordnance expert. But this was the problem at these small bases, he concluded, especially when there was no international military presence. There was no one to do such dangerous work and no one to complain to when Nepali workers were forced to do work they had not signed up for. Eventually Bashu's boss backed down and he kept his job, but he remained wary of those above him and realized that he had to watch out for himself. Bashu's experiences in Afghanistan were harrowing, and in all of his tales, he seemed far more concerned about getting injured due to Supreme's lackadaisical approach to security and the company's constant corner cutting than by a Taliban attack.

It was even worse for those who made the least and were viewed as the most replaceable, like cooks.

THE HAPPY COOK

Perhaps my most perplexing interview in Nepal was with Kusang, a slight young man who seemed to smile even when talking about some of the more harrowing circumstances during the three years he worked in Afghanistan. We sat on the porch of his friend's house, who, after hearing of my project, had introduced us. I don't think I asked more than three or four questions as the afternoon slowly turned into evening. There was no stopping Kusang after he started his emotional tale. As he described some of the hardships he faced, instead of growing sad or depressed, he seemed more spirited, as if reminding himself of what he had been through renewed his self-confidence further.

Nine years before, Kusang began, his family was facing tough times. Not sure they could afford for Kusang to keep studying, they decided it would be better for him to look for work abroad. His cousin had found work in Afghanistan as a cook, and Kusang had previous experience training in a kitchen in a hotel in southern Nepal, so the family called the broker who first brought his cousin to Delhi. Following the instructions of the broker, he and his father traveled to Delhi where his father left him in the care of another broker. Kusang was twenty-three at the time.

Things did not go well at first in Delhi. The broker put him in a dingy hotel filled with other Nepalis looking for work, he said. He was left to wait with unemployed men and women being trafficked to the Middle East. Each day he called the broker, who always said that something would come through "tomorrow." After twenty-two days Kusang recalled precisely, he considered returning home but realized this would bring shame to his family, not to mention that they had already paid a few thousand dollars to the broker who promised him the job. Finally, the broker told him that he had found him a job as a cook for Supreme that would pay him $500 a month in Kabul.

When he arrived in Kabul after two further delays, the job he found was very different from the one the broker had described to him. Kusang would not be a cook, merely a kitchen worker, mostly cleaning dishes and carrying supplies from the storage area at the large base located adjacent to Kabul Airport. What's more, his salary would not be $500 as he had been told, but $350 a month. This was not much more than he could make in Nepal, Kusang told his new boss. His boss replied that if didn't like it, he was welcome to leave; there were plenty of others who would take the work. With no contacts in Kabul other than his cousin who was working at a distant camp, little money, and no return ticket, walking out the front door of the compound was not an option. He resigned himself to the work.[15] Many other Nepalis had it much worse than he did, he told me. Some had to stay in the work camps for a year before they found any work. At least he had a job.

As the days passed, he was most annoyed that he had extensive cooking experience and several of those above him in the kitchen hierarchy had no experience at all, and yet he was left to do the cleaning up. Most of the other kitchen workers were fairly unpleasant to those below them, particularly if they were from different countries. One of his Bengali bosses liked to berate him and beat him once when he tried defend himself. Since the other Indians in the kitchen took his boss's side, Kusang was the only one reprimanded, his boss telling him that another such incident and he would be fired.

The hours were long and uncomfortable. The day shift wasn't too bad, from 5:00 a.m. to 3:00 p.m., with a thirty-minute break for lunch. But the night shift was truly dreadful, stretching all through the night. He had to clean the kitchen and make sure it was prepared in case a late-night flight of hungry soldiers arrived.

This was also the time when many of the Afghan workers who had other jobs on the base would try to steal supplies from the kitchen stores, and Kusang constantly had to chase them off. Once, on his birthday, he decided he had enough and when he caught one worker stealing a wheel of cheese, he reported the theft. After filing a report with the military police, the Afghan worker sought him out and told him that nothing would happen to him since he was on the base, but that as soon as he returned to Kathmandu, the man's cousins would be waiting for him and would cut off his head. While Kusang wasn't convinced the man's cousins could actually make it to Nepal, he kept looking over his shoulder for weeks after in fear of some other sort of reprisal.

Instead of dwelling on these hardships, Kusang preferred to talk about the parts he enjoyed most. The base had a dance once a month for the workers, and coming from a more rural area in Nepal, he had never before attended such an event. At once he fell in love with the music. Most of the other Nepalis, he said, liked these events because each person was given two tokens for alcoholic beverages, which they quickly drank and left. But Kusang danced for hours and eventually became friends with the deejay who would play Nepali songs for him once most of the rest of the audience had cleared out. These moments seemed to help dull the monotony and constant stress of being harangued and threatened by those working above him.

Often homesick and sharing a room with three other workers most of the time, Kusang would quietly listen to Nepali music and think about his family. He didn't feel that he could return home until he had earned enough money to repay the broker and provide something for his family. Since his wages were so low, he was stuck. As a kitchen assistant, he was prohibited from having access to the Internet, since perhaps, he was told, he would send Afghan insurgents details about the base. For Kusang, this was ridiculous. He was allowed in only a very limited area within the base, so had no useful information. Besides, the only Afghans he knew worked on the base and tended to be rude to him.

He was not allowed access to a cell phone either. To get around some of these restrictions, there was a thriving black market on the base, and Kusang could occasionally borrow a cell phone and purchase phone credit from workers who had smuggled in phone cards. Each time he did this, he was risking his job.

After two years of work as an assistant, Kusang was promoted to the status of a cook and his pay increased to $500. What made Kusang even happier was that with his promotion, he was finally able to get a phone and a computer. He could now call his family on Skype, though he still had to rely on Afghan workers to smuggle Internet cards onto the base, which he bought for 5 euros each.

As the Americans withdrew troops, Kusang's dining hall closed. He was considered a good worker, however, and instead of being fired, he shifted from working at the Kabul airport to International Security Assistance Force (ISAF) headquarters in the center of the city. He continued living at the Kabul Airport compound, and the trip into ISAF each day was harrowing, he said. He had heard tales of the Taliban overrunning Kabul and was concerned about kidnapping, but just as worrying were the Afghan police. Since Supreme did not get new visas for its employees, Kusang and his colleagues had questionable legal status.

One day, about halfway to ISAF, the bus driving the Supreme employees was stopped at an Afghan police checkpoint. The policemen checked the passports of everyone on the bus and found that almost none of them had visas. They pulled the Afghan driver off the bus, and Kusang and his friends waited for several tormenting minutes while the Afghans negotiated outside and the driver made some frantic calls to the office. He had heard of Nepalis, like Teer, sent off to Afghan prison. Luckily, Kusang said, it seemed that there were so many of them on the bus that Supreme felt it could not afford to lose so many employees at once and perhaps the police were reluctant to have to deal with an entire bus full of illegal workers. So instead of heading to the police station, the driver paid the police a $500 bribe and they were allowed to continue on their way. This episode, Kusang said, made him hold his breath even more on his twice-daily bus rides.

As the evening wore on, the conversation drifted, and we talked about love and music and other things less related to Afghanistan. Toward the end, Kusang returned to his time working for Supreme, re-flecting on how challenging it had been and how alone he had felt, but how it had also been "one of the best moments of my life." These challenges, he said, were simply the types of things we all must face. They had made him who he was, he decided.

Still, even good-humored Kusang was frustrated by the lack of support he had received from the Nepali government. As companies lied about wages, foreign governments denied them visas, and brokers exploited them, who was looking out for these workers? Particularly after the twelve Nepali workers were killed in Iraq in 2004, public pressure increased for support for these migrant workers, but those like Kusang hoping for help were to remain disappointed.

CHAPTER 10

A PROTECTIVE GOVERNMENT?

GETTING A PASSPORT

While conducting interviews in Nepal, I began to prepare for my later research in India and elsewhere. Part of this included filling out vast piles of paperwork (seven copies of each form!) and standing in line at the Indian embassy while waiting to secure an Indian research visa. Forced to leave my passport at the embassy for three weeks, I was in the tricky position of not being able to leave Nepal since I had no passport. I found out during this period, however, that I could go to the U.S. embassy and apply for a second U.S. passport. All I had to do was submit a letter explaining why a second one was necessary. It was a strikingly easy process, and my new passport arrived in a couple of weeks. This contrasted sharply with the bureaucracy that Nepalis looking for work abroad endured.

To get a better sense of how the onerous process worked and how workers often tried to get around it, I began visiting the government offices that these workers relied on between interviews with contractors. As I visited the various government departments in search of some answers about the immigration process, what I found was a system that seemed almost more confusing with each step. The murkiness of securing passports and permits seemed to hang over those I interviewed, adding one more layer of uncertainty, particularly for those considering work in conflict zones. It also drove many to consult with the brokers and agents who were there, offering their services at each twisty turn.

I started at the Kaski Administrative Center in Pokhara, where those living in west-central Nepal could apply for passports. Much of the time, I spent sitting with Naresh, who ran the help desk, and chatting with those who stopped by his office on their way into the compound. On my first day there, I was a little surprised to find that although he ran the information desk in the government office, he actually worked for the Swedish Development Agency, not the Nepali government. As was the case with several government offices, the Nepali government was using this international nongovernmental organization not so much to support its work as to allow it to avoid the cost of setting up its own office. Naresh's salary, the colorful posters on the wall, and the new computers were all donations from the government of Sweden. As a result, his office was also significantly better equipped than the offices of those actually working for the Nepali government just across the compound, something that must have generated a certain resentment.

The government compound held several buildings with a courtyard in the middle and Naresh's small office off to the side. It was a good place to observe some of the comings and goings of the place. Applicants with folders of documents lined up at a series of windows at the main building, others sat under one of the shady trees, and on two occasions, police led shackled prisoners past to the police station in the rear. Just outside the compound, a couple of older men with makeshift desks helped applicants fill out forms for a small fee, and near them lounged a group of younger men. It took only a few minutes of watching applicants arrive to realize most were not interested in using the information center but instead relied on these young men to help them navigate the process as quickly as possible.

The young brokers had a tacit agreement with the guards of the compound as well as the officials. So, for example, when an dirty old motorbike pulled up and a man looking as if he had just ridden down from a village dismounted and started rummaging through his backpack for documents, two of the brokers casually wandered over to him. They began chatting, asking if he needed any help. He pointed to a couple of his papers, and after a minute or two, an agreement was reached and one of the brokers walked him over to the head of one of the lines to process his paper without waiting.

One broker pointed me in the direction of one of the shorter, stern-looking brokers. I was told his name was Hemanta Gurung, and after serving two decades in the Indian Army, he had been in Afghanistan he said. Hemanta was busy with one of his clients, so I went back to talking to some of the other applicants in Naresh's office. About thirty minutes later, however, Hemanta beckoned me over to a cement ledge away from the bustle. Although there was nothing illicit about our discussions, I found that these brokers preferred discussing matters privately, where they could not be overheard. It also gave them a certain air of mystery.

Before going to Afghanistan, Hemanta said, he had worked for three years in Iraq, mostly driving an armored truck in convoys. In the dust on the concrete step, he took a couple of my business cards and demonstrated the various evasive maneuvers they employed when under attack and how each vehicle in the convoy was supposed to respond. After working in Iraq for a year, Hemanta heard that there were even better-paying jobs in Afghanistan. He returned to Delhi and, using the Iraqi visa in his passport, convinced the Afghan embassy in Delhi to give him a visit visa for Afghanistan. Once he had the visa and the name of a broker in Afghanistan, he set off.

Work in Afghanistan was not as satisfying as it had been in Iraq. The broker he spoke with initially promised him $2,800 a month, but when he arrived at the security firm, it turned out that his contract was for only $900 a month. Furthermore, he was a static security guard, which was not as interesting as the work he did driving in Iraq, he said. Feeling overqualified, bored, and underpaid, he quit after just four months and returned to Nepal. Back home he bought a pig farm with some of his savings, but the farm didn't require much work and was not paying as much as he wanted, so he decided to use some of his contacts to work at the district center. Here, the money was much better, he said. It wasn't as good as Iraq or Afghanistan, but he was pretty good at his job, and some weeks, even after paying off the police and officials, the money could be close. Getting a passport wasn't easy, and Hemanta was good at facilitating.

Trying to get a complex document like a passport at the administrative center required waiting in multiple lines to get a series of signatures, even if the person could find out which windows were the correct ones and which stamps were necessary for which documents. An alternative was to pay Hemanta or one of his colleagues 2,000 to 3,000 rupees ($20 to $30, a significant amount for a Nepali worker). Hemanta would then take the person through the process, walking him to the head of each line. Here he would slip the file onto the desk opened to the correct page; the official would ignore the broker, looking disinterestedly at the other documents from the person who had actually waited in line, but once done would quickly stamp the broker's documents before moving onto the next petitioner. Hemanta would then move onto the next line and instead of staying there all day, he could get the petitioner through in twenty minutes or so, he explained. He would then take the money from the petitioner, and at the end of the week, in some private place, divide a little less than 2,000 rupees or so per transaction among all the officials who had facilitated the document. This way, no money actually changed hands inside the government compound and the allusion of fairness was maintained, despite the fact that almost everyone was paying.

Sometimes, Hemanta said, he missed driving in Iraq, but he was getting older, he said, and this provided a steady income.

WAS IT LEGAL?

Everyone who wanted to work in Afghanistan needed a passport, but beyond this, it was not always clear to potential workers what other paperwork or documents were necessary. Many who went through Nepali manpower firms also had their companies secure them Nepali labor permits, but others skipped this. Somewhat surprisingly, I noticed as I collected interviews that those in better-paying jobs, like a couple of the Nepali hydroengineers I spoke with, rarely got Nepali work permits and those in lower-paying, labor-intensive jobs were more likely to have secured them. The process appeared to be taking advantage of those with lower status and fewer connections. As I asked other workers about permits, it became clear that it was the ambiguity around the process that made it difficult for those who were poor and not well connected to navigate the system.

For Nepali laborers, one of the most confusing aspects of talking about work in Afghanistan was the assumption by many that working there was illegal. However, it turned out that getting answers about whether work there was legal or not was difficult.

When a job was arranged through legitimate channels, it was called *ramrari*, or done properly. Most Nepali workers didn't use the words *legal* or *illegal* and tended to refer to the work as being "set," as in arranged, which implied some sort of working around the law. In many cases, it wasn't clear whether work that had been set was necessary illegal. Did they mean it was illegal under Nepali law for Nepalis to work in Afghanistan, or was it illegal under Afghan law?[1] In most cases, contracting companies were more concerned with conforming to American law if they were American based. There were also questions about whether the workers were transiting legally through India or the United Arab Emirates.

The more I looked at it, the clearer it became that whether it was legal was a complex, troubling question for potential contractors.

The Nepali Department of Foreign Employment's website listed countries where Nepalis could work, and Afghanistan was on the list. The only two countries that were noted as banned were Libya and Iraq. However, other countries were completely left off the list, such as Guatemala, while Costa Rica and Panama were there. Congo was listed as okay, but Zimbabwe was not on the sheet at all as either approved or restricted.[2] The annual report from the Department of Foreign Employment also listed workers heading to Afghanistan as receiving 605 permits the previous year (598 for men and 7 for women).[3] This number was thousands lower than the number of Nepalis actually working there, suggesting that work in Afghanistan might be legal with the correct permits and that some people were actually applying for these permits. But in interviews, workers said again and again that they thought their work in Afghanistan was illegal.

I also discovered there was some precedent for the shifting regulations around Nepali migration. Female migrants in particular faced a series of challenges over the past three decades as laws changed. Until 1998, women were allowed to migrate to the Gulf only if they were accompanied by a male relative. From 1998 to 2003, there was a complete ban on female labor migration, which was then partially lifted and replaced with a partial ban on domestic labor for women under twenty-seven.[4] While many of these laws were described as protecting women from exploitation in foreign countries, some women's rights groups argued that the laws reinforced Nepal's patriarchal structures. The confusion around the restrictions on labor migration for women, however, paralleled some of the ambiguity around migration to Afghanistan, particularly poorer, less well-connected laborers looking for work abroad.

To try to answer some of the questions about the legal status of migrants, Dawa and I joined the hundreds of young Nepalis waiting outside the Department of Foreign Employment's offices.

FINDING THE RIGHT WINDOW

The Department of Foreign Employment labor permit office was bustling every day we visited. In the alleyways around the compound, small storefronts were set up where brokers could help with paperwork. Lines of mostly young men and the occasional woman formed as they waited for copies to be made or photos printed. A handful of men lounged on benches near the entrance. Some of them, slightly better dressed than the young applicants trying to get their papers in order, were clearly brokers offering their services.

Inside, a line on one side of the courtyard snaked around and back onto itself, with at least two hundred men standing patiently in it, waiting to submit their documents. To the right, another fifty or so men pushed up around a window where the processed passports were being distributed. Dawa and I stopped first at the information window in a small, recently constructed outbuilding. Its new furniture and bright walls stood out compared to the dullness of the government buildings, and it was not surprising to find out that the information office wasn't built by the government, but with European Union (EU) funds—a larger version of the office that Naresh worked in.

The pleasant, well-dressed woman inside who technically worked for a Western international nongovernmental organization was also far more helpful than any government employee I spoke with. Only a few people stopped by this office, and she seemed more than happy to look up from her Facebook page to chat with us for a while.

When I asked if I could see a pamphlet or two, she eagerly began handing me a copy of each, reaching behind her chair to open a new box. The shiny magazines all had the EU flag stamped prominently on them, alongside the seal of the International Labour Organization. They focused on types of exploitation and other useful themes, but none of the material explained how the labor process worked, how the workers could get their labor permits or visas, or how, exactly, they were supposed to protect themselves. There seemed to be plenty of information on what people shouldn't do but nothing on what they should do.

The young men we interviewed who were eagerly lined up outside the office were not interested in hearing more advice about avoiding exploitation. They wanted to know how to get abroad and find work. I was not surprised that few seemed interested in the materials the office was distributing.

When I asked about whether it was legal to work in Afghanistan, she took out a sheet of paper and studied it. Yes, she said; Nepalis could go to work there. I looked at the list, which was different from the one I had seen on the website. Scanning through quickly, I noticed that Russia was missing and asked why it was not included. She said that currently Nepalis were not allowed to work there, but she thought that it was primarily a "misunderstanding." She said that a similar thing had happened with Uzbekistan, but it was now back on the list of countries they could go to. She had no idea how the decisions were made about which countries were included and which were not.[5] She suggested we interview the chief of the department and offered to help us make an appointment.

Dawa and I returned the following week to meet with the chief. After we passed through two police checkpoints, the inside the chief's large office felt disconcertingly sparse compared to the crush of people outside. A heavy ledger sat on his desk that looked better used than the new computer alongside it. On the wall was a large flat-screen television with feed from eight surveillance cameras, though the fact that another four spaces were blank or flickering suggested the system was in need of some maintenance. We each sat on overstuffed sofas as I described my project.

Once I had finished, the chief explained that Nepali immigration for labor purposes was strictly banned in Afghanistan, Iraq, Libya, and Ghana. While my focus was on Afghanistan, I couldn't help asking about Ghana first, which seemed to me to be more stable and prosperous than the other banned countries. "The situation is not good there," he responded cryptically.

Since he did not seem interested in elaborating, I moved the conversation back to Afghanistan. Careful not to get any of my contacts in trouble, I mentioned that I had been in Afghanistan for five years and had seen a good many Nepalis there. Had the country been banned recently? I asked, No, he responded, but a few exceptions have been made in the case of Nepali security contractors working for the United

Nations. "Just the United Nations?" I followed up. "Well," he said, "the U.N. or the Afghan government."

I mentioned that I thought I might know some Nepalis there who were not working for either of these groups. Ah, he said, there were also some exceptions made for individuals who worked for companies that had U.N. contracts. If a company had a U.N. contract, a Nepali could work for it even if the Nepali was doing work that was not directly on that U.N. contract. Since the United Nations had far fewer contracts than the U.S. government, this did not account for the thousands of Nepalis working for Supreme, DynCorp, or dozens of other firms. Either the chief had little understanding of what most Nepalis working in Afghanistan were doing there or he was deliberately trying to mislead us. In order to keep the conversation pleasant, I asked some more questions about the general procedures of his department.

While the chief was more than happy to chat with us, he had few details about the work his office was doing. He shuffled through some papers on his desk a couple of times, but couldn't find the official figures for how many permits they had given out in the past weeks. Toward the end of our conversation, we were interrupted when one of the men who had been sitting outside the office in the courtyard knocked on the window next to his sofa. He said a few words through the window quietly to the chief and tried to pass some files through to him. Looking only slightly embarrassed, the chief waved the man away, telling him to come back later in the day.

He then turned and laughed to us about how it was important to sometimes do friends favors. Clearly this was how the system really worked. It was informal relationships, probably facilitated by a little money that was never visible but was always there, making it difficult for laborers to navigate the system. The bureaucracy remained intentionally confusing and opaque. And once the workers left Nepal, things only got worse.

IN AFGHANISTAN

After spending time in these various Nepali government offices, I compared them to the Afghan government offices I had visited. The Department of Foreign Employment in Kathmandu looked much like the foreign registry office in Kabul. This was where Nepali laborers who did not have the support of a large company had to go to get their visas renewed. It was located to the north of the city center in a compound with a slightly crumbling Soviet-era building at its center. Similar to the office in Kathmandu, it was surrounded by other smaller buildings and offices and a courtyard filled with applicants and brokers, moving to and fro with files in their hands.

Beginning in 2006, I had been to the offices on several occasions to get my own visa renewed, though in each case I had a friend or colleague along to assist me. The first time I went, I was able to get everything done by going to just one office. As the government "improved" after international funds started to flow into the country, however, getting the visa soon required filling out more forms and moving between three or four offices (it seemed to vary every time I went) on different floors of the building. None were marked, and all were filled with individuals waving forms. With the help of a friend and my Dari, I was usually able to navigate the process after a stressful hour or so.

For a Nepali laborer, however, particularly one with limited English and papers from a not very well-known company, the worker was likely to struggle with the process, and he was also likely to be a target for Afghan officials seeking bribes. Unsurprisingly, the Nepalis workers I interviewed almost all came to rely on their companies if they were in more legitimate, higher-paying jobs or Afghan brokers, if they were not. Most of these brokers they found through the original brokers who had helped the workers secure visas and contracts in Kathmandu, Delhi, and Kabul, and these networks seemed to grow and become more complicated as the war went on.

In the earlier years after the U.S. invasion between 2003 and 2005, almost all the Nepalis going to Afghanistan whom I interviewed arrived with a contract already set up. Their company would pick them up at the airport or even fly them in on a chartered plane if they were bringing in enough workers simultaneously. As the surge brought more money and jobs to the country, a process that particularly accelerated in 2009, rumors of the opportunities available there led more and more Nepalis to set out for Kabul without a job set up before hand. These numbers increased as they were joined by Nepalis who had been laid off from jobs in Iraq as the American efforts there waned. These workers, who already had experience in conflict zones, were also more likely to simply set off even if their documents were not in order. In Kabul, these groups mingled, creating a community of both employed and unemployed laborers from Nepal, Sri Lanka, Bangladesh, and India, and the brokers from Afghanistan, Nepal, India, and elsewhere who got them there. This created great opportunities but also significant chances for exploitation.

Pratik's experience highlights some of these risks. He had been in Kabul for almost ten years, moving higher and higher up the administrative ladder at the logistics company he worked at. I interviewed him in a coffee shop in Kathmandu while he was on a one-month leave back in Nepal to hear about his experience and how other Nepalis were navigating the system.

Before Afghanistan, Pratik had worked in Dubai, so he had some contacts who put him in touch with a broker. When he arrived in Kabul, he was met at the airport by a man working for this broker and taken directly to the compound, which was run by a Nepali broker but owned by an Afghan who collected rent and other kickbacks. The Nepali broker had originally come to Kabul as a security contractor, and while working for the company, he had helped some friends back home find jobs, taking a small payment each time. Pretty soon, he discovered, he could make more doing this than working as a contractor, so he set up recruiting firm that became a lucrative business.

The broker's compound was one of a handful scattered across the city at that time, though the number would climb to at least twenty as the surge continued.[6] Many were in the middle of the city, a couple even in the neighborhood I lived in, though I never noticed them when I was there, which was not surprising considering brokers' desire to keep low profiles due to the questionable nature of their work. While the brokers often called them "hotels," the high walls made them feel prison-like, Pratik said. Dirty and at times overcrowded, they were difficult places to spend time.

Pratik's compound charged $5 a day, which was cheaper than some of the others, which charged around $10. These fees were often deferred until the worker found a job, which meant debts would mount as the days went by. Some of the camps combined other illicit activities and housed brothels and illegal bars as well. The brokers made their money by collecting additional fees for helping the workers arrange their visas or other papers and, once they found a job, their initial month of salary.

Pratik and other contractors I interviewed explained that Afghan brokers who owned the compounds had connections, or in some cases, more formal partnerships with both Nepali and Indian brokers. In one case I heard of, a female Nepali broker even married an Afghan broker. I did my best to track down some of these brokers, but they were elusive. Two Nepali reporters I interviewed who had been in Afghanistan managed to confirm the stories about the brokers, but as one of them said, these people were "ghosts," using falsified documents, bribes, and informal connections to move people from country to country and find them work. They had no interest in speaking to journalists or anthropologists who might disrupt their lucrative business.

The difficulty finding work was something of a shock for Pratik, but he had some money saved and could continue to stay with other Nepalis at the compound. But as the days wore on, he grew bored and increasingly nervous. Most Nepali job seekers in Kabul rarely left their compounds, waiting instead for brokers to come to them with job offers. Some with expired visas were nervous about being arrested (a fear that increased as police enforced this more starting in 2009), but others simply said that there wasn't much to do in Kabul, so better to stay in the compound and play cards or watch TV if you were in one of the nicer ones. Finally, after seven months, just before he was considering giving up, Pratik found a job at Global Service and Trade, a logistics company based in Virginia.

His luck turned. His boss there was an American who wanted to set up his own high-end housing and logistics complex mostly for American and European contractors. He partnered with an Afghan to fund the company, and since he needed private security to guard his compound, he asked Pratik to join him, along with two other young Nepalis. Pratik was hired to help coordinate the Nepali security contractors they were planning on employing. He did not have as much experience as some of the Nepalis who had served in the British Army, but he spoke good English, and in a job like this, his flexibility and willingness to do any type of work made him ideal for a young business in the rough, fast-moving world of serving contractors.

The timing of the opening of the compound with its thorough security was fortuitous. After attacks on several smaller, private guesthouses in the heart of the city in 2009 and 2010, the United Nations and other international organizations started demanding that their employees live in larger compounds that met certain security standards. (Journalists and more independent internationals like myself continued to rent private houses in the city.) Some of the new restrictions, such as that the residential building could not directly abut the wall of the property (which made it easier for attackers to enter, something that was common in more urban neighborhoods in Kabul), meant that the eligible compounds were often located on the more remote and open Jalalabad Road to the east of the city. The compound, similar to the Pinnacle but catering to people from more companies, was soon filled with almost 2,000 contractors and other international workers.[7]

Eventually, like the Pinnacle, the compound became a target itself, and in 2009, when protests about rumors of U.S. troops burning Korans at Bagram Airbase spread across the country, two thousand to three thousand angry Afghans assembled outside throwing stones.[8] Pratik, who was in charge of security that day because his boss was on vacation, had to convince his guards not to fire on protesters, eventually getting permission from the Ministry of the Interior to shoot rounds into the air to scare them back. The compound continued to draw attention, and in 2014, a Taliban suicide attack targeted the compound, with Zabihullah Mujahid, the Taliban spokesman, calling it a foreign "safe house."[9]

While the American eventually sold the company, Pratik stayed on and nine years later had risen up through the ranks and was working as the coordinator for the company's almost five hundred Nepalis by the time I spoke with him in fall 2015. Pratik was an outlier: a young worker who had gone over with no connections and not been brought down by the various schemes and tricks of the brokers in the camps. Many more went over, failed to find work, and sank deeper into debt. In some cases, they had to contact family members back in Nepal to send them money.

There was almost no support for these workers. Caught without a visa, they could be thrown into jail, though that was better than what happened if they were caught by the Taliban. The nearest Nepali embassy, in Islamabad, Pakistan, was of little help. Where else could these workers turn? I managed to track down a few of the other more extreme cases, but by this point, I had been in Nepal for six months, and I was anxious to make sure that I had researched comparative cases as well.

I had been collecting contacts from other countries through the Nepalis whom Dawa and I interviewed, and I was particularly interested in contractors from Georgia and Turkey, who generally worked a step or two up the contracting hierarchy. Were the Nepali contractors experiencing the same things that these other contractors were? How did differences in race and religion shape their experience? Were they open to the same vulnerabilities and potential exploitation that young Nepalis were? Were there ways to make the system more transparent, to prevent some of these worst cases of exploitation?[10]

So while Dawa stayed in Nepal to conduct more interviews with young people considering work abroad, I planned my travel to Georgia, then Turkey, and finally India to look at comparative cases.

CHAPTER 11

OF ROSES AND REVOLUTIONS

TBILISI IN WINTER

This was my second trip to the former Soviet Republic of Georgia. I had first come as a student in winter 2000, when the country was still reeling from a decade of conflict driven by instability in three separatist regions (South Ossetia in the north, Abkhazia in the northwest, and Adjaria in the southwest), a newly assertive Russia in its "near abroad," areas that had formerly been under the influence of the Soviet Union, and uncertain support from the West, anxious to get oil out of the Caspian region to the east. This led to a series of short, internal wars that left the country much poorer than it had been when it was one of the more favored republics in the Soviet Union. It was now a divided country with few economic opportunities and corrupt officials. In Tbilisi, the Georgian capital, the crime rate soared.

Tbilisi that January was cold and bleak with only occasional electricity. My best friend and I lived with a tiny elderly Georgian woman, who cooked us an unending supply of rich Georgian delicacies, like khachapuri, a loaf of bread with cheese and egg cooked into it. Most of my memories of that time are of the three of us huddling around the inadequate ceramic heater, playing chess. The decaying grandeur of the city made an impression on me as a twenty-year-old college student. Rustaveli, the broad central avenue with stately architecture, still had the burned-out buildings from the brief civil war in the early 1990s. Surrounding it were neighborhoods of Soviet apartment buildings. The center is distinctly European despite the fact that Tbilisi is farther east than Baghdad and Georgians go out of their way to highlight their European ties, in part to assert themselves in what has historically been a dangerous neighborhood. It was a welcoming place for young Americans abroad.

Surrounded by Iran, the Chechen region in Russia, Turkey, Armenia, and Azerbaijan, Georgia has long been pulled in different directions by conquering empires ranging from the Mongols to the Arabs. It was under the influence of both the Ottoman and Persian empires in the nineteenth century before ultimately falling under the sway of Russians late in the century. Despite the European feel of its capital, the precarious nature of Georgia's position has also dragged it into America's wars, first in Iraq and later Afghanistan, with Georgian contractors and soldiers playing an important role, though one very different from that of their Nepali counterparts. In many ways, I came to find the lesson that Georgia taught was that countries could better organize to protect their contractors, but that race and claims of Europeanness helped almost as much as its more organized foreign policy. This helped contractors secure better positions while still facing difficulties that their American colleagues did not.

Throughout its history, Georgia often existed in a semiautonomous state, and the waves of immigration that followed various invasions made it both diverse and cosmopolitan. As late as the beginning of the twentieth century, Georgians were a minority in Tbilisi, which was dominated by Armenian merchants but also had sizable communities of Germans, Jews, and Azeris.[1] In 2000, when I explored the back alleys of the old town, which had not yet been renovated (UNESCO would do this five years later), the Turkish baths and neat brick mosque and the mixture of Georgian, Arabic, and Russian scripts within the cathedral were all evidence of its multicultural past.

During my first visit, Georgia was perhaps best thought of as in between. The government, led by former Soviet Foreign Minister Eduard Shevardnadze, was corrupt and elections were highly manipulated, but there were still many more basic freedoms than there had been under the Soviet period. It was exciting to sit in basement bars and drink Georgian wine while listening to Georgian youth attack the government and their parents' generation, which ran it.

Fifteen years later, when I arrived for the second time, the city was far more prosperous. The electricity was on full time, new buildings were springing up along Rustaveli, and restaurants aimed at foreigners, serving Georgia's delightful cuisine, were everywhere. I met Nino, a former contractor for Chemonics, at one of the brand-new cafés, a glass building with a giant metal bicycle outside it. In her mid-forties, with dark hair, she dressed in stylish European clothes. I asked her what she thought the bicycle was supposed to represent, and she laughed and shrugged. It was just another of the glitzy new structures lining Rustaveli.

Led by some of those disenchanted youths I had met in 2000, Georgia underwent the so-called Rose Revolution in 2003. Throwing off of the Shevardnadze government was the first of the so-called Colored Revolutions in the former Soviet Union (the Orange Revolution in the Ukraine and the Tulip Revolution in Kyrgyzstan followed).[2] This brought Mikheil Saakashvili to power, riding in on a wave of Western goodwill and optimism about the supposedly inevitable nature of democracy. As a graduate student at Columbia, where he had gone to law school, I heard him speak to a packed hall about the promise of the transformation of former Soviet political systems. His charisma and fluent English helped him play to the international media.

With this international attention came an increase in international support and money. The European Union provided funds for governance and civil society projects, and new government buildings went up in the city. The interest, particularly of the European Union in Christian Georgians, despite that it did not actually border on any member countries, was in stark contrast to the relatively limited amount of support Nepal received when it abolished the monarchy and went through its own democratization process three years later.

Many Georgians embraced these supposedly European ties after the revolution, and when I returned to the apartment I had lived in fifteen years before, I found a new municipal building in its place, complete with both a Georgian and an EU flag flying outside it (despite the fact Georgia is not a European Union member and not likely to be one soon). Amid a fit of police reform, the government decided more transparency was needed in both a literal and figurative sense, and new police stations built across the country were encased with glass, making the police literally more transparent.

The U.S. changed its approach to Georgia dramatically following the September 2001 attacks, as it did with many of its allies, and counterterrorism funds became increasingly available for smaller countries that could make it look as if they were helping President Bush's War on Terror. As a result, Georgia received some $65 million in military aid in 2002, leading its Ministry of Defense to focus on building counterterrorism programs. An official I interviewed described how the Georgia Train and Equip Program brought U.S. advisors to the country to train troops at the Krtsanisis training range south of Tbilisi, but also included equipment, such as a dozen UH-1 Huey helicopters.[3] As I'd seen before with American funds, international companies chasing contracts came too.

Nino began as a translator at the American embassy, but in a post–September 11 world with the Americans now funding a Georgian counterterrorism unit and bringing in Americans to train with the Georgian army, her boss suggested she apply for a job as a military liaison at the contractor KBR. KBR, a Houston-based contractor, was a subsidiary of Halliburton and, in addition to work in Afghanistan and Iraq, had contracts with the U.S. military in a number of countries. KBR was overseeing much of the logistics of the training on a base just outside Tbilisi starting in 2002. Nino worked closely with the trainers, who were primarily former U.S. military now working as contractors for Cubic, a U.S.-based firm.[4]

In later years, other firms such as MPRI gained additional U.S. contracts in Georgia.[5] One reporter recalled that at trendy boutique hotels catering to foreigners, "gossip over Saperavi wine [was that] well-paid jobs out at Krtsanisis were there for the asking."[6] Those who took these jobs were soon pulled deeper into the international webs that linked various contractors across the globe.

As the war in Afghanistan expanded in 2005, KBR and other contractors needed an increasing number of experienced employees in Kabul and other parts of Afghanistan. Those working in Tbilisi were ideal because they had experience working within KBR's structures, but they could be paid less than most Europeans or Americans since the standard of living in Georgia was lower than that of most European countries (a large number of similar contractors in similar positions came from Bosnia-Herzegovina, Serbia, and Romania for similar reasons). Unlike the Nepali government, with its ambiguous relationship with contractors, the Georgian government was more than happy to have its citizens working abroad, increasing the international visibility of Georgia's ties with Europe and America, and sending money home. So with a passport in her hand and an Afghan visa waiting to be picked up in Dubai, Nino set off.

"By nature," she said, "I am fairly brave." But work in Afghanistan was not what she had expected, and KBR had not given her much of a briefing before she set out. She was surprised at how dangerous it seemed once she arrived. She was required to wear Kevlar vests and a helmet whenever she traveled. A slight woman, it was hard to maneuver in all that gear designed for male soldiers. While she was never near an attack, the regular air raid sirens were stressful, she said.

Contractors were just starting to expand their roles in Afghanistan, and the base she was living on had not kept up with the increase in personnel. She and other contractors lived in drafty tents despite the extreme temperatures. Everything had a ramshackle feel to it, as if the mission had just been thrown together. The only good thing about working twelve hours a day was that she did not have time to dwell on the conditions, she said. Luckily, KBR in Georgia had just received another sizable contract for work with the Georgian military, and five months after her arrival in Afghanistan, Nino was able to apply for a higher position at KBR but based back at home in Tbilisi.

While she certainly did not look back on her time in Afghanistan fondly, Nino admitted it was beneficial to her career. After working for a while for KBR in Georgia, she found another position abroad, this time in Uzbekistan, which was far removed from the conflict and more comfortable. She also liked that she could communicate with the local Uzbek employees in Russian. Later, when she applied for a new World Bank–funded position, the woman interviewing her told her that she liked her courage for taking the job in Afghanistan and then offered her the position.

Like Nino, most other Georgian contractors I interviewed went over with firms on logistics or similar administration contracts. At least initially, there were fewer Georgian security contractors. The Georgian government's quick and strategic decision to commit troops to America's efforts in Iraq and Afghanistan would change this.

A BEACON OF LIBERTY

During the early Saakashvili years, there was much talk about Georgia's commitment to democracy in comparison with more authoritarian regimes in the area. President George Bush, when on a visit to Tbilisi in 2005, called Georgia "a beacon of liberty." As a result, one of the main thoroughfares in Tbilisi had been renamed George Bush Avenue.[7] The reality of Bush's visit had more to do with the solidifying of coalition allies for the wars in Iraq and Afghanistan than democracy promotion. This coalition relied on both troops and contractors and reveals the international nature of the outsourced military industrial complex.

Georgia's contribution of troops to the ISAF effort was remarkable. At one point, Georgia had more troops in Afghanistan than any non-NATO country, and by 2012, when it had around sixteen hundred troops, it was, per capita, the greatest contributor to the effort.[8] Explaining Georgia's commitment to the intervention, in a letter to the *Washington Post*, Mikheil Saakashvili wrote, "Georgia has been grateful for the extent to which the U.S. and Europe have stood alongside us over recent years. Now we are proud to stand—and fight—alongside you."[9] At the crossroads of empires, Georgian adventurers and soldiers were not often inclined to stay at home. "Georgia is not big enough for Georgians," noted Georgian author Archil Kikodze told me, and prominent Georgians had served in the armies of other countries and empires for centuries, often with distinction. In the war against Napoleon, the Russian Imperial Army had twenty-five Georgian officers.[10]

This was not the first time Georgian soldiers had been in Afghanistan. They made up a large contingent of the army of Shah Abbas, ruler of the Persian Empire in the seventeenth century, which fought across much of what is now Afghanistan. Four centuries later, there were legends about Georgian villages still in Afghanistan. While more myth than anything else, Georgian journalist Lazare Orbeli, who wrote pieces on Georgians in Afghanistan, told me that the memory of Georgian troops in the Soviet army that had fought in Afghanistan did remain. He and his Georgian photographer had even been greeted in a bazaar in Afghanistan in Georgian, he claimed. In Georgian narratives, this was not a new war.

Georgia's presence in Afghanistan, however, had more to do with its current geopolitical reality than these historical connections. Concerned about the threat of Russia and hoping for more international support for its claims on its breakaway territories, Georgia had rarely been shy about its desire to join NATO and the European Union. But to what extent had the strategy of supplying troops been successful in bringing Georgia closer to Europe and the United States?

Sitting in his office in an American-sponsored think tank, the former deputy minister under Saakashvili, who had overseen these programs and the later Georgian deployment to Afghanistan and Iraq, reflected on the increase in cooperation with the American military. Giorgi, now in his mid-forties at the most, was typical of Saakashvili's preference for a new generation of bureaucrats: young, sophisticated, and fluent in English and the bureaucratic jargon of international development. As part of Georgia's attempt to integrate itself more fully with NATO structures, he had gone to officer training school in Ankara and held a Turkish officer's rank as well.

George Bush had trouble gaining the same international support for the invasion of Iraq that he had in Afghanistan. In Georgia, he found himself a willing partner, and Saakashvili experienced little domestic resistance as the initial deployments received general public support. Georgia's 850 troops in Iraq in 2005 made it the sixth highest contributor to the international coalition, and by December 2007 that number had climbed to over 2,000 troops.[11] Only the United States and the United Kingdom had more.

The aim, Giorgi told me, was to show the American government that if you give us something, you will get something in return and that Georgia was not just consuming U.S. security but contributing to it. When I asked him if he thought this strategy worked, he wavered. It was not clear that Georgia was any closer to joining the EU than it had been a decade ago. However, in Washington, it certainly helped him gain access to members of Congress who voted on appropriations bills, which ensured continued financial assistance to the Georgians.

U.S. senators knew that the Georgian presence essentially meant two thousand fewer American troops in Iraq, and assistance programs in Georgia, run by Chemonics, MPRI, and other contractors that worked in Afghanistan and Iraq continued to grow. As a result, U.S. assistance to Georgia, which hovered around $100 million a year from 2000 to 2007, jumped in 2008 to $400 million, the largest category of which was "Peace and Security," the goal of which was in part to "reform, train and equip the Georgian military to meet NATO standards and to support contributions to international peacekeeping and security operations."[12]

At least initially, the Georgian soldiers were primarily serving in support and noncombat roles. This meant that many of them were performing guard duty at American installations in Iraq. The work they did was little different from that of many of the Nepali security contractors I had interviewed. The difference was that by sending them over in Georgian military uniforms, the U.S. government could claim to have another ally involved in the mission and Georgia could negotiate for various forms of aid. While the Georgians I interviewed claimed there was also a cultural distinction that made the Georgians bound to support their European allies, the explanation was more instrumental: Nepalis were going over individually with no government support, while the Georgian government had essentially been able to negotiate a more collective bargain. In a sense, the United States was just outsourcing one more aspect of the conflict to the Georgian military instead of a private company.

While Georgia might have been making progress reinforcing its alliances internationally, renewed trouble in the separatist regions of the country threatened the administration. Once in office, Saakashvili proved to be more nationalistic and aggressive toward the autonomous territories than he had indicated initially, and Georgian troops entered the disputed territory of South Ossetia in early August 2008. While many of the details are still unclear, most agree that the Georgians were the provocateurs and initially pushed the South Ossetian irregular troops back from a key border town.[13] The Russian military responded in assistance of the South Ossetians, using immense, and disproportionate, force. Tanks rolled south, jets bombed the Georgian city of Gori where Stalin was born, and quickly most of the Georgian army was routed. Russian soldiers were on Georgian soil, not far from Tbilisi. On August 11, 2008, four days into the South Ossetian war, Georgia called back the First Battalion, its most effective fighting unit, from Iraq.[14]

A tenuous cease-fire was reached, but the effects of the brief war were disastrous for the small republic, with 850 dead and 138,000 forced from their homes.[15] The diplomatic repercussions were serious, with Europe and the United States rattled by the recklessness of the Georgian government and at Putin's renewed interest in the region. While Georgia's allies in the west half-heartedly expressed solidarity with Georgia, the fact that Georgia had initiated the conflict provided an excuse for the United States and EU to stay out of the military fray.

The experience of Georgian troops in Iraq did not translate to the home front the way the Georgian military hoped it would. In Iraq, Georgians primarily played support roles for the Americans, whose firepower was vastly superior to their Iraqi (and later Afghan) opponents. In the war of South Ossetia, the roles were reversed, Giorgi said; it was like "we were the Taliban and Russia was the U.S." Troops who had been in Iraq were accustomed to calling in for rapid air support, but these resources simply weren't there in the fight against the Russians. Looking ahead, the Georgian military decided something had to change.

ONTO LEATHERNECK

With these lessons in mind, when President Obama sought additional international troops for the surge in troops to Afghanistan in 2009, the Georgian military responded quickly. Aware of their spotty record in the conflict in South Ossetia, the Ministry of Defense wanted to make sure that its troops "smelled gunpowder," Giorgi said. It was not enough for Georgians to serve as security guards in Afghanistan; the government was adamant that they take on a lead role, as opposed to many of the other small countries that volunteered their contingents with the understanding that they would be far from the frontline.

Initially, in 2009, Georgia had 170 troops under French command based in the relative safety of one of the International Security Assistance Force's (ISAF) large Kabul base.[16] Coinciding with the American surge, they increased this to one, and later two, battalions, bringing their numbers close to 1,600 troops on the ground at the height of their involvement in 2011.[17] The vast majority of troops were assigned to

Leatherneck, the massive base in the desert in Helmand, one of the centers of U.S. operations at the time. They also manned several smaller camps around the base. Of his visits to these desert camps, Giorgi said, between the heat and the dust, it was "hell on earth."

I interviewed Lazare, a veteran war reporter who was embedded with the Georgian troops in Helmand, at a French café in the upscale neighborhood that the Soviet elites had lived in. He looked like a war reporter, with a dark beard, chaotic hair, and a cigarette constantly in his hand. Lazare covered Georgia's civil war in the divided region of Abkhazia and had some of the best connections among Chechens in Georgia, which gave him a reputation for rugged reporting, earning him the opportunity to be the first reporter to accompany the Georgian troops to Afghanistan. Lazare's time with the Georgian troops in Helmand coincided with several attacks that led to a decline in public support for the mission, despite the pride regarding Georgia's contribution that came through in most of the interviews I conducted.[18]

Since the Georgian soldiers were all military, they were treated more as part of the mission than Nepali contractors were. That said, as an initially small contingent of soldiers at an international base, the Georgians were aware of some of the differences between them and soldiers from other countries. The Georgians I interviewed liked to brag that they were the most courageous soldiers, and when it came to real fighting, there were none better. Yes, the Americans and Russians tended to have better weapons, but the mostly young soldiers from rural areas embraced their frontline assignments, Lazare said, disparaging countries that sent their troops to safer support bases.

Lazare said the Georgian troops appeared to thrive in the American system because the Americans were said to be the most disciplined of the troops in Afghanistan and had transparent systems with clear chains of command. The British were too slovenly, allowing their soldiers to grow beards, and the French refused to go to the frontline, Georgian soldiers complained. This American approach allowed the passions and bravery of the Georgians to work in a more constructive way, according to Lazare. The problem of the Georgian wars in the 1990s and 2000s, he continued, was that they were simply too chaotic, without clear chains of command or communication. As the Russians attacked and the Georgians fell back, there was panic in the ranks and they had no contingency plans. The clarity of the American military structure allowed Georgian soldiers to focus their energies on fighting.

Attempts at integrating the different militaries did not always work. At one point, a female Georgian soldier, whose brother had been killed previously in the South Ossetian war, wanted to go out on patrol. The Georgian officers, taking a traditional patriarchal view of protecting their "sister," refused. She managed to advocate with some American officers who allowed her "outside the wire," overriding the orders of the Georgian officers and increasing the tension between the two groups. In a more severe case, a male Georgian soldier thought a female American soldier was making sexual advances when she showed him her tattoo, which was in a risqué area. The result was not rape, an official told me, but it was not good, and she had pulled a weapon on the Georgian soldier. Officers scrambled to reduce tensions between the two groups.

Listening to Lazare explain the position of the Georgians in relation to the American military, it was hard not to draw parallels with the role of Nepali Gurkhas in the British Army. In this postcolonial period, enlisting foreign troops is a practice frowned on by the American military establishment,[19] but the Georgians were clearly there to support an American effort. None of the Georgians I interviewed who had been in Afghanistan thought that Georgia had the slightest interest in Afghanistan itself. The Americans even bought their uniforms for them, one pointed out. The cultural claims of brave Georgians who were, if anything, too exuberant in conflict and needed both American weapons and structure to help them focus their efforts was eerily similar to the ways in which the British reminisce about Nepali Gurkhas, who simply needed some British discipline in order to be effective troops.

Contributing to the sense that the Georgian soldiers were being used by America was that unlike most other soldiers deploying there, Georgian soldiers had to volunteer to go. Serving in Afghanistan brought a jump in salary, from approximately $500 a month for those serving in Georgia to $1,500 a month for those in Afghanistan. With most of the soldiers in the Georgian army coming from rural areas, this was a real economic opportunity.

Fifteen hundred dollars, however, was still a good deal less than they could earn as private security contractors, and increasingly for many Georgian soldiers, going to Afghanistan for a tour was only the first step. As more rotated through, many of those who returned left the army to join private security contractors like DynCorp and Global, where they could earn even more. Another contingent went to work for the HALO Trust, a demining organization that had done work in Georgia and Abkhazia, as well as Afghanistan and Iraq, and relied on former military personnel to staff their operations.

For international contractors looking to maximize their profits, Georgian soldiers were appealing. Paying Americans six-figure salaries was cutting into their profits. Since Georgian soldiers were considered white, this allowed the contractors to charge more than they charged for an Afghan or Nepali guard. Yet with Georgia's lower living costs, former Georgian soldiers were far less expensive than other European or American security contractors. The Georgian Ministry of Defense quickly realized the dangers of the draw of contracting as more soldiers returned to Georgia and immediately left the army, taking jobs they had either set up or at least heard of while deployed. Those at Leatherneck, which at the time was awash with contractors from all sorts of companies, had a particularly easy time making contacts that helped them find new work.

By 2015, 11,500 Georgians had served in Afghanistan, a significant percentage of the armed forces, and retention became an issue.[20] To try to prevent defections of the soldiers who were now among Georgia's best trained and most experienced, the army started requiring soldiers to sign a four-year contract before deployments to Afghanistan, which would guarantee at least three more years of service after their return.[21]

This didn't stop many from leaving and accepting the potential legal consequences, however. As a result, the number of Georgians in a range of contracting positions grew, and the Georgian presence increasingly resembled the presence of Nepalis, albeit with better salaries and fewer questions around legality.

The growing Georgian contingent also meant that Georgia became a target for various insurgent groups. Lazare arrived in Helmand not long after a Taliban attack killed three Georgian soldiers. Many of the Georgian soldiers thought that some local villagers helped in the attack, and Giorgi, who was in the Georgian Department of Defense at the time, said that the soldiers went out "looking for revenge." It wasn't clear whether they had done much more than threaten local villagers, but this permanently soured their relations with the Afghans around the base.

Just a week after Lazare left, an Afghan employee on the same base allegedly helped the Taliban orchestrate another suicide attack that killed seven Georgians and wounded nine others.[22] The attack coincided with the uploading of a video on YouTube, "Taliban Jihad Against Georgian Troops in Afghanistan," allegedly by a Taliban spokesperson who threatened attacks on these "Georgian crusaders" if they continued their presence in Afghanistan.[23] A national day of mourning was declared after each incident, with flags across the city lowered.[24]

Following this incident that more than doubled the number of Georgians killed in Afghanistan, public protests broke out questioning the continued military involvement. There was an increasing sentiment that Afghanistan was not Georgia's fight, and the benefits of participation were no longer outweighing the costs. By the time of my visit to Georgia in January 2016, the number of Georgian troops had been cut in half and they had returned to providing only security forces and some rapid reaction teams.[25] Despite Georgia's contributions to U.S. efforts, membership to both NATO and the EU felt further off than before. Russia's pressure on the Ukraine, which led to the annexation of Crimea, received a muted response from the EU and the United States, leading Georgians to doubt whether either would come to their aid if Putin turned his eyes south again.

Much of the political uncertainty from the 1990s and the threat of Russia remained in the minds particularly of Georgia's youth. With over 1 million Georgians living outside the country (out of a population of only 6 million), much of the talk among young people in the bars and cafés on Rustaveli was of emigrating. The war in Afghanistan for many young men with limited skills provided an ideal opportunity, and soldiers continued leaving the military. Georgian contractors for companies like DynCorp began looking elsewhere for employment using the credentials they had built up in Afghanistan. One of Georgia's commanding officers in Afghanistan left the military to become an advisor for a high-ranking military official from Abu Dhabi he met while training in the United States, Giorgi said, and most of Nino's Georgian friends who had worked with her at KBR in both Georgia and Afghanistan were now working elsewhere in Europe. These trends all suggested that Georgians were likely to continue to appear in farther afield places working security and in other contracting positions.

For those who remained in Georgia, economic conditions were certainly better than on my first visit in 2000, yet there was also a sense that opportunities had been lost over the past decade. Just outside Tbilisi, a refugee camp was built for those made homeless by the 2008 war in South Ossetia. These identical cheerless temporary homes stretched for almost a mile just south of one of the main highways. They could have put the camp somewhere else more secluded, Archil Kikodze told me as we drove past. Here, he pointed out, the refugee housing allowed the government to remind those driving by of the continued threat presented by Russia to the north.

For Georgia and the soldiers and contractors who went to Afghanistan, the experience of the war provided some American military aid and sent some remittances home. But the cost was high, with thirty soldiers killed in Afghanistan and little sense that this had done much to make Georgia's place in the world any more stable.[26] For both soldiers and contractors taking part in conflicts far removed from their homes, the consequences of these flows of military personnel and money across the globe were not always what they seemed initially.

While contractors continued to send money home, the political impact of Georgia's presence in Afghanistan was largely contained. This was not the case in Georgia's larger neighbor, Turkey, where the military and contractors mixed far more rarely and where Turkey's shifting politics did much to shape the fate of Turkey's contractors abroad.

CHAPTER 12

ECONOMIC OTTOMANS

OUR MUSLIM BROTHERS

Seven hundred miles to the west, I arrived in Ankara late on a cold January night in 2016. I was expecting a ride from Aykut Zik, a retired Turkish general I had been introduced to through a string of mutual acquaintances. But I couldn't find him, despite the almost empty airport arrivals hall.

Ankara, Turkey's capital and administrative center, never draws the same number of travelers and tourists as the more vibrant cultural hub of Istanbul, but on my visit, it was even quieter than normal. A recent spate of bombings by the Islamic State angered over Turkey's support of the U.S. campaign in Syria had at first primarily targeted Western tourists to the country. But a bombing of a peace rally three months before had killed almost a hundred Turkish protesters in Ankara.[1] That January was also the coldest in a decade, and most of my fellow passengers were bundled up tightly, heading quickly out into the night.

Finally, I got through to the general on his cell, and he walked across the empty arrivals hall, apologizing and remarking that I had been hard to spot since I was younger than he was expecting. Driving into the sprawling city, we chatted about his time in Afghanistan. On the way into the city, it was difficult not to notice the new presidential palace, or Ak Saray (the white palace), which dominated the skyline, as well as many conversations I had over the next week. The one-thousand-room complex cost more than $600 million to build. The architectural style was meant to remind viewers of Turkey's Seljuk past as warrior horsemen who had ridden into Anatolia in the eleventh century, before the founding of the Ottoman Empire, which would eventually include most of the Middle East, as well much of the North Africa, the Balkans, and Caucasus.[2] Six months later, the palace would be attacked during the failed military coup in July 2016 that attempted to overthrow the government of President Recep Tayyip Erdoğan.

The new palace was a sensitive topic of conversation. Depending on whom I asked, it was a sign of the economic growth and renewed Turkish pride of the past fifteen years generated by the rule of the Justice and Development Party, more commonly known as the AKP, or a sign of the waste and corruption of the increasingly authoritarian regime. Erdoğan, the AKP head, had moved in Putinesque fashion from prime minister to president while strengthening the Islamist message of the AKP, which appealed particularly to those in rural areas. In response to criticism around the palace and his increasing cult of personality, he declared that the palace "belongs to the nation, not the president."[3] Still, many had their doubts about Erdoğan's commitment to democracy.

The palace represented the Turkish government's move away from its secular roots, from its violent postwar World I struggle for independence toward a more conservative Islamic government that the president argued was more reflective of the values of Turkey's majority. The AKP-led government had also deemphasized the role of the Turkish military, which was the dominant institution in the government founded by Mustafa Kemal Atatürk and the secular regimes that followed largely in his footsteps. Not everyone agreed with this drift of the government and the military, which had previously overthrown governments in 1960, 1971, and 1980, often positioned itself as the defenders of Atatürk's secular legacy. The military, with its history of NATO engagement and cooperation with the United States during the Cold War, was also perceived as the most pro-American of Turkish government institutions.

I was interested in what many of these changes meant for Turkish foreign policy and the role of their soldiers and civilians in Afghanistan. On a personal level, since I had lived in Turkey a decade before and my grandparents had met in Istanbul, I wanted to know what this might all mean for the future of the country. How had the contributions of both soldiers and civilians affected Turkey? What did the war in Afghanistan look like from the perspective of a country that is both Muslim and a member of NATO? With the Turkish military playing a major role in Afghanistan, it seemed necessary to me to look at the experience of Turkish soldiers in the country to understand what the conflict meant for Turkish contractors.

As we drove into Ankara, General Zik chose his words carefully when talking about the government and its relationship with the military. Zik and the other soldiers and contractors I spoke with were far more comfortable discussing the role of Turkey in Afghanistan, which was much less politically controversial. Both sides of the political spectrum seemed to agree that Turkey's involvement was a good idea. For the more Islamist, this reflected a reaching out by Turkey's Muslim brothers, while for the more secular, this was an international humanitarian mission that reinforced the bonds between Afghanistan and Turkey that Atatürk had established with the Afghan king, Amanullah Khan, in the 1920s.[4]

Many in Turkey were happy to discuss how the Afghan and Turkish states shared initially parallel and then drastically divergent histories in the twentieth century. It was an opportunity for Turks to reflect on their success over the past century in comparison with many of their neighboring countries. After seizing power, Atatürk successfully pushed through a radical set of reforms that, at times brutally, transformed the remnants of the Ottoman Empire in the Anatolian region into a secular republic based on a Turkish ethnic identity. This helped link Turkey more closely to Europe and stimulated economic growth, but it also left unresolved questions about non-Turkish minorities in the country who were often either killed or oppressed and about the unresolved relationship with Islam. It also left Turkey at odds with almost all of its immediate neighbors and much of the Muslim world.

After meeting with Atatürk, Amanullah Khan pushed a similar set of reforms in Afghanistan. His government was not as strong as Atatürk's was at the time, and Afghanistan was more rural and difficult to control. His reforms were too rapid and met with strong resistance from religious leaders and tribes in the rural areas.[5] A chaotic civil war had ended these reforms in 1929, and similar reforms were not attempted again until the rise of the Soviet-backed government in the 1980s. During these years, the tracks of the two countries separated: Afghanistan's economy stagnated, while Turkey, despite some economic ups and downs and several military-backed coups, reasserted its geostrategic position between Europe and the Middle East.

Zik and other Turkish officials I interviewed were more interested in some of these political and historical trends than other military and contractors I interviewed, perhaps because of this shared history or, perhaps more simply, as members of the International Security Assistance Force (ISAF) coalition, they were constantly talking about ways to "fix" Afghanistan. They often suggested that the problem in Afghanistan was tribal, with the ethnic divides that persisted, and this explained why Afghanistan had never really developed as a nation. While I didn't always agree with the idea that Afghans did not possess a national identity, there was still a clear contrast with the Turkish trajectories over the past eighty years, despite some of their cultural similarities. Turkish military officials were particularly well positioned to assess the difference between the Turkish strong-state model and the weaker Afghan case, even if this was a way of reflecting on their own success over the past century.

As we chatted and ate dinner later that night, General Zik's stylish dress and dignified smile made me think of photos of Atatürk that still dominate many of Turkey's public buildings despite the AKP's attempts to move away from such symbols of the secular nationalism of Turkey's founding. Zik seemed to have an almost endless supply of energy. He was officially retired, but was the editor of a Turkish military journal and was working on his third book all while finishing course work for his PhD. He confided that his dissertation advisor was significantly younger than he was, and he found this a little strange, but was amused by this role reversal after decades in the military giving orders. His friends and subordinates obviously thrived off his energy and leadership, and after he called several colleagues and friends who served with him in Afghanistan, nine of them gathered at his pleasant suburban apartment on a bitterly cold Sunday morning, all happy to be interviewed. And so my research in Turkey began.

In 2009, Turkey was the lead nation for ISAF, coordinating its work around the country. At various points, Turkey has also been in charge of two of the so-called provincial reconstruction teams, located in Jowzjan in the north of the country and Wardak in the center.[6]

My interviews with the Turkish officers and retired military personnel made it clear that as a member of NATO since 1952, Turkey simply did not have as much to prove as Georgia did. Instead, its military was a well-established part of various international efforts. While Turkish-American military relations had weakened somewhat following Turkey's refusal to support George Bush's invasion of Iraq in 2003, the person-to-person ties that remained between the two militaries still mattered more, and these came up in interviews again and again.

Almost all the former Turkish military personnel I interviewed had trained at some point in the United States or with NATO troops at various bases in Europe, and they often had fond memories of these experiences. One artillery officer reminisced fondly about a sports car he bought while attending the U.S. Naval Postgraduate School in Monterey. In California, he said, it seemed that everyone had a sports car, so he got one too. The one thing he marveled at in particular was the Seventeen Mile Drive road through Pebble Beach, where drivers had to pay a toll to rich people so that they could drive on a road and see how rich the residents were.

Still, as the only Muslim country in NATO and with deep historical ties in the region, the experience of Turkish soldiers and contractors was bound to differ from their other NATO colleagues.

THE BEAT OF THE DRUM

The Turkish military museum in Istanbul puts on a daily show where soldiers in Ottoman dress sing traditional military songs while surrounding a drummer playing the *mehter*, large stand-alone drums that the leader hits energetically with two mallets for which the style of music is named. Before the show, a narrated slide show on the history of the mehter and the other traditional instruments used reenactors to demonstrate its use among Turkic horsemen as they rode across the plains, conquering what is now Central Asia and the north of Afghanistan. Afterward, a group of musicians in historical Ottoman costumes came out with their instruments and formed a semicircle. For the following thirty minutes, each mesmerizing drumbeat reverberated through my body. The little girl in the row in front of me actually started marching in the middle of the performance.

On our way out, Asil, a recently retired captain, explained that the Turkish military also had a more modern military band, but that he preferred this one. There's something about the drums, he told me as we went off to explore other exhibits. After spending ten days conducting interviews in Ankara, I had flown to Istanbul, and Asil volunteered to take me around the museum, which was housed in the former military academy where Mustafa Kemal, who later took the name Atatürk, studied. There's even a classroom where mannequins of his classmates sit with their hands raised. One seat is empty so that visitors can have their picture taken participating in the lesson.

The museum is a reminder of how much of the debates over Turkey's past are actually debates about its present; even seemingly far-off issues such as questions about Turks in Afghanistan tend to reflect back on the more pressing current political issues. As we walked through the museum, the installations told a very specific story, one that avoided some of Turkey's history of belligerence in the region.[7] The Turkish invasion of Cyprus, largely perceived internationally as an act of aggression, was referred to as a "peace operation" to defend the Turkish population in the north of the country. This militaristic approach, however, had shifted in recent years.

Over the past decade, the AKP attempted to revive Turkey's influence among its neighbors, while, not coincidentally, emphasizing its Ottoman history, when the Turks ruled from Vienna to Yemen. The promotion of this imperial past could be found in foreign policy attempts reaching out to their "Turkish brothers" in the Central Asian -stans and on the billboards in Istanbul, advertising soap operas based on the stories of Ottoman soldiers and their families. This was in contrast to the Cold War focus on its relations with United States and Soviet Union and, most notable, a shift away from the EU, membership to which appeared to be more distant than it had been a decade before.[8]

The AKP's regional approach was also notable in its turn away from military force toward more subtle forms of soft power, relying on business and trade. The strategy was referred to as an approach of "zero problems" with its neighbors, drawing on Turkey's growing economic influence.[9]

As the Cold War ended, the opening of Eastern Europe and the Central Asian Republics (where primarily Turkic languages are spoken), the privatization of industry in Russia, and the collapse of Yugoslavia opened new markets for Turkey, which had been relatively cut off from its neighbors during the post–World War II period. Turkish Airlines flights to neighboring countries increased from nine a day at the end of the Cold War to twenty-nine a day by 2010.[10] Previous governments tried to expand Turkey's commercial and economic interests in the region, but these attempts were more successful in the past decade.[11] Simultaneously, Erdoğan's government relaxed visa restrictions, encouraging visitors from neighboring countries and the movement of Turks abroad.[12] Erdoğan and other Turkish diplomats took Turkish businessmen with them when they went on trips abroad, and Turkish trade was a key topic of discussion.

Perhaps the most extreme example of Turkey's move away from military approaches to security concerns and toward a more economically driven approach to its neighbors was the engagement of the Turkish government with the Kurdistan Regional Government (KRG) in northern Iraq. This was not simply about political ties, and when Erdoğan made the first trip of a Turkish premier to Iraqi Kurdistan , he was there in part to inaugurate the Erbil airport, which had been built by a Turkish joint venture.[13] These diplomatic ties encouraged rapid Turkish investment in the area. Between 2007 and 2013, Turkish exports to Iraq climbed from $2.8 billion to $11.9, and the number of Turkish firms working in northern Iraq went from 485 in 2009 to three times that number in 2013.[14]

Toward the end of my walk through the museum, I entered a small room featuring more recent NATO operations and the role of Turkey in Bosnia, Somalia, and Kosovo. There were photographs, equipment, and a few flags the Turkish forces had flown there, but nothing on Turkish operations in Afghanistan. This was in part because Turkish operations in Afghanistan were ongoing. It also reflected something about the complex nature of Turkey's unique position in Afghanistan.

NATO PARTNERS OR FELLOW MUSLIMS?

As the only Muslim majority member of NATO and by far the largest contributor of Muslim troops to the international effort, Turkey's role in Afghanistan was different from that of other foreign troops there. After finishing our tour of the museum, we went to a café down the street for coffee, and Asil discussed how his colleagues in Afghanistan were constantly aware of their image in other Muslim countries, echoing much of what I had heard from General Zik and his former colleagues. The war in Syria, remaining tensions over Cyprus, and continued Kurdish separatist activity all gave Turkey reason to avoid being

seen as simply America's lackey in Afghanistan, something that was
much less concerning for some other allies like Georgia. The Turkish
military and diplomats attempted to carve out their own unique position in Afghanistan.

All the Turkish officers I spoke with went out of their way to cul-
tivate relationships with Afghan officials while in Afghanistan and felt
that they were much better at this than many of their ISAF counterparts. Part of this was emphasizing cultural similarities. While Turks,
particularly in the military, are generally relaxed in terms of their practice of Islam, their joint celebration with Afghans of religious holidays
and other aspects of their shared Muslim identity certainly helped them
build relationships, Asil said (though it probably also helped that the
members of the Afghan military I knew also tended to be some of the
least religious of the Afghans I met). One atheist Turkish officer I interviewed said that he faked his belief to help encourage friendships with
Afghans, while another told me that his wife would never wear the veil
in Turkey, but that didn't stop him from letting the Afghans he worked
with believe that she did.

In addition to these religious ties, several of the smaller minority
groups in the north of Afghanistan were Turkic, and Uzbek and Turkmen, languages spoken by some communities in the north, are often
mutually intelligible for Turkish speakers. This led to some interesting
dynamics since the Turks often preferred to work with these minority
groups, particularly as opposed to Pashtuns, whose language is much
more distant and difficult for Turkish speakers to pick up. Those who
probably benefited most from the perceived close relations between
Turks and Afghans were the numerous Turkish officers who took part
in some of the training and partnering programs set up by NATO that
provided foreign mentors to key Afghan civilian and military officials.
These "mentors" did not always fit the typical master-apprentice mold.

Asil said that he was a twenty-seven-year-old lieutenant when he was
assigned to mentor a general in his mid-fifties. The general had fought
against the Taliban but was old enough to have served in the Afghan

military during the Soviet period, before Asil was born. At first, Asil felt there was almost nothing he could say to help the general, telling him that he should consider him a younger brother. The general, however, brought him to all of his high-level meetings, in part because within the Afghan military, it was something of a status marker to have an international mentor. Eventually they grew closer, and the general began seeking out his advice regularly. Those who were probably least comfortable with the relationship, he continued, were the U.S. Army Rangers who were also working with the general and were a little shocked by the lieutenant's young age and the amount of influence he had.

Repeatedly in interviews, members of the Turkish military expressed how close they felt to the Afghans they worked with, a sentiment that several felt even more strongly after watching some of their NATO allies treat Afghans with more distance and less sensitivity. While some officers expressed cynicism when Afghans called them "brothers," both the military and civilians I interviewed felt a common history and religious identity gave them a starting point for conversations and relationships that other NATO members, and particularly Americans, often lacked.

These ties could make things much easier for members of the Turkish military. One officer said that in 2009, when he was in charge of the Turkish contingent at the joint international base at Kabul Airport, at that time under the command of the French, there were riots across Kabul following the burning of several Korans at Bagram Airbase—the same incident that had led to the attack on Pratik's compound. As these riots grew, the officer said, there were concerns that protesters might breach some of the compounds near the airport. The French commander of the base came into the Turkish officer's office and asked if it would be all right with the Turks if they took down all the flags except the Turkish one (normally all the flags of the various militaries stationed there hung together). The officer replied that it wouldn't bother him, but what mattered more was whether it was okay with the leaders of the other coalition countries. The French commander went ahead and took down the other flags, and a few minutes later, the officer's American counterpart barged into his office, demanding an explanation for why the American flag had been taken down. In the meantime, the rioters had moved on, he said.

Such stories about Afghan tolerance or even embracing of the Turkish presence may have been exaggerated, but they were common among both military and civilians I interviewed. One of the favorite and most common tales, several interviewees said, was that if you had a Turkish flag on your shoulder, you were certain to avoid Taliban attacks. Rumors went around that American troops and others were putting these flags on their uniforms and on their vehicles to avoid attacks. I had trouble finding any verified examples of this, but it was clear that the Turkish approach to the intervention in Afghanistan was different from that of most of the other ISAF countries.

IN IT FOR THE LONG HAUL

One of the central questions for all the members of the Turkish military I spoke with was how the particular relationship between Afghanistan and Turkey might play out in the future. This longer-term thinking was one of the key differences between how Turks and some of their NATO partners approached their roles in Afghanistan.

Turkey, like many NATO members in the country, contributed to provincial reconstruction teams, which were located around the country and helped coordinate development and relief efforts, backed by a military force.[15] The key difference was that the two Turkish provincial reconstruction teams (PRTs) located in Wardak and Jowzjan were led by civilian diplomats, not military officers, as was the case of other NATO PRTs.[16] With twenty-nine civilians and seventy-nine military personal, the civilian-to-military ratio was much higher than the U.S.-led PRTs I had visited, which had hundreds of troops and fewer than ten civilians from the United States Agency for International Development (USAID) and the State Department.[17] Working with the Turkish International

Cooperation and Development Agency (Turkey's version of USAID), the Turks I interviewed who had worked on these PRTs felt that Turkey's contributions had been far more effective than the work done by other PRTs, despite not having near the same level of funding. All told, the spending was not inconsequential; led by these PRTs, the Turkish government had spent $2 billion on humanitarian investment by 2014.[18]

The difference in the Turkish approach did create some tension, the former public affairs officer from the Wardak PRT told me. While relations between the Turks and the nearby American base were good, it was still clear that the approaches did not always complement each other. He compared the infrastructure and development projects that the Turks ran with those run out of the nearby U.S. base. The Turks built four elementary schools, a vocational high school, and several infrastructure projects. When they found out that many of the apples grown in Wardak were being exported to Pakistan, where they were stored and then reimported to Afghanistan after the growing season was over, they built two cold-storage facilities. This allowed the farmers to keep the apples in Afghanistan and avoid reimporting them.

In contrast, the U.S. programs were smaller. They included a fountain in the town center and the rehabilitation of a park. Most of these were so-called Commander Emergency Response Programs, a set of programs that essentially gave U.S. military officers money to do development or respond to other community needs immediately.[19] These quick impact projects were not meant to lead to real development, the public affairs officer said, just to win hearts and minds in the counterinsurgency effort. But still, these things were more related than the Americans understood, and it was clear that he was not the only one who saw a glaring difference between the long-term effectiveness of the Turkish and American projects; the Afghans did as well. The public affairs officer said that he hoped this would be the basis for a longer-term Afghan-Turkish partnership.

Turkey's president agreed, declaring that "Turkey will stay in Afghanistan even after all the other forces have left, and will leave only when our Afghan brothers and sisters tell us, 'Thank you, now you can go home.'"[20] But as security deteriorated in Kabul and other parts of the country, it was difficult for the Turkish military to continue promoting some of these ideas of brotherhood. But what would become of Turkish-Afghan brotherhood if they were not treating each other like brothers, some of the Turks wondered.

For example, Asil described driving down some of Kabul's heavily fortified roads and realizing that almost all of the compounds he was driving past were completely inaccessible to Afghans even though it was their country. They see these convoys, he said, and yet they can't approach them. In the end, the best the Afghans could hope for was a job cleaning toilets on one of the bases, he said. All this made life cheap and made it easier for him to understand how young Afghans would want to join the insurgency.

Ultimately, the longer-lasting impact of the Turkish presence might have much less to do with the military. There were also Turkish-funded schools in Afghanistan, and in the wake of the failed coup in July 2016, the Afghan minister of education was quick to reassure people that the Turkish government was not going to cut off funds for the eight thousand Afghans in these Turkish-sponsored schools.[21] In an effort to build ties with some of the potential future leaders of the country, there were also a significant number of scholarship programs aimed at young Afghans, like my previous research assistant Sediq, who had received his BA, MA, and a free flight home on a Turkish military plane each year, all courtesy of the Turkish government. Sediq was now fluent in Turkish and, unsurprisingly, a great fan of the country.

Afghanistan was not left out of the AKP's push to make Turkish business interests an increasingly important aspect of their foreign policy. Doing this in a war zone was trickier than expanding trade with its more stable neighbors, so their policy had to use some creative approaches. To this end, Turkey took advantage of the fact that they had one of the largest and most centrally located embassies in Kabul, soliciting donations from Turkish firms to build a hotel on its grounds for future Turkish businessmen abroad.

Still, this connection between Turkey's military and private business efforts was not always as smooth as it could have been. I found in my interviews of former Turkish military personnel in Ankara and Istanbul that they, similar to their American counterparts, often knew little about the growing role of their fellow countrymen working as contractors in the country.

CHAPTER 13

TURKISH ENGINEERS AND OTHER HEROES OF THE INTERVENTION

THE WOOD MERCHANT LEADS THE CHARGE

"There's a story about this that we like to tell each other," the deputy head of the South Asia Desk at the Turkish Ministry of Foreign Affairs told me while we were meeting at his office in the Foreign Ministry. His assistant served us Turkish coffee, and his chief of staff was sitting nearby, listening. "The story," he continued, "is about a Turkish foreign affairs bureaucrat who dealt primarily in oil."

In Istanbul he met an old Russian, who said that he could take him and his team to an untapped oil field in Russia, ripe for development by Turkish firms. So the man and his team went with the Russian. First, they flew to Moscow. Then they took a smaller plane to a city in Siberia. Here they got on a train and traveled for a few more days into a remote area, and, indeed, there in the middle of the Russian tundra was a rich oil field.

On their way back home to tell their colleagues of their discovery, they stopped in a small neighboring town first to buy some souvenirs. When they arrived, however, walking down the street in this town in the middle of nowhere was a man who appeared to be Turkish. Shocked at seeing another Turk thousands of miles from home, they asked him what he was doing there, and he replied that for years, he had imported wood from the region and sent it to a small town in Turkey's southeast. He was there organizing some trucks to transport more wood and was about to send another shipment home. Far from being the first Turks there, this simple wood seller had beaten the oilmen by years. "This," the official at the Ministry of Foreign Affairs said, "is how it is with Turkish businessmen: wherever you go, you are sure to find one, long before the government knows anything about it."

While he told the story in a farcical manner, there were clear parallels with the Turkish presence in Afghanistan. The Turks sent troops to Afghanistan, at first a few hundred, and later, during the surge, more than a thousand. The numbers hovered around 1,600 for several years, before dropping down to 500 in 2016.[1] Beyond this, several large Turkish firms oversaw construction of some American bases, and some of the Nepalis I interviewed had done security for these firms or worked on nearby projects. There were few precise figures, however, about the extent of the Turkish civilian presence.

A useful comparison with more public data is the Kurdish region of northern Iraq, where construction business for Turks had boomed in recent years. In describing this growth, Christina Bache Fidan, a policy analyst I met with at a think tank in Istanbul, argued that Turkish firms in northern Iraq were particularly well positioned due to their "depth of experience; proven track record of performance in Russia, Central Asia, the Middle East and North Africa; contiguous land borders with the KRI (Kurdish Region of Iraq) for low-cost transport of equipment and materials; availability of Turkish Kurds to act as intermediaries with Iraqi Kurds; and similar cultural values."[2] While Afghanistan might not have a contiguous border, it was still easier for Turkey to reach it than almost any other industrialized country except Iran and Pakistan, and the Turkish Kurds who served as intermediaries in Iraq were replaced with Afghans who spoke Uzbek or Turkmen, both closely related to Turkish.

Yet as the official's story about the wood merchant suggested, government policy was still far behind the businesses already abroad, and once I started sifting through lists of smaller international construction firms, it was striking how many were Turkish. Like the Nepali government, the Turkish government appeared to have a rather limited view of the role of their citizens in Afghanistan. At the same time, Turkish contractors took advantage of their status, as both Muslims and members of a NATO country, to move in the higher levels of the contracting hierarchy. This gave them certain opportunities, but still left them exposed to some of the vulnerabilities of contracting.

When we discussed the number of civilian Turks working in Afghanistan, the deputy head of the South Asia Desk estimated that there were about a thousand.[3] Analysts from one Ankara-based think tank put the number at 3,000.[4] As I surveyed Turkish companies working in Afghanistan and individual contractors working for American or international firms, it became clear that both numbers were underestimates. When I asked the Turkish contractors themselves, their estimations tended to be between 4,000 and 8,000 depending on the year. According to a list from the Turkish embassy in Kabul, sixty-two Turkish firms were working in Afghanistan, but I quickly found contractors for companies that were not included on the list. Turkish officials seemed to be familiar only with those working for large firms or directly with the Turkish embassy.[5] As I tracked the number of Turkish companies in Afghanistan and got better estimates of their number of employees, I was increasingly convinced that the contractors were far more accurate in their estimates than the government officials were.

Clearly the Turkish government was having trouble keeping up with all the Turkish wood sellers who made their way to Afghanistan.

AN ENGINEER ABROAD

As I spoke with Mehmet in the dim lobby bar of a hotel in Ankara's busy shopping district, it was difficult not to be infected with some of his enthusiasm. As he spoke, he lowered his voice, leaned his head down toward me conspiratorially, and then laughed, stabbing with his finger to make a point. He had a twinkle in his eye that might have just been from the candles, but it still gave the impression he had played dice with the devil and won. And in a sense, perhaps he had.

Mehmet was not inexperienced when he went to Afghanistan in 2002 at the age of fifty-five. Like many of the other Turkish contractors I interviewed, he had worked abroad before, but his experience was more varied and colorful than that of most others. He had jobs previously in Europe, the Middle East, and all across Turkey. His background was in engineering, but he did contract work in a variety of roles as he moved from project to project. He did some construction management and some program planning and eventually came to focus primarily on design—the field that he thought Turkey had the biggest advantage in in terms of skill-to-cost ratio.

I tried to keep up with him as he rattled off the names of the various companies he had worked for in countries across the region. He jumped between a series of engaging tales from his younger years of crooked deals, bribe-seeking officials, Israeli security restrictions, and eventual deportation from Russia. (He had left all his things there and spent several months trying different airports and land crossings to retrieve them, but was barred at each one.) He started his own firm in Turkey, but as the economy struggled, he followed the large Turkish firms that began looking more regionally for work.

In the 1990s, the First Gulf War and the economic shock of increasing integration with world markets had slowed the Turkish economy. Turkey's engineers, architects, and other skilled workers like Mehmet found themselves struggling to find jobs at home. New government policy meant they now had an easier time working abroad than in the past. Their salaries were lower than those of their Western European counterparts, but Turkey's strong technical schools meant that they had skills that were lacking in many of their less industrialized neighbors. As a result, through the 1990s Turkey looked abroad to help drive economic growth, and trade with those states bordering on Turkey increased from $4 billion in 1992 to $82 billion in 2008.[6] Mehmet joined this outflow and spent time building skyscrapers in Israeli, worked on power plants in Russia, and eventually did rehabilitation work on the Stari Most bridge, a sixteenth-century Ottoman landmark in Bosnia-Herzegovina that had been destroyed in 1993 by Croat troops.

Mehmet said that the U.S. invasion of Afghanistan was, at first, just one more opportunity. His peripatetic lifestyle until that point suggested that when he arrived in Kabul to work as a subcontractor on the construction of the new U.S. embassy, he would not be there for long before moving on to a new country and a new project. But he found that his experience dealing with crooked bureaucrats, cultural differences, and the shifting demands of construction projects in conflict zones, all while attempting to maintain engineering standards, made him suited for the undeveloped and unregulated construction boom in Kabul. Thirteen years, later he still had an office there.

Mehmet was initially hired by a Turkish firm that had a contract with KBR, which won a $100 million contract to build the new U.S. embassy.[7] The firm he was working for ended up in a dispute with KBR and never completed its work on the project. This was not an entirely uncommon or unexpected occurrence in the contracting world, but it meant Mehmet found himself unemployed in Kabul, but with $45,000 in cash for the work that he had done, along with some important initial contacts.

Instead of returning immediately to Turkey, Mehmet used this money to set up an office with an Afghan finance manager who had been let go from the same Turkish contractor he had been working with. After a couple of months Mehmet found the manager's brother an even savvier business partner and brought him aboard. The brother spoke Dari, Pashto, English, Turkish, Urdu, and German. He had important contacts in the Afghan business community and could also easily converse with international officials and businesspeople. This was exactly the type of partner that Mehmet needed for his work in a multicultural environment. His partner joined him for meetings at military bases to show that his company had an Afghan face and could also help oversee the Afghan construction teams working for the company, which were as large as a hundred at several times.

Mehmet's first big break came in the form of a contract from a mid-sized American public health nongovernmental organization (NGO) to design a series of hospitals the NGO was hoping to build across Afghanistan. The contract was just for designing the hospital plans, but after the construction contractor who was supposed to be building the hospitals proved unable to do this at the costs they had estimated, the NGO had Mehmet take over construction too. This evidence of his ability to deliver a tangible product despite the political and economic uncertainty in Afghanistan at the time gave him a reputation in the construction community. Soon he was jumping back and forth between a series of small and midsized projects, designing and building everything from blood banks to media centers for Afghan universities.

There were plenty of economic reasons for Mehmet and other Turkish contractors' involvement in Afghanistan beyond the lower costs Turkish firms charged in comparison with its Western neighbors. There were deep connections between various Turkish construction, transport, and logistics firms and their larger American counterparts. In a similar manner that most of the Turkish officers I had spoken with had trained in America at some point, the Cold War partnership had encouraged a close working relationship between the Turkish and American military-industrial complexes. These men had already worked together on projects in Iraq, Bosnia, Germany, and elsewhere. The companies for the most were second-tier contractors that subcontracted from American or multinational companies with direct contracts with the U.S. Department of Defense or the United States Agency for International Development. This meant a smaller but still sizable piece of these contracts for figures like Mehmet who were willing to embrace the dangers and uncertainty of working in Afghanistan.

One of the reasons that Mehmet had been successful was that his was a family business, and after his initial success, his wife, an architect, came to Kabul to work with him. One of his sons was also an architect and the other a civil engineer. Both worked for the family business at various points over the past decade, and his office expanded and contracted based on how many projects he was working on, usually having between seven and twelve staff. Most were Turkish or Afghan, but he was not averse to hiring others, and the office was a diverse place.

As the surge continued and funds kept pouring in, Mehmet said he was amazed at how much money was available and how much of it was being wasted. There seemed to be almost no logic to it. At one point, he said, he had a proposal to renovate the epidemic hospital in Kabul, which had been badly damaged in a fire.

The situation was dire, he said. Since the structure was unsound, all the patients were in tents in the hospital garden. His proposal for renovations, which he estimated at only $75,000, was under review by the U.S. embassy. During the contract negotiations, U.S. officials visited the site several times, Mehmet said. Each time, Mehmet would wait for them at the hospital grounds, and they would arrive late in long convoys, since each vehicle had only one official in it, the other seats taken up by their armed escorts. Because traffic was bad, this meant the embassy was paying for a convoy with a dozen security contractors to cross the city, taking hours, burning gas, and using security guards who were probably needed elsewhere. Within a few visits, he was convinced that the embassy had already spent enough to rehab the entire hospital just driving officials out to visit it and quibble about a couple thousand dollars more or less in construction costs.

Other cases were more egregious. Large U.S. and international firms would collude on bids to drive up the prices of contracts, he said. American firms often had "secret partners," who would put in fake bids to drive up prices. These were accusations I had heard before from many Afghans, and Mehmet tended to agree with the claims that social ties were allowing both certain contracting companies and a handful of well-connected rich Afghans to dominate the contracts the United States was offering.[8] Turkish contractors like Mehmet seemed more aware of some of the regional politics and social ties that facilitated much of this nepotism and corruption than their American counterparts.

Mehmet gave as an example the case of Mansur Maqsudi, an Afghan of Uzbek origin who had become a U.S. citizen. Maqsudi had first made headlines by marrying the daughter of Uzbek president Islam Karimov and becoming a major force in Uzbekistan's corrupt business world. In his eventual bitter divorce from the president's daughter, it had come out that Coca Cola had colluded with the Uzbek president to strip Maqsudi of his business interests in the country in retaliation.[9] Mehmet said that Maqsudi had moved south and was now working with members of the Karzai family in various legal and illegal businesses in Afghanistan. Businessmen with ties in both America and Afghanistan were, after the big companies, the ones who were able to manipulate the system the best.

All of this made the business climate in Afghanistan hostile to outsiders like Mehmet; he was able to secure small contracts that did not attract much competition, but he could never rise above a certain level on the contracting hierarchy. He was also frustrated that the Turkish government had not done more to advocate for Turkish businesses in the country. The Afghan government owed him $300,000 for a building project inside the presidential compound that his company had completed in 2014. The problem, he said, was that he signed the contract under the Karzai government, and when Ashraf Ghani came to power, he shifted the contact from the president's office to the Kabul city government. Later, for a reason he did not understand, the contract was shifted again, this time to the Ministry of Urban Development, which now refused to pay it because they were not been the ones who signed it originally. He had no idea what to do next. Although his dealings had been clean on the contract thus far, he was now thinking that it would be necessary to pay a bribe just to get it paid. Paying $10,000 in cash to an official to finalize the payment was certainly better than losing the entire $300,000. This was the problem of being a Turkish company: the Turkish embassy would do nothing to help. If the same thing were to happen to an American company, he said, you could be sure that the American embassy would be able to exert pressure to get the contact paid.

Mehmet felt that there were other risks too, considering that the Turkish embassy provided Turks in Afghanistan less diplomatic support than some other countries. For example, his friend had been with several Americans at an international restaurant when one of the rather infamous alcohol raids took place. These rarely seemed to be designed to actually stop the drinking of alcohol but were thought to be thinly veiled messages by the Afghan government to the international community.[10] After the arrests, Mehmet said, American soldiers showed up within an hour at the police station to pick up the American citizens, while his Turkish friend remained in prison for six months.[11]

Now, at age sixty-seven, Mehmet concluded that he was getting older and the contracts were drying up, which meant that it was probably a good time to be winding down business. He still had an office in Kabul, and his employees were wrapping up a couple of open contracts, but he was no longer actively looking for more work in the country. Instead, he had taken $1.5 million of the money he earned in the country and bought a small cement factory in Turkey, which would pay him a steady return without too much work or risk. "I turned $45,000 into $1.5 million," he said, "and all in all, that's not too bad."

Despite this, as we finished our tea, I noticed that as he described his new factory, some of the enthusiasm and the sparkle from his eye faded in comparison with how he had described his early, risky days in Kabul.

A BUSINESSMAN

Mehmet and other Turkish contractors faced hurdles in Afghanistan that were different from the challenges a Nepali might face and demonstrate some of the ways that nationality shaped how contractors were treated in Afghanistan. While by far the majority of Nepalis in Afghanistan were in security or more basic labor, I had interviewed a handful of more business oriented-Nepalis, and Mehmet's case had many parallels with Akash, a Nepali businessman I met in Kathmandu earlier in the fall, who also happened to get caught up in an alcohol shakedown by the Afghan police.

Akash was from a Newari family, a traditionally business-oriented caste in Kathmandu, so business was in his blood, he said. He laughingly told me that his four-year-old son had set up a coffee shop in his kitchen that morning and tried to charge him for bringing him his coffee. So perhaps it was not surprising that in Afghanistan, Akash would be particularly attuned to business opportunities.

We met in a trendy café in November 2015, three months before my time in Turkey, in the trendy Lazimpat neighborhood of Kathmandu, home of the Shangri La Hotel and several other five-star facilities. It was the type of place that businessmen and diplomats working at the nearby embassies frequented. While we seemed to be interrupted every two or three minutes by a text or call on one of Akash's two cell phones, these intrusions did little to slow him. He had some business experience before going to Afghanistan, he said, and earned an MBA in Nepal, but not unlike Mehmet, he mostly wanted me to understand the strides that he made through wheeling and dealing among contractors in Kabul.

He got his start using the Nepali connection with security companies and first went over in 2005 on a three-month contract to assist with an audit of the British security company ArmorGroup's work in Afghanistan. Part of this involved checking prices in the bazaar against prices listed on receipts, and slowly he began to get a better idea of the complexities and opportunities of the market in Afghanistan.

Akash did not like the idea of simply working for other people, so when his contract was up, he planned on returning home to be with his family. In the meantime, he had become close to one of the senior British employees of the company. The Brit had been in and out of Afghanistan for twenty years and was planning on setting up his own guesthouse, which would subcontract out from ArmorGroup, and he wanted Akash to join him. This type of arrangement was common among security firms and other contractors, where employees would leave the company with handshake deals and set up subcontracting companies taking business from their former employer. This enabled the firm to maintain a 15 percent cut, or "administration fee," that most contracts had and do less work, while allowing the former employee to greatly increase his income.[12]

The guesthouse they set up was a more upscale version of the camp that Pratik worked for. Akash made it clear that he did not just want to work for the guesthouse; he wanted in on the financing and a percentage of the profits. His British partner agreed and promised him a place big enough for him to bring his family and the potential for more business opportunities. So he joined the venture while still working for ArmorGroup on reduced hours.

Together Akash and his partner opened a guesthouse, along with a restaurant and bar, that became popular among the security ex-pats in Kabul. It was open during my time in Kabul, though was a little loud for my taste. On the other side of several layers of security, the bar had strobe lights and five-dollar beers, which the mostly male patrons were guzzling quickly.

But not everything about the setup went smoothly.

While foreign non-Muslims were technically allowed to drink alcohol, there were rumors of other illegal activities there, in part fueled by the presence of pretty Thai bartenders. Regardless of whether there were illegal activities going on, the rumors were enough of a pretext, and the police raided the compound, hauling eight of Akash's Nepali staff off to prison and threating Akash and his wife at gunpoint.

These problems seemed minor six months later, when Akash's partner was shot and killed while driving. Akash was required to go down to the station to identify his mangled body. He was told that it was a random robbery, but he later heard that the shooters had been wearing Afghan police uniforms. That his partner had been coming back from the bank with pay for the staff also suggested that it might have been an inside job and whoever arranged the killing could still have been working for Akash.

As Akash began rethinking his business options in Afghanistan, ArmorGroup was acquired by G4S, a much larger security company.[13] The new arrangement meant new contracts for procurement of food and other essential supplies in particular, creating an opening for Akash. Previously much of the procurement had been done internationally, with the U.S. government requiring many contractors under them to import all food and other supplies, including even water in some instances.[14] Akash said that ArmorGroup had been importing rice from India and bananas from the Philippines, all of it gathered by a subcontractor in Dubai who took a cut in addition to inflating prices.

Now the U.S. government was demanding that their contracting companies "Afghanize" their approach and buy products locally. The hope was that this would stimulate economic growth while also contributing to the counterinsurgency efforts by winning local hearts and minds. So Akash, a Nepali, registered an "Afghan" company, called Afghan Services Ltd., to help the military Afghanize.[15] He relied on the contacts he already had from his work with the guesthouse to start procuring food and other goods for G4S. While G4S remained his primary customer over the next five years, as in Mehmet's case, one contract led to another, and soon he was supplying a variety of goods for international companies across Kabul, relying on a team of Nepali and Afghan employees that eventually grew to fifty.

Despite his success, that there was no Nepali embassy in Kabul meant that there was no one to offer him support, and business became increasingly challenging. Nepalis were considered particularly easy targets, and as he had seen when his guesthouse was raided, it was difficult for Nepalis to get out of jail once they were arrested. As the years went by, he became increasingly upset with the corruption and roadblocks thrown up by the Afghan government. Americans, he said, could show their passport and just walk into any government office. For Nepalis, all the Afghan officials did was see the opportunity for bribes; he concluded that the immigration and visa office treated Nepalis like an ATM. Even when he knew that he filed the correct paperwork, officials would deny him whatever license or permission he needed, holding out for a cash payment. Akash knew his rights, but he had no way of making sure anyone protected them.

Mehmet had some similar experiences, but both Afghan and American officials treated Turks differently. As a Nepali, Akash was never able to get some of the larger contracts that Mehmet could and was considered an easier target. As long as Akash paid the right bribes and stayed in his "correct" place on the food chain, his business could grow slowly, but there were limits. Like Mehmet, he also knew that international contracts were likely to dry up in Afghanistan, so he started reinvesting back at home, buying a small hydroelectric plant and even setting up an indoor soccer facility that Kathmandu's young people could rent out by the hour.

Back in Kathmandu, Akash's major complaint was that after spending so many years working in Afghanistan, he felt that he was not appreciated enough by some of the other internationals he had worked with. While those working for DynCorp or other companies that were funded directly from the United States often had an easier time securing American visas, Akash had already been rejected twice already. He thought it was probably because they were suspicious of the Afghan visas in his passport. He tried applying for visas from Australia and Hong Kong as well and had run into issues there too. "With a million dollars in the bank, I can live like a king in Nepal," he said, but he was not able to visit his sister in San Francisco or take his son to Disney World. While the unregulated business of the war economy allowed Akash to make some good money, it would never be the money that Westerns could get and it did not lead to access to a more international, legitimate business world. There was a reward, but the risk that predatory Afghan officials would take over his businesses was eventually too much.

Nepalis in Afghanistan trying to set up their own businesses were left in this no-man's-land where they struggled. The long history of Turkish investment abroad and the at least minimal support of the Turkish government gave Turkish businessmen an advantage. But I would come to see that other, simpler variables eventually did even more to shape one's experience and seemed to constantly conspire to ensure that only the big firms made the real money.

CHAPTER 14

BUILDING AN EMPIRE?

SIZE MATTERS

While both Mehmet and Akash set up small firms that rarely employed more than a hundred people, other international subcontracting firms were enormous. Large Turkish companies like Yüksel and EMTA had a dozen construction sites across the country, some of which employed hundreds of employees each. Large bases like Bagram would generate a surrounding constellation of compounds where these companies worked and, for example, the gas for the base was stored. While not as large as some of the U.S. firms they were subcontracting from, these Turkish companies still had such a substantial presence that it was possible to work in these contractor compounds, occasionally commuting to other international bases and feel far removed from the Afghan world outside the compound walls. Comparing the various types of Turkish businesses contracting in Afghanistan shows some of the diverse ways in which companies became embedded in both Afghanistan's warlord-based economy and more global economic flows, and the consequences for individual contractors.

While the laborers working for most Turkish firms were Afghan or Pakistani, the majority of the administration and engineers on these projects were Turkish.[1] These workers tended to be either experienced skilled workers who made a living working abroad or were recent university graduates who could be hired cheaply. When I asked Onur whether his work in Afghanistan was his first job abroad, he laughed and said that it was his first job ever. Growing up in Ankara, the son of a schoolteacher and a forest engineer, he was interested in the preservation of historical monuments. Upon graduation, however, he found jobs were scarce in the preservation world and that his engineering degree was sought after if he was willing to travel abroad. Luckily, after graduation he found that Yüksel was hiring.

By the time Onur arrived in Afghanistan in 2011, Yüksel was one of the largest of the Turkish firms active in Afghanistan and had moved from being a secondary subcontracting company to receiving funds directly from the U.S. government for many jobs. Much of the success of large Turkish companies like this was due to their connections to both the American and Turkish militaries. Yüksel was closely tied to Turkey's right-wing Nationalist Movement Party and according to its annual report was the sixty-sixth biggest business in Turkey, doing work in "Libya, Jordan, Saudi Arabia, Qatar, the United Arab Emirates, Iraq, Afghanistan, Uzbekistan, Kazakhstan, Georgia, Ukraine, and Romania, and [with] company representatives in Ghana, South Sudan, Ethiopia, Algeria, Oman, Pakistan, Turkmenistan, Azerbaijan and Russia."[2]

The Yüksel Group is divided into twelve subsidiaries, a common strategy in the contracting world. Breaking a company into pieces like this has several advantages in terms of taxation, but it also enabled contracting companies to subcontract to different parts of their holdings. This practice is common in defense contracting and makes spending more difficult to regulate.[3] These firms were often run by former employees and had compatible goals. Even better-known, supposedly more transparent companies like DynCorp had spun off companies like DynCorp Technical Solutions and DynCorp Systems and Solutions, which had their own U.S. government contracts too.[4]

Onur admitted that he did not have much of a sense of what was going on in Afghanistan before taking the job. Life on the vast Bagram Airbase was a shock for the young man. It was a strange mixture of both extreme ease mixed with periods of great discomfort. Onur enjoyed that all of his meals were provided for him, his laundry done, and there was little he needed to do in order to go about his day-to-day business. In his free time, he went to the base's gym and played video games, mostly with other Turkish engineers. You could almost completely forget that there was a war going on, he said.

In other ways, Bagram felt much like a jail. Over the course of the year he worked there, he left its walls only a handful of times, mostly for meetings at the main office in Kabul and once to get his passport renewed. You could only move around the base with a badge, he complained, and before he got his security clearance from the Americans running the base, he was required to check in every morning, getting his eyes scanned to confirm his identity. This process was more onerous for the Afghan laborers who were not allowed to live on the base. At times of heightened security, the process of checking in and out of the base could take three or four hours, depending on the length of the line. This meant that for a 7:00 a.m. start time, some Afghan workers were required to line up by 3:00 a.m., which caused logistical and morale issues on the work site.

The work itself was not so different from some of the tasks he would have performed if he were working in Turkey, he said. It was the conditions and variables that were so different. One day a project would be moving along smoothly, and the next day a bombing or a protest near the gate to the base would mean that none of his workers would show up.

While the differences between the experiences of Turkish engineers and Nepali security contractors were pronounced, there were similarities in how both groups had to navigate the conflict settings. The Turks generally had more secure paperwork and visas allowing them to travel around Afghanistan and Iraq, but since most were not contracting directly from the U.S. Department of Defense, they were only entitled to certain support (like base access or use of the dining facilities), while figuring out much of the rest on their own.

Another Turkish construction manager I interviewed who worked for a much smaller firm told me that when contracting from the Army Corps of Engineers in Iraq, the company had to scout the site where the construction was to be done before submitting a proposal. In this case, the contract was for building at a small Marine outpost in Fallujah. Larger companies would normally be able to arrange site visits complete with security teams. As a smaller firm with no contacts among the Marines and only a limited amount of experience in Fallujah, the construction manager decided the only way to get to the camp was by taxi.

Once in the taxi, however, he realized the United States usually used different names for the camp site than the locals did, and the driver had trouble finding the base. So the construction manager stopped and asked an Iraqi policeman, but he didn't know either. Then the manager remembered seeing a Marine roadblock up the road; since the camp he was looking for was going to be a Marine camp, surely they would know. The taxi driver would not go near the three Humvees at the block, so the contractor got out and approached slowly on foot. When he was about fifty feet away, every Marine from the three Humvees leveled their weapons at him. He shouted that he was Turkish and was just looking for directions, lifting his shirt to show that he had no gun, but they just kept their guns pointed at him, shouting for him to get down on the ground. For a terrifying moment, he was sure he was going to be shot by the very people who were paying his company. After a moment, the Marines approached, frisked him, and let him go when they saw his Turkish passport but refused to help him with directions. He eventually made it to the camp, but later, on his way back to his office, he heard shooting and found out that the roadblock he had stopped at had become a part of a firefight. The driver took the long way around to get back.

Not having the established relations of the bigger firms made smaller firms struggle and take extraordinary risks. It also made work in Afghanistan a game of chance. Small companies that secured a large contract could be set for years, but failure to secure one would bankrupt them or worse. For Onur, Afghanistan gave him some good experience, and he was able to jump to other jobs in Libya, Iraq, and, eventually, by the time that I interviewed him, he found a job that he was enjoying in Turkmenistan.

In Turkmenistan, security was more relaxed, which made life far more bearable. Onur was still hoping to return to Turkey sometime in the not-too-distant future or perhaps pursue an advanced degree in historical conservation, which he still studied on quiet nights on construction sites. It had been difficult thus far to translate some of his international experience onto a résumé that would get him an equivalent job in Turkey. So for the time being, he said, it seemed to make more sense to keep moving among international contractors where he had good connections.

FORWARDING FREIGHT AND FREIGHT FORWARDING

While larger Turkish firms tended to be well integrated into the military-industrial complex that established itself in Afghanistan, smaller firms often had to take more risks and be more flexible. For some firms, this meant not having personnel on the ground and relying primarily on Afghan contacts.

A few days after talking with Onur, I drove away from the historical monuments on the European side, over the Bosphorus, into the heart of Istanbul's industrial growth area. After passing from Europe into Asia, the views of the Bosphorus Strait gave way to high rises crowding the highway. There was little of Istanbul's charm in the warehouses, shopping malls, and highways, yet when it came to shipping between Europe and Asia, the offices in these neighborhoods controlled a vast amount of the commerce that crossed from one continent to the next.

Materials generally came by train or ship into a warehouse on the European side, before being taken across to their primary warehouse on the Asian side where they were then picked up by trucks taking them to destinations all over the Middle East. The fighting in Syria had slowed business, several of those people working in the shipping industry I interviewed said, but the recent U.S. dealings with Iran gave these businessmen hope that there would be more economic opportunities to the east to make up for the instability to Turkey's south.

I met Özgen in a trendy café with glass cases of delicate desserts. He arrived before me, and when he stood, he towered above me, his hand enveloping my own when we shook. A mountain of a man, he managed to immediately give off the sense that he would get freight to wherever it needed to go without the client worrying about it. He had spent his entire life working in shipping and ran Freight Logistics, a firm he had taken over from his father, who had founded the company forty years before. He didn't speak English very well, but brought along a young woman who worked in his office to do the translating for us, and so, in a subdued manner, he began to explain how his business operated, waiting every few minutes for the young woman to translate.

Shipping to Afghanistan was only part of the shipping that Freight Logistics did, Özgen explained. It also did a lot of shipping to Iran, Turkmenistan, Iraq, and other parts of the Middle East. You couldn't rely on shipping to Afghanistan exclusively as a business, he said, since things had been so up and down there. In general, when dealing with unstable areas, you needed to spread your risk around. Those that were most successful had a variety of contracts moving goods to a variety of destinations, to make sure that if one conflict got significantly worse, it would not bankrupt the entire company.

In 2014, Özgen sent six hundred trucks to Afghanistan, mostly transiting through Iran and Turkmenistan. In 2015, as the Afghan economy stagnated in response to the international troop withdrawal and the protracted political negotiations around the Afghan presidential election, the number of trucks his company had sent dropped to just sixty. This year, as of late February 2016, things were looking better, and they had sent ninety trucks so far.

Most of his customers were Turkish or Afghan firms that had some sort of subcontract from the U.S. military, but he was open to working with others as well, he said. He made many of his contacts in Dubai, and this was where most of the transactions took place and the contracts were signed. Other companies had offices in Turkey, and he'd arrange the transport directly with them.

On occasion, he said, he'd get an order from an Afghan company that didn't have access to the banking system (or, he implied, might be trying to avoid government detection and taxation). In cases like these, the owner of the company came to Turkey or sent a representative with cash in a belt or hidden in a vest. Özgen chuckled as he described this, demonstrating how these Afghan businessmen would pull money out of their belts, pockets, and other various spots all around their bodies until there was an imaginary stack on the table in front of him.

Once he had a deal with the business, they would agree on a pickup point (usually in Turkey), a drop-off point in Afghanistan, and a time frame. Özgen relied on Afghan partners who did most of the border paperwork in exchange for cash payments. These "agents" had relationships with militias across the north of Afghanistan and would supply armed escorts for convoys heading into more dangerous areas. The guards on his convoys were all Afghans, but larger firms he knew hired Nepalis and other international security contractors.

When he started contracting out trucks for delivery in Afghanistan five years before, he said he did not understand the geographic or political layout of the country. Later he decided it was best to think of Afghanistan as divided into two zones: a somewhat safe northern one and a dangerous, Taliban-controlled southern one. Shipments to the north were usually easy; it was the ones in the south that he had to worry about. Just a couple of months ago, the Taliban captured two of Özgen's fuel trucks. His Afghan partners were able to get the shipments released after a couple of weeks, but it took a payment of $35,000 to make this happen. It was a headache, but this was how business worked in Afghanistan, he sighed. More common, Özgen said, was theft. Whenever he shipped cable through Turkmenistan, it always seemed that each roll was missing a few meters once it was delivered. His assumption was that the border guards were cutting off a portion for themselves, though it was also possible that the drivers had something to do with it. As a result, he spoke of a "tolerance ratio," the cost of doing business as usual.

The bureaucracy along the border could also be problematic. One of the issues, he said, was that compared to other countries he shipped to, there is no customs law in Afghanistan. This was not entirely true; there was a law someplace, but it was not regularly enforced or transparent. So for businessmen in Turkey with effective Afghan agents, there was no sense in understanding customs law because they were always working outside of these agents.

At this point, Özgen took out his phone to show me photos of two flatbed trucks stranded at the Afghan border with Turkmenistan. They were donations from UNICEF that he had been contracted to ship to Kabul. It was not clear why the customs agents were refusing to process the paperwork, he told me, but the driver wanted to show him that there were no problems with the cargo itself so had sent him the photos. Probably they were just waiting for the agents to arrange a bribe, he concluded.

Overall, he said, he was not too worried. His agents had good connections in the Afghan government. I asked if ever worried about losing a shipment or payment from these agents who were so far off; it seemed that they could have disappeared easily. He shrugged and said he didn't have any choice except to trust them; besides, they all came from good families.

This phrase "good families" was interesting, because it suggested the ways in which the Afghan economy had increasingly grown connected to a mafia of political and economic elites who were involved in industries like transportation and construction. His company, one of several Turkish transport companies shipping to Afghanistan, was the point at which money was transferred from the more transparent world of contracts and bank transfers to cash exchanges and bribes. The way the various pieces came together into a global web was striking to think about while sitting in this quiet Turkish café: the U.S. Department of Defense contracted KBR to build a base, KBR contracted out some of the building of that base to a Turkish construction company, the Turkish construction company contracted a logistics firm, that firm paid Özgen to transport its materials, and then Özgen paid cash to Afghan agents, who paid local militias to run protection rackets. In some cases, those local militias then paid the Taliban not to attack them, meaning that U.S. dollars were flowing straight to the enemy.[5]

Özgen continued to describe how his business represented the thin veil between companies that claimed to be a legitimate part of the global business work and the patronage-based economics of militia and tribes in Afghanistan.

Taking a sip of my tea, I could envision the money the U.S. military was spending in Afghanistan slipping away.

WAR IN A GLOBAL WORLD

Listening to Özgen describe the ways in which he arranged for convoys to cross borders, how contractors and subcontractors and sub-subcontractors were linked together, it was easy to see how the global economy shaped pathways across Asia in support of the U.S. war in Afghanistan. Everything from trucks of fuel to Nepali laborers took these routes and were intertwined.

Still, there were other times when nationality and the quirks of individual countries mattered. Turkey's major role in subcontracting in Afghanistan suffered a rather severe setback in 2011 when Afghanistan started more rigorously enforcing some of its own tax laws on international companies working in the country. The U.S. embassy interceded on behalf on some American companies, but Turkish companies, particularly smaller ones like Özgen's, were left with less direct diplomatic support. Rather than pay back taxes, some of the largest Turkish firms simply sold all the equipment they had brought to the country and left. Others made deals with local partners. As Özgen found with his associates, Afghan businessmen were quick learners, and some of those who had been working for Turkish firms had set out on their own.

The attempted coup in Turkey in July 2016, six months after my interview with Özgen, also raised questions about Turkey's position in global politics. Erdoğan's anger toward the United States for not doing more to condemn the coup, along with his rapprochement with Russia, shifted Turkey's orientation. As conversations with General Zik and others indicate, it is often the personal ties that are the most long-lasting and meaningful.[6] After decades as a member of NATO, Turkey's officers, as well as members of its military-industrial complex, have deep connections with their NATO and American counterparts that makes it likely that Turkish firms protected by Nepali security contractors will continue to follow behind American troops, wherever they happen to intervene next.

For Turkish firms, big and small, their complaints were more over how they were treated by the Turkish government than their roles internationally. Several contractors complained to me that American dollars tended to go to American firms, yet there did not seem to be a real preference for Turkish firms when the Turkish government was looking for contracts abroad.[7] Turkish businessmen I interviewed felt they were much more successful seeking out American dollars than they were in securing contracts from their own government to support the Turkish development projects across Afghanistan. The Turkish contractors I interviewed found the authoritarian drift of Erdoğan's government worrying, but better support for Turkish businesses abroad would go a long way toward making the government palatable. In a few cases, contractors felt the Turkish embassy was helpful in giving them assistance, but most thought the embassy did not help them enough.

While international politics shaped the experience of Turks in the war generally, in a subtle way, there were moments when the experiences seemed particularly vivid. It was in a series of interviews among a certain subset of contractors, those that had been detained by the Afghan police or those kidnapped by the Taliban or other groups, that the importance of citizenship and nationality stood out. As I found a growing number of contractors who had been through such ordeals, I started comparing their experiences with each other and with my own to look at the effect of nationality on contractors in this outsourced war.

CHAPTER 15

DETAINED

ROADBLOCKS AND SPEED BUMPS

The Afghan police station hallway was dimly lit, but it was difficult to tell whether this was simply a lack of electricity or was done on purpose to make the place feel more menacing. The glow from the office at the end of the dark hallway should have been inviting, but more so, it suggested that all those entering should abandon hope.

As I crossed international borders tracing the routes of the contractors and soldiers, the contrasts among countries became more striking. In some ways the borders mattered little, but in other ways, they greatly shaped the experience that these individuals had in Afghanistan. Citizenship was particularly important when it came to determining the extent to which your home country might (or might not) help you in moments of crisis.

For soldiers, it was almost always clear what role their governments had in protecting them legally within the country, which was done almost entirely through the military bureaucracy.[1] Staff Sergeant Robert Bales, for example, who went on a rampage killing sixteen Afghan civilians near Kandahar, was sentenced to life in prison in a U.S. military court after being brought back home, despite calls by Afghan officials to have him tried in an Afghan court.[2] Other cases involving U.S. soldiers were similarly adjudicated in the American legal system (usually in military courts). For civilians, their legal protections were less clear and less predictable.[3]

This was particularly true when contractors ran into problems with the Afghan police. The extent to which a country assisted a citizen who was detained really depended on the country and the person who was detained and if that person had any connections. Your experience of war depended on who you are and who you know.

As an American citizen, I was warned by the State Department that "no part of Afghanistan should be considered immune from violence, and the potential exists throughout the country for hostile acts, either targeted or random, against American and other western nationals at any time. . . . There is an on-going threat to kidnap and assassinate U.S. citizens and Non-Governmental Organization (NGO) workers throughout the country. Afghan authorities have a limited ability to maintain order and ensure the security of citizens and visitors."[4] I was encouraged to register with the American embassy if I insisted on working in Afghanistan, but beyond this, there was no promise of assistance other than being provided a list of lawyers if anything happened to me.[5] In reality, there was much more help behind the scenes, but this was never transparent and it was difficult to tell how to access it. Americans did enjoy certain privileges that others did not, which become clear one of the times I was detained.

That early evening in 2006, I was driving with a group of international friends to a restaurant in the center of Kabul. Not long before, a couple of Westerners working for the United Nations had been accused by the Afghan government of illegally communicating with the Taliban. This was one of the first major disputes between President Karzai and the international community over the tendency of the United States to ignore the sovereignty of the Afghan president.[6]

When my friends and I were stopped at a checkpoint, two rather enthusiastic policemen asked us for our passports. Until that point, the Afghan police had a fairly laissez-faire attitude toward internationals in the country and tended to wave them quickly through checkpoints. Two of us had our passports with visas in them, but the others had left their passports at home. Our driver protested for a while, trying to convince them to let us pass, but the policemen were clearly pleased with their catch and they instructed the driver of our van to follow their car to the central police station. There, they paraded us in through the dark lot of the police station, down two long hallways, and directly into the office of the police chief.

The police chief himself had little interest in us when it became clear that we were civilians working for nongovernmental organizations who just did not have their IDs with them. But to dismiss us immediately would make his officers and, by extension, himself look bad, so he had us all take seats on the plush sofas in his office. While we sat there, he went through several files on his desk, made a couple of phone calls, and turned back to the files on his desk. While he took his time, I began looking through my phone, wondering who I could call to extricate us from this fastest.

I had the U.S. embassy consular number in my phone, but I had little hope that anyone there would be helpful. I knew several Americans working in the embassy, but they were not the types of officials who would have pull in a situation like this. It did not seem like a good idea to call them unless things were really dire. Several of my Afghan friends worked in the government or had government connections, but it was not clear which of these would be most helpful. In some cases, I could imagine these connections doing more harm than good if they were someone the chief was not aligned with. This was common in the complicated and turbulent world of Afghan politics. Luckily for us, after what felt like an hour but was probably only ten minutes, the chief turned to us and in Dari, with the occasional English word thrown in, gave us a brief lecture on how we should carry our documents at all times.

He stopped talking abruptly, and we were dismissed.

While the police had little interest in detaining us, this was certainly not true of all contractors, particularly when they ran afoul of

Afghanistan's complex networks of corruption and patronage. These moments of crisis bring up questions about citizenship and rights that were central issues for many of the international contractors working in Afghanistan.

TURNING A BLIND EYE

Samet was a Turkish contractor I interviewed in Ankara. Slight and dressed in khakis and a button-down shirt, he seemed more of a businessman than a retired military officer. He had served in the Turkish NATO forces in Afghanistan for a tour, working on logistics and transportation.

Through this work, he became close to several high-ranking Afghans, including the chief of police of Kabul, whose position was extremely influential and coveted among Afghan officials. The two had taken part in officer training and exchange programs together. After returning to Turkey, he had at one point escorted the chief of police on tours of Istanbul and Ankara. He even introduced him to his wife, he noted, something that was remarkable for Afghans, who rarely introduced women to foreigners.

Not long after his Afghan tour, Samet retired. At that point, in his early forties, he was still fairly young and found himself bored. When a friend told him about a well-paid position doing similar work for a company that arranged International Security Assistance Force (ISAF) support for the Afghan National Police (ANP), he applied. The firm had good relations with the Turkish military and hired retired Turks in particular. With his connections with the chief of police, he was an appealing candidate and was quickly selected for the position. Upon arriving back in Kabul, he fell back into some of the same routines and in general enjoyed both the work and the relatively freer lifestyle that contractors were afforded from his time as a soldier. Their accommodations were nicer and while he still worked long hours, the company's laxer policies meant that there was less oversight or need to constantly report to those above him. Of course, for some, this created an opportunity.

After several months on the job, as he became more involved in the project, he realized a rather large amount of money was missing from the company's ledgers. It seemed that the Afghan contractors, who did most of the actual work supplying and transporting the goods, had overbilled by $12 million. With so much missing, those working in his company must have been aware, he concluded. He had seen similar cases of overbilling before, which was usually swept under the rug, but that had been a few thousand dollars perhaps. This amount of money was simply too much to be ignored.

At first, he kept quiet, he said, but he realized that if he did not report this, he would be part of the corruption as well. So in a meeting with several ISAF officials, he carefully brought up that there might be some issues with the contracts he was overseeing, particularly that money was missing. He was hoping they would go after the case themselves, but instead they instructed him to "find the leakage points." As he dug, he realized it was not several points, but just one point and that most of the overbilling had been done through what were essentially fake contracts to an Afghan firm owned by a key ally of President Karzai.

After following up with his bosses, he went to meet with the three-star Afghan general who oversaw the distribution of the contracts with his translator. The general was coy at first, Samet said, suggesting that if he spoke with some people, the general could get him most of the money back, but he told Samet that the company with the fake contracts should be allowed to hold on to at least some of the money. Giving it all back would be too insulting to them, the general said. Samet protested that the work had never been done. As he continued, the general grew agitated, cutting off Samet's translator. Things escalated quickly. Soon the general was yelling not just at the translator, but at other men who worked for the general who were in the room. Amid the confusion, Samet said he had little idea what was going at all.

Then two guards hurried into the room and most of the other men started looking down. The guards walked over to Samet, motioned for him to rise. They pulled him out of the room and dragged him to a small cell in the adjacent building. "You need to know when to resist and when not to resist," Samet said, and it was clear it was best to go quietly, at least until he could figure out how serious the situation was. They did not take his phone, and in the dark cell, his first call was to his old friend, the police chief, who did not officially outrank the three-star general who had just detained him but was in what was generally considered a more influential position. But there was no answer. He tried another few times, and it became clear that the police chief had turned his phone off.

Unable to call colleagues from his company who could very well have been involved in the corruption as well, there was not much for Samet to do but sit in his cell. I asked whether he considered calling the Turkish embassy, and he said no; there was nothing that they could have done. In a situation like this one, it was personal connections that were most valuable; he needed someone who could talk to the general reasonably.

A couple hours passed before Samet's phone rang. It was the police chief's assistant, calling him back. Samet told him about the detention, but the assistant seemed to know what had happened already and told him there was nothing the police chief could do. It was a sign in itself that the chief himself would not come to the phone. Apparently Samet had angered people high up the ladder, close to the presidential palace, which even the police chief could not trump. At this point, Samet said, "I lost all hope. He was like an uncle to me." Sitting in the cell, he was convinced there was nothing that could be done about the corruption in Kabul. Several hours later, the guards came and opened the cell door, leading him out of the compound and onto the street. Perhaps some pressure had been applied to the general, or perhaps this was only a warning to Samet. There was no explanation.

The general had clearly sent his message: Don't go digging around in corruption cases. Not long after, Samet handed in his resignation and went back to Ankara. He was scared he had angered the wrong people and might end up killed. With the $12 million still missing, it was the lack of clear rules in the corrupt world that contractors and officials mingled in that Samet said worried him the most.

And, in such a system, what about those with no connections?

LOCKED AWAY

From Samet's story I was convinced that because he was a Turk, the Afghan police felt that they could push him around far more than they could an American. The general consensus among the Turks I interviewed was that most of those at the embassy had little sense of what Turkish contractors were doing or what support they needed. But the influence of the embassy, the presence of which was perhaps simply an indirect form of reassurance, was far greater than it was for the Nepalis, who had no diplomatic support whatsoever in the country. Teer Magar, whose story I discussed in the Prologue, knew this well.

As Teer described during the afternoon I had spent with him in Bandipur in central Nepal, he had been in a similar situation as Samet, but instead of an afternoon in a cramped cell, he had ended up spending almost three years in a prison built primarily for insurgents. Teer had gone to Kabul with six friends, believing they had jobs lined up working for a catering company. When it turned out the vacancies had already been filled, they were forced to stay for a while with a Nepali broker in Kabul while he waited to hear if there might be another opening. Finally, a job did open up and a broker contacted him about working for Contrack International, an international construction firm that worked on several international bases.

Though not as well known as some of the larger primary contractors, through 2011, Contrack had received over $1.5 billion for "fast-track, design-build projects" in Afghanistan funded by the U.S. Army Corps of Engineers, according to testimony by the company's CEO before a congressional hearing.[7] Company literature explained this included a $32 million contract to build Afghan National Army (ANA) Camp Scorpion SOF (Special Operations Forces) Training Faculty and another for $127.6 million to build the runway and "aircraft apron" at Camp Bastion.[8] The company was large enough to draw explicit attention from the Taliban, and in December 2012, a blast tore through one of its offices located close to several key international bases on the Jalalabad Road in Kabul, not far from the Pinnacle. In response to the attack, Taliban spokesperson Zabihullah Mujahid stated that "the company was under our surveillance for a long time."[9] In general, however, it was difficult to find much information about the company, and Teer was surprised that he had been hired since he had no construction experience.

This lack of experience, it turned out, did not matter much, since the management of the firm had other work in mind for him. When Teer was hired, the United States was simultaneously tightening security on various bases in part as a reaction to a rise in attacks on American soldiers by Afghans working for or partnering with the coalition, while also funding more and more building projects.[10] It was becoming harder for companies with no security clearances to get workers on the base. Procedures varied at different bases, but some had such backlogs for security check requests that even if a company had been approved to work on the base, that did not mean all the workers for the company would be approved in time to work on the project.

Often these workers could come onto the base, but only if "escorted" by someone who had already gone through the clearance process. This was also typical at places like the U.S. embassy where I could visit with a visitor's badge, as long as I was in the constant presence of a designated escort. If, on one of these bases, you had someone with a security clearance who was allowed to escort five workers on the base each day, you could charge other firms to escort their employees onto the base, provided you had an employee essentially spend the day with them. And so once Teer Magar got his security clearance, this was what he was hired to do. Since it really didn't mean much other than picking a few people up at the gate each morning and sitting around while they worked, he did not complain. Besides, both the food and accommodation were good, he said. Initially he worked at a base in Kabul, but then he was shifted to a base farther south where tensions over who controlled contracts and certain projects were more complicated. There were rumors of local contracting firms, in particular, threatening each other with violence and the kidnapping of family members, particularly during the bidding on contracts.[11]

Teer said he was not sure who exactly made the phone call, but he thought it was probably some rival of his construction firm. He remembered the exact date and time that the National Directorate of Security (NDS), Afghanistan's intelligence service, descended on his office. NDS had been tipped off that someone at Contrack was spying for the Pakistani's intelligence agency (ISI). When NDS raided Contrack's offices, the plans of American and Afghan bases that they found there seemed to confirm the accusations. Despite Contrack's pleas that they had had these plans since they were building on these bases, Teer, who just happened to be in the office at that time, was arrested with three other Afghans who were there with him.[12] At first Teer and his coworkers thought Contrack would pay a bribe and they would be released, so he was not concerned that he was not given a lawyer and that he had not spoken with anyone from the company. But a week passed and then another. Still no one from Contrack contacted him. During this time, he was also unable to contact his family, who was growing concerned about him.

Later he heard the rumor that Contrack had considered paying a bribe for his release, but the head of NDS became involved in the case, which made the price of the bribe too high. Finally, he got to speak to a lawyer on a mobile phone that a guard handed him, but the lawyer didn't seem to have much information either. The lawyer was in Kabul and was not willing to go down to Helmand, where Teer was being held. It was too dangerous, he told Teer, and hung up. Later, when Teer tried to call the lawyer back, the lawyer said that he was no longer on the case. Teer suspected that he had been threatened by the NDS. That was the end of his legal support.

By the time his trial arrived, Teer had no one and was unable to understand much of the courtroom proceedings; he was supplied no translator and did not speak Pashto.[13] At the end of the trial, he was sentenced to eighteen years in prison. After the group appealed, the sentence was raised to twenty years. This was how Teer found himself living in the British-built maximum-security prison in Helmand.

A LONG ROAD HOME

More than anything, prison was boring and lonely, Teer said. He had always been skilled with his hands, so he passed the time by making small objects out of tiny beads and glue. Other prisoners noticed him doing this, and eventually he set up a business, where he would buy beads the guards smuggled in; make airplanes, chairs, and other small toys; and then sell these to the other prisoners to give to their children on visiting days. Teer himself received no visitors.

The only person Teer spoke to from outside prison was a French member of the International Committee of the Red Cross, who checked on prisoners in the facility.[14] He brought Teer books, and finally he was able to send a letter home to his wife. For twenty-five months, Teer's only means of communication were his limited English and the few words of Pashto he had been able to pick up. He gradually grew more despondent.

Teer Magar's luck finally improved purely by chance when a compassionate Nepali journalist based in Kabul heard about his detention. Subel Bhandari was the bureau chief in Kabul for Deutsche Presse-Agentur, a German news agency. He had filed a variety of stories on the conflict since 2010, before that working for the AFP in Kathmandu. Informally, he also took an interest in the fates of his various fellow countrymen in the country; another Nepali security contractor I interviewed, marveled at Subel's wide array of connections and the way he got around town on his motorbike. It was a level of mobility that no other Nepali seemed to have. Working in Kabul, Subel had heard rumors that a Nepali had been in prison in the south of the country for several months, but his inquiries with the Afghan government led nowhere. It is not easy to find prisoners in the Afghan system, and he was not sure where the prisoner was being held or what exactly he had been charged with. As it happened, one day he was shopping in one of the Western supermarkets in the center of Kabul when he overheard two other Nepalis who had worked with Teer before his arrest talking about Teer's case.

With their help, he got some more specific information about the case and was able to get in touch with Teer, but it was not clear there was much he could do to help. As a reporter, he did not have much influence within the Afghan government, and without a Nepali embassy in Kabul, Subel was forced to rely on personal contacts and the limited support of the few Nepali diplomats who passed through the country. One of these diplomats was the Nepali ambassador to Pakistan, Bharat Raj Paudyal, who had worked hard to get Nepalis released from similar situations, but the ambassador was limited in terms of both his ability to get information about Nepalis in such situations and influencing the Afghan legal system. Eventually Subel managed to contact an old school friend in Kathmandu whose father had recently been elected prime minister of Nepal. He convinced his friend to lobby his father to write a letter and send it to Bharat Raj Paudyal to use to pressure the Afghan government.

These diplomatic wheels began slowly turning, and after twenty-five months in Helmand, thanks to Subel and Ambassador Paudyal, Teer was shifted to a prison in Kabul. He enjoyed visits by Subel and finally being able to speak to a fellow Nepali. Teer's initial optimism that came with his transfer was dulled when he realized the new prison was much rougher and his sentence had not yet been reduced. Almost immediately after he moved in, another prisoner stole all of his personal belongings, and still there was no official word from the Nepali embassy.

While all this was going on, Ambassador Paudyal had a rare meeting with President Karzai on a diplomatic visit. At this meeting, among other discussions, the new ambassador also finally delivered the letter from the Nepali prime minister requesting leniency for Teer. Karzai read it and issued a decree. A few days later, Teer was on a plane headed home.

Throughout his story, I was surprised that Teer was not very critical of either the company that he worked for when he was arrested or the Nepali government. He thought that Contrack had made an effort to get him a lawyer, even if the lawyer never showed up. Moreover, he was pleased that they had continued paying 50 percent of his salary for the first twenty-two months he was imprisoned.[15] As for the Nepali government, he concluded that it didn't have the resources to do much.[16] He was more frustrated at the Afghans in the competing firm: he felt they had manipulated the security forces to ensure that his company would be raided.

More than in other interviews, Teer Magar pointed to the similarities between Nepalis and Afghans. Both countries, he said, were places that had gone through much hardship, and the ordinary people were at the mercy of those in power. Perhaps his time in prison had helped him sympathize with Afghans in a position similar to his own. I asked him if he would consider working abroad again, and he quickly said he would, though not in Afghanistan, since "I don't think they will let me back in."[17]

I noticed toward the end of the interview that beneath his vest, he was wearing a shalwar kamees, the long robe traditionally worn by Afghan men. The one he had on was rather stylish, the type he might wear to a wedding. I asked him about it, and he said that they had been required to wear such outfits in prison and that they were comfortable, so he brought this one home with him. When our conversation ended and I watched him head back to the poultry farm that he now co-owned with his brother, I wondered how the experience hadn't left him more resentful. Was he still in shock from the years of detention? Or was I simply too caught up in Western assumptions about justice and individual rights?

There was a terrible gap between the support that American and other Western civilians in Afghanistan received and the abandonment of those from Nepal and other poor Asian countries. In Kathmandu, thinking about the experience of Teer Magar's wife and what it must have been like for her not to be able to contact her husband, I began to look into the types of assistance that were available to these migrant workers.

MIGRATION ASSISTANCE

If the Nepali government was not going to help Teer Magar and his wife, who should? In Nepal, a country highly reliant on international development aid and with a government that only lightly regulated the presence of nongovernmental organizations (NGOs) and civic groups, it made sense that there would be a series of NGOs or international nongovernmental organizations (INGOs) devoted to the protection of the 500,000 Nepalis who annually went abroad for work. And in Kathmandu, there were a number of organizations large and small, such as the International Organization for Migration (IOM) that facilitated migration and worked with economic refugees. Most of these organizations were focused on the migration of Nepalis to the Gulf or sometimes Europe. I began my research at one of the NGOs that was known for being dynamic and taking on more difficult cases.

"Maiti Nepal" is a reference to the *maiti*, or the family that a woman is born into (as opposed to the family of the husband that she eventually moves to with marriage as is the practice in Nepali's patriarchal society). A maiti is a safe place thought of nostalgically, and Maiti Nepal was set up to help Nepali women who had been trafficked for sex or otherwise taken advantage of abroad to return to Nepal. On our way to meet with the information officer, Dawa pointed out the director who was just leaving the office. Anuradha Koirala was a winner of a CNN Heroes award in 2010 and was something of a Nepali celebrity. She graciously welcomed us before continuing on her way to other meetings.

Maiti Nepal was doing important work assisting Nepali women who had been exploited abroad and working to help them reintegrate themselves into a patriarchal society that more often than not blamed them for being abused and exploited. The compound had a dormitory where women could stay if their families would not let them return home. They even had a school for their children on the premises.

When I asked the information officer about women getting trafficked to Afghanistan, he said that they had heard rumors about this but had not yet seen any cases. He was hesitant to make any conclusions without any evidence. There were men who were being trafficked, he said, but as we knew already, almost all of these cases were of voluntary trafficking, and public education campaigns about the dangers of such work were their only real response.

Maiti Nepal did, he continued, have a recent case of a woman who had been rescued after being trafficked to Syria. I asked how she had gotten there, since it seemed a little surprising that there would be much profit in trafficking women to a place where a conflict was so active when they could be sent to more profitable and stable places in the Gulf. The officer wasn't sure, but perhaps she had been brought there as domestic help before the fighting started, and once the war got too dangerous, whoever had been holding her had simply let her go.[18]

Maiti Nepal's spokesman said they did not have any plans to do follow-up work in Syria. More generally, when they recently had cases in Kenya and Tanzania where there was no Nepali government presence, there was not much they could do. He said that while they often spoke of rescuing women, what the organization did primarily was provide women with resources back in Kathmandu after they were brought back to Nepal. Little was being done to help those outside the country. The organization helped the government carry out much of the work that it should have been doing to protect its citizens, but it did not have the resources to do much in places where Nepal did not have a diplomatic presence, which made Afghanistan, Iraq, and most other conflict zones out of reach.

I visited more of these organizations, and it became apparent that none of them were focused on the plight of Nepalis in places like Afghanistan. Some were simply understaffed and overwhelmed by the issues that they were dealing with. Others felt that they should focus their efforts on Gulf countries where more workers were being exploited as domestic laborers or in dangerous construction jobs, or in Malaysia, which had the greatest number of Nepali workers in it. Afghanistan and Iraq were far more difficult to work in. Besides, some reasoned, most of those in places like Afghanistan had made the conscious choice to go to a war zone. What did they expect would happen?

I asked one lawyer working for a legal aid foundation what he would do if was contacted by someone like Teer who was arrested in Afghanistan. He said that he would call the nearest Nepali embassy and have them help. He echoed the sentiment of the information officer at Maiti Nepal suggesting "the NGO's role is to support the government." So it seemed that in a case like Teer's, the informal assistance of a kind-hearted journalist like Subel was genuinely his best way to seek assistance.

It seemed from the perception of the contractors that the Nepali government was not interested in doing much more than extracting bribes from workers as they returned home. Companies had little concern for the welfare of their workers, and the United States was more concerned with conducting audits of the money being spent than paying attention to whether companies were protecting their workers. Even supposedly independent NGOs felt that they did not have the authority or ability to help these migrants in need.

Teer was released (and then only after almost three years) only because a Nepali reporter overheard a conversation in a grocery store and had taken it upon himself to help. It was a sad reflection on the unequal protections that foreigners of different nationalities had in Afghanistan. These imbalances could be even more extreme and distorted in some surprising ways, when it came to how insurgents and other criminals treated foreigners that they had kidnapped in Afghanistan.

CHAPTER 16

KIDNAPPED

ANTALYA

My last stop in Turkey was on the southern coast. Here where the mountains drop dramatically into the eastern Mediterranean sits Antalya, an ancient Roman city that is now home to a little over a million inhabitants, most of whom cater to the tourists who come down to lie on the beaches and take in the landscape. When the weather is right, Arda told me, you can ski and swim on the same day, "like in California."

The city, though, was glum when I arrived in February 2016. While it was still too early for the spring tourists to start arriving, there was increasing concern that many would not come this year, scared off by the war in Syria. Although Antalya is geographically much closer to the Syrian border than Ankara and Istanbul are, there were fewer Syrian refugees in Antalya than there had been in the larger Turkish cities, since it was not seen as a helpful transit point for those headed to Europe. What really damaged southern Turkey's tourist industry was the recent shooting down of a Russian jet by the Turkish Air Force for violating its airspace. Russia had largely gone on its own by supporting the Asad regime in Syria and had frozen relations with Turkey after the incident.[1] Just the day before my arrival, Turkey claimed its airspace was violated again, calling the Russian ambassador to Turkey to apologize. While many of the tourists who normally came to Antalya were European, in recent years, a growing percentage were Russian, and with new direct flights to Russia, they now made up perhaps 50 percent of Antalya's visitors.

Arda, who was a Century 21 real estate broker and relied largely on income from vacation properties, joked that the economic downturn was going to bankrupt him and that he was going to start looking for a new job. This would not be the first time his career path had taken a sharp turn. He and Hakan, who had worked for him in Afghanistan, sat in Arda's bright office telling stories about working as contractors in the construction business. The longer I listened to their stories of narrow escapes and business deals gone sideways, the less reason I had to doubt that Arda would find some way to land on his feet even if Antalya's economy went south. For me, his story was compelling because it highlighted the different ways that the Turkish government approached the war in Afghanistan, but it also suggested how the Taliban and other insurgent groups saw those contractors who came from various countries to take part in the war.

Arda was the type of savvy businessman who thrived in the chaotic world of Afghanistan and Iraq during the wars.[2] Before heading to Iraq, he had been unemployed in Turkey. One day when his sister-in-law and her husband were visiting, her husband received a call and stepped away for a moment. When he returned, he asked Arda, "How's your English?" Two days later, Arda was on a plane headed to Iraq for a temporary two-month position doing some fill-in translation work for a construction company that his brother-in-law was part owner of. Once there, the company realized that Arda was useful for more than translation, and after a couple of more senior people in the Iraq office decided they had enough of the dangers of living in Baghdad, Arda found himself promoted to the position of country manager.

Working in Iraq was something like working in the Wild West, he said, and he told stories about rocket attacks and dealing simultaneously with insurgent bombings that were disrupting his supply line while he was trying to finesse angry U.S. Army engineers who did not understand the delay. At one point, just to get gravel, they sent an armored escort to accompany trucks all the way from Jordan because they could not get rocks locally. "We were paying $120 a ton!" he laughed.

Once a week, he said, he would go off to the bank, always at a slightly different time, and pick up $300,000 in funds transferred to him by the head office in Ankara. He would go alone and tell absolutely no one. He'd ride through Baghdad with an AK-47 in the foot well, an MP 5 in the door pocket, and a 9-millimeter Berretta in his belt. Sitting at his office desk, he mimicked having one hand on the grip of his pistol while steering with the other. He decided if something were to happen, he was going down shooting. The difference between Iraq and Afghanistan, he explained, was that in Iraq, you knew that even if you were a Turk, they would kill you, probably in some nasty way once they had taken an electric drill to your knee or ankle. Looking back on it, however, he concluded, "I'm not a psychologist, but we were all kind of crazy to be over there."

In 2009 he had been in Iraq for six years, and with American-funded contracts starting to dry up, he pressured the company to start looking at Afghanistan, where he correctly guessed U.S. funds would start flowing. They agreed to send him over to Kabul to test the waters. This appealed to him because he heard that Afghanistan was far safer, and even though some Turkish contractors had been kidnapped there, they were all safely released.[3] He arrived with few contacts and headed over to the Turkish embassy, where they recommended a translator to him. He set up an office in Wazir Akbar Khan, one of the diplomatic neighborhoods in Kabul, and started knocking on doors.

Arda used some of his former connections from Iraq to set up a meeting with the U.S. Army Corps of Engineers and eventually received a small contract to build two incinerators for a U.S. base. From here, his business quickly expanded. Soon he was building a variety of buildings and camps, most with American funds, and his bosses in Ankara sat back and watched the money roll in. In need of good project managers, he called his old friend Hakan to oversee his biggest project, an $18 million Afghan national Army (ANA) camp just below the mountains in Khost, not far from the border with Pakistan.

INTO THE MOUNTAINS

A common superstition among many of the internationals in Afghanistan was that attacks, kidnappings, and other misfortunes always happened just before you were going to leave. Arda's kidnapping was no exception. In April 2013, he was overdue for a break, but his boss from Ankara wanted to come to the country to inspect some of the construction sites. Arda agreed to delay his vacation, coming up with a week-long itinerary for his boss to tour the company's key projects. The last of them would be their largest project in Khost, which Hakan was supervising.

Since the roads in this area were considered unsafe, they chartered a helicopter from an Afghan company that primarily provided air transport to contractors like Arda when they could not get rides on U.S. or NATO military flights. They were scheduled to spend April 21 in Khost and fly to Kabul. Two days later, Arda would leave to meet his wife in Southeast Asia, where he had prepaid the hotels for their vacation. "I never got that money back," he grumbled to me as he began his account.

On the morning they were to leave the camp in Khost, the weather report called for some afternoon thunderstorms. Arda asked the Russian pilot about the storms, but the Russian assured him they were leaving early enough to miss the bad weather. Arda had used this helicopter company before, so let the issue go. In addition to Arda and his boss, ten others were on the flight: six other Turkish employees going on leave, the Russian pilot, his Afghan copilot, a Kyrgyz navigator, and an Afghan police officer, hitching a ride back to Kabul.

As they took off, the storm clouds had begun to roll in, but the pilot assured them that by flying north and slightly east toward Jalalabad and then turning directly north, they could avoid the storm. The helicopter had another client they were supposed to pick up that afternoon, so the pilot wanted to make sure they made it back to Kabul. As he often did, Arda said, he was lulled by the heavy, rhythmic engines and feel asleep almost as soon as they took off.

Half an hour later, he was woken with a jolt as the helicopter began to shake.

Looking out the window, he said, he could tell the pilot had taken them up a valley, trying to cross over the mountains without getting too high up into the rain clouds. The pass, however, was higher than the Russian pilot had expected and they were in the process of turning back as the weather continued to worsen. As they wound their way back down in the direction they had come, it became clear that they would have to land until the weather improved. The pilot first tried to set them down on a hillside, but the incline was too steep and they had to pull up. The copilot pointed farther down the valley at a dry riverbed, and they landed there. Once on the ground, the group piled out of the helicopter, shaken, but happy to be on solid ground. Immediately they were greeted by a group of Afghans standing above them from a village just above where they had landed, watching them.[4]

At first, Arda thought things might be okay. The Afghan copilot had gone to speak to the villagers who had gathered on the hill above, returning to tell them that they were friendly and there would be no problems. Arda and a few others walked around and snapped some photos. "This is not the type of adventure you can find anywhere else," he said while leaning back in his office chair laughing at their naiveté. "We didn't realize the adventure hadn't started!"

While they were taking selfies, a couple of men from the village wandered down to the helicopter, chatting with the copilot and took a look around. Arda realized later they must have been checking to see if they had any weapons on the helicopter. A few minutes later, he also noticed that the crowd on the hill, which seemed to have been the entire village out in the excitement of the helicopter landing, was now only men. The women and children had all been taken away.

Many of the men also had guns. This was when Arda said he started to worry.

Trying not to panic the group, he hustled them back toward the helicopter. The pilot started the engine up, but as he did, one of the young men from the group of villagers came running down the hill, shooting his Kalashnikov. Three bullets went into the pilot's window, though amazingly they did not hit him. He turned off the engine, and the group realized they were there to stay.

The twelve were pulled out of the helicopter and quickly searched. They weren't kept in the village for long, and a group of local men, armed with assault rifles and rocket-propelled grenades (RPGs), marched them up into the barren mountains. They hiked without stopping for six hours. He later heard that to throw the authorities off the scent, the villagers burned the helicopter so that it might look as if it had crashed. Hakan, who remained at the base in Khost, was shown satellite images of the wreck of the helicopter. Sitting at his desk, Arda opened his laptop and showed me some grainy video footage. It was a group of teenage boys throwing rocks at the burning helicopter while men cheered in the background. They received this later from some of their security contacts, Hakan explained.

The ruse did not work for long. When helicopters crash, they tend to throw off debris in all directions, and a security analyst looking at images of the site realized that the wreckage was in a confined area, so there must have been no impact. By this point, Arda and the other hostages were long gone, and the authorities were in no position to track them through hostile Taliban territory.

Over the first few days, their captors were clearly trying to move them quickly away from the site of their abduction. Some men in the village took their passports and information and apparently sent it on to some of their associates higher up in the ranks. Eventually the twelve were split into two groups, with the pilots and police officer going off in the other group, and a couple days after that, two of the other Turks were taken off in yet another group. This left only Arda, his boss, and an older Turkish worker and his son together. Over the next days, they were marched from one compound to another, higher and higher into the mountains.

Arda said that he was always thinking of escape, but in a place like that, where could he escape to? The mountains were desolate, and there was not much vegetation to hide in. The weather was good, but he had little faith that he would survive long out in the open. There was an occasional settlement and they all seemed to welcome their captors, so even if they did run, there was no one to help them.

Despite the futility of escape, there were only a few times when he thought they might be killed. One of these was around the fourth day, when they stayed for the night at an isolated house in the mountains that had a low pit inside it, covered with fine sand. One of the other captives said that he had heard that the Taliban liked to execute people in such places. They had given them all some medicine earlier that day, and the other captive speculated that they gave people drugs to calm them before cutting their heads off. Looking around, Arda said, the setting was suitable: they were far off in the mountains with no one to disturb them.

For the most part, he clung to his conviction that the Taliban were not known for killing Muslims and that most Turks who had been captured had been released. Still, as a fairly secular Turk, he realized he needed to convince them that he was actually a good Muslim. He never prayed when in Turkey, but he told me that once he was captured, "I was the fastest man to ever learn how to pray." Initially, he would stand next to the other captives he knew were the most devote and copy their motions. He was worried they might be tested on their religious devotion and he had the others teach him the precise wording of the prayers when their captors were not around.

In the meantime, Hakan was working to try to locate Arda's group and figure out who was responsible for the incident. He contacted the Turkish embassy, but it was not clear how much they knew or whether they were going to be able to be much help.[5] They were looking into it, they told him.

Simultaneously, Hakkan relied on the contacts the security firm that they employed had. The firm itself was a Turkish subsidiary of Yüksel, but they partnered with well-connected Tajik Afghans from the Panjshir Valley to provide both their guards and their intelligence. Having been kidnapped in a Pashtun part of the country, these Tajik Panjshiris had less influence than they did around the capital, but they were still able to produce more information than the embassy did. Their conclusion was that the kidnapping had been opportunistic, which was good news, and that the villagers were in the process of trying to sell the Turks to an insurgent group. The question was who would end up buying them.

On the twelfth day of their captivity, they were told at 8:00 p.m. to start walking. The first sign that something was different was that their guards had changed and the number of fighters escorting them increased. The fifteen new guards were a more rugged bunch. One of them had a large 50-caliber machine gun, and another carried multiple RPGs. There was even a medic to attend to any injuries the group might sustain if they got into a firefight.

These were the highest mountains they had crossed yet, and with places where the path was only a foot wide with a drop off the side that seemed endless. The slope was steep, and the footing was rough, causing pain in Arda's foot. When he returned to Turkey, he learned that the walking had led to a stress fracture.

At one point in the night, their guards made them stop when they heard another group approaching on the narrow path. They cocked their weapons and prepared an ambush. It was only after their cry of "Taliban" was responded to with another confirming cry that they realized the group were also insurgents. Still, as Arda said, there are many different Talibans. Their captors were afraid that another group would either kill the captives or steal them for profit, so they told them to act as if they too were insurgents and part of the group. Arda shook the hands of the fighters in the other group, told them all "Salam Aleykium," and the other group continued into Afghanistan to conduct attacks. While he was unsure of their destination, his current captors had not killed him yet, which made him think it was better to stay with them than to try to alert the other group, who might be even less friendly.

Finally, in the morning, heading down to a lower altitude, the fields seemed greener, and in the distance, Arda saw a school. The flag flying in the schoolyard confirmed what he had already suspected: they were in Pakistan.

"I TAUGHT THE TALIBAN HOW TO MAKE MENEMEN"

The four Turks who remained in the group were picked up by the side of the road, and the group spent six hours in an SUV blindfolded, bumping down the road. The road eventually smoothed out, and when they slowed, he could hear children playing outside and the call to prayer from six or seven mosques simultaneously. They were taken to a safe house in a fairly urban area, which Arda latter found out was in Waziristan.

On the first night, they slept on the floor. By the second day, their captors bought the Turks beds, with canopies of mosquito netting. They remained chained to each other, to slow any escape attempts, but for the most part, he was surprised at how well they were treated. After spending days walking through the mountains and losing a lot of weight, things were at least a little more predictable in the safe house.

There were always at least three guards in the house armed with Kalashnikovs. The men took their guns with them wherever they went, even when they were just going out for bread. There were other captives in the compound, who were Afghan, and Arda assumed were members of the ANA or Afghan National Police. They were not allowed to speak to them and were kept in a locked room and hooded when they were taken to the bathroom outside, something they did not do to the Turks. Perhaps this was because their captors were more concerned about the Afghans being able to recognize them, but this was never clear to Arda.

At first, it was a little like being in a zoo, Arda said. People kept stopping by to look at them. Clearly their captors were gaining a certain sta-

tus from their presence. Visitors always asked first if they were Muslims
and then if they were Turks. Immediately in response, the older Turkish
captive would start reciting Koranic verses to prove their piety. Often
the visitors would respond with other verses, and Arda was particularly
impressed when the older captive corrected the wording of one of the
visitors. Their most disturbing visitor was a young man who came by
and gestured at his chest mimicking an explosive vest to let them know
he was preparing for a suicide mission and would soon be traveling to
Afghanistan. He was visiting so that he could be more familiar with
what the enemy looked like, they were told.

Arda had more experience in conflict zones like Afghanistan and
Iraq than the other captives, and as a result, they tended to look to him
for leadership. He worked to keep their spirits up and to keep them
from panicking. They collectively decided that trying to escape was a
bad idea. Still, as the days dragged by, it was difficult to remain optimistic. Without a positive mind-set, Arda said, they would not survive. Your
mind in a situation like that is like a cell phone that has returned itself
to the factory settings: everything was blank, and all they could think
about was survival.

One of the things that particularly came to depress them was the
food. For breakfast, they and their guards ate only bread and a sort
of artificial cream, popular in Afghanistan, that Arda said gave them
all diarrhea. Finally, he said to his friends, "Wouldn't it be great if
we could have *menemen* for breakfast!" referring to the Turkish-style
omelet of eggs, onions, peppers, and tomatoes. Everyone agreed, and
they started reminiscing about different foods from home. Arda then
stopped them and said, "Remember this moment because I will make
you menemen."

"Next, I said, 'Taliban come here!'" He continued his story, now
laughing. He knew the Pashto word for onion and was able to describe
peppers and tomatoes with some gestures. For eggs, he put his hand
around by his rear trying to mime a chicken laying an egg, which ended
up causing them all to laugh. The guards, bemused by these antics,

brought the ingredients the next time they went out to the bazaar and then let Arda sit with the cook while he prepared menemen.

I asked if the menemen tasted like home, and he responded, "Of course it was not good!" But it was better than what we had." He paused triumphantly, "And I taught the Taliban to make menemen!"

And so this went on for twenty days.

Finally, the "big boss," who had been to see them only on the first day, came to visit again, bringing them new clothes and new shoes, promising them that they would soon be free. Arda and the others were skeptical and did not want to get their hopes up. The next day, their captors loaded them in two pickup trucks and they started hiking again back into Afghanistan. Two days later, they arrived at a larger village and were told that a helicopter would land and pick them up at 4:00 p.m. Arda thought perhaps that it was a trick so the men could shoot down the helicopter.

He wasn't surprised when no helicopter arrived.

They were then loaded into another car and left the village. Now the roads they were driving on were a little better traveled than the mountain paths they had mostly been on. Some ways down the road, they were stopped at an ANA checkpoint, the first they had seen since being taken captive.

Arda was torn: Should they run from the car? Things seemed to be looking up, but a failed escape attempt could ruin everything. But what if this was their one chance? As the commander of the post walked over, he seemed to notice that those in the back seat were foreign, so he leaned in and in English asked where they were going. Arda was overjoyed. This had to be their opportunity to ditch the driver and be rescued by the ANA. He rapidly explained that they had been kidnapped, but instead of looking shocked, the ANA commander did not seem particularly bothered by this. The driver handed the commander a sheet of paper. The commander scanned it, handed it back to the driver, stepped back from the car, and waved them through the checkpoint.

Looking back, Arda said, it must have been some kind of letter from the Taliban, but at that moment he was shocked and disheartened. Maybe it was all a trick and they weren't going to be freed. The ANA seemed to have little interest, and they were again driving down the road with their captors.

Just when Arda thought that there would be no end to this, one of his captors handed him a phone with a call from an official with the Turkish embassy speaking Turkish and telling them not to worry: they were released and a helicopter would be there soon. They were brought to another house where they killed a sheep and had the best meal they had eaten in weeks.

The helicopter still did not come.

The emotions of thinking they were free and then realizing they were not began to wear on the group, Arda said. Their captors also seemed to be growing agitated. The next morning, the captors seemed to have had enough and shoved the group roughly into another car and drove them down the road. The driver then stopped the car and shoved them out. "Walk toward that wall" he said. Arda was terrified they were about to get shot in the back.

Tentatively they walked toward a nondescript looming wall outside a large compound. Then an American soldier appeared and began walking toward them, and Arda realized perhaps they really were free.

IT MATTERS WHERE YOU COME FROM

Arda and Hakan still did not know many of the details surrounding the entire episode. It was only afterward that Arda found out that it was the Haqqani network, an insurgency group and tribal criminal network allied with the Taliban, who ran the safe house where he was held most of the time. It was still not clear why there was so much confusion around their release and why the helicopters kept being delayed. Later, a colleague told Arda that the Turkish military couldn't find a free helicopter to pick them up. This would have been a rather shocking coordination failure, and the military officials I questioned on the matter felt that miscommunication was far more likely. Still, ANA helicopters eventually picked them up from the American base, not a Turkish one, Arda said.

As I gathered more accounts of contractors, it was clear that nationality shaped the experience of war or even of being kidnapped. As Turks, it did seem, Arda thought, that even while the experience was terrifying, they were treated much better than they would have been if they hadn't been Turks. The Afghan prisoners, for example, were treated much worse, hooded and confined. While all of the Turkish prisoners were released after a month, the Kyrgyz navigator was detained another two months and the Kyrgyz government had had to pay a ransom, Arda heard.[6] The Turkish government denied paying any money for their release, as did Arda's company, but it was not clear if this denial was done just for publicity's sake or not.[7]

Arda still wasn't sure what happened to the Russian pilot, though several months after he had been released, he received an intelligence report claiming he was still being held. He also heard a rumor that the Afghan policeman who had been with them had been killed. Perhaps this difference in treatment was that the Afghan policeman was seen as more of a traitor than the Turks were, or perhaps he was simply less valuable to the insurgents and not worth the risk.

In following up on other kidnapping cases, I saw a striking spectrum of difference in the lengths a country would go to in order to help its citizens. The Americans had real firepower and teams trained to negotiate and rescue hostages, though they also were reluctant to use any means that had the appearance of negotiating with terrorists, which made them more hesitant to act. The consensus among many in Kabul was that British citizens received more support of this type from their embassy, which was bolder in such situations than their American counterparts. Some other countries were less hesitant to negotiate. For example, the Taliban claimed it secured more than $20 million and a promise from South Korea to withdraw all of their troops by the end of the year in exchange for the release of twenty-one South Korean missionaries who were kidnapped in 2007 (two had been killed while captive).[8]

Countries with smaller diplomatic missions also faced serious logistical challenges. Indian reporter V. Sudarshan provided a detailed
account of the negotiations around the release of three Indian truck
drivers who were kidnapped in Iraq in 2004 in his book *Anatomy of an
Abduction*.[9] For the Indian diplomats involved, there were numerous hurdles, and even the most basic questions were challenging: How do you
contact the group that has taken the hostages? Where do you meet?
How does money get exchanged? Particularly important for the Indian
diplomats was the need to distance themselves as much as possible from
the Americans in Iraq. While the drivers where transporting supplies
that were ultimately going to be used by the international presence
there, India emphasized that their workers were not occupiers but in-
nocent workers.

One of the problems that the Indian mission faced in the case of
the three Indian truck drivers who were kidnapped was the reluctance
of the firm that employed the drivers to accept any responsibility for the
situation. The Indian government did not want to organize the payment
to the kidnappers, so they could continue to claim that they did not pay
the ransoms to terrorists. They were, however, willing to facilitate exchange between the company and the kidnappers. The company, which
did not have an office in Iraq, would not even send a representative to
Iraq at first to take part in the negotiations, and the Indian diplomats
had to exert a good deal of pressure in order to get them to participate.
The diplomatic cajoling involved showed how difficult it is to hold contracting companies responsible for their employees. This was similar
to the case of Teer Magar, who essentially appeared to have been dis-
owned by his company as soon as he was taken into police custody.

The Nepali government faced similar public criticism of their failure to save twelve Nepali contractors who were kidnapped in 2004 and
later killed in Iraq after crossing the border in a truck convoy from Jordan. The Nepali government had made appeals through Al Jazeera and
a Sunni cleric to the kidnappers to release the Nepalis. They did not,
however, send a government delegation to Iraq or elsewhere in the re-

gion. They were also later criticized in the Nepali press for not "being in touch with the relevant entities and agencies, and particularly not reaching out to Iraqi tribal heads, who had more influence in such a situation.[10]

This was not the case of all companies and organizations. In Afghanistan, the Aga Khan Foundation had a good reputation for protecting its employees. Led by the head of the Ismaili Shia sect, the international nongovernmental organization did work across the globe, but particularly among Ismaili communities. Seen by most as a Muslim organization, the foundation had connections inside and outside the Afghan government that served them well in a crisis.[11]

In India I spoke with one administrator from the Aga Khan Foundation who had been involved in hostage negotiations after two of his colleagues had been kidnapped from a convoy that they had all been driving in. In contrast with the Indian truckers in Iraq, in this case Aga Khan handled negotiations directly, though clearly with the support of various governments, and the British eventually sent a helicopter to pick up the hostages once they were released. During my research in India, a similar case occurred when another Indian working for Aga Khan was kidnapped. She too was released.

All of these examples suggest that war is not experienced evenly. Country of origin, class, religion, and race all shaped what a contractor might experience in Afghanistan and the extent of the support the contractor could expect.

Hakan and Arda traveled to the Turkish embassy after the ANA brought them back to Kabul. The chaos around their release and the fact that the helicopter did not arrive when they were originally told it would was worrying, and Arda did not have much respect for the Turkish Foreign Ministry after spending years in Iraq and Afghanistan witnessing their bumbling. In his case, there was also the awkward fact that Erdoğan had actually announced their release several days before they arrived at the American base, when their freedom was still in question.[12] This was exactly the type of move that could have upset the entire negotiation, Arda said. The Turkish embassy kept them in Kabul for three days of debriefing. They were instructed not to speak to the press under any circumstances, and the embassy refused a request from the CIA to allow their agents to sit in on the debriefings. They heard later that their captors had been asking for the release of some of their friends and relatives from Bagram.[13]

After almost six hours of speaking with both Arda and Hakan, we went down to the shoreline and looked back on the city, perched on a cliff over the sea. Both men had married since leaving Afghanistan and had settled down from what Arda called his "crazy days." He had put back on the twenty pounds he had lost while in captivity.

While Arda and Hakan remained critical of the Turkish government, as we lingered over kebabs and beers at a restaurant across the street from the sea, I thought of the Russian pilot who was supposedly still being held and compared their experience with what happens when one's government was not there to facilitate negotiations or provide much support at all. What about the contractors with no place to turn in a war zone?

CHAPTER 17

HOM BAHADUR

VISIT VISAS

Hom Bahadur wanted to make it clear from the beginning. If I was going to write his story, he wanted to be sure I did not use a pseudonym. He wanted his name written down so that people would know what had happened to him.

Dawa and I been told stories during interviews of several groups of Nepalis kidnapped by the Taliban or other groups. Often with no help from the Nepali government and only a little more from the companies they worked for, other Nepalis working in Afghanistan had all pitched in to secure the release of their countrymen. The lack of Nepali diplomatic presence and the fact that companies viewed these workers as expendable contributed to situations where workers could be exploited almost endlessly by those who had better understandings of migration and Afghan law. The workers themselves rarely had anything other than personal connections and the goodwill of strangers to fall back on.

Several of the contractors from Matta Thirta described a DynCorp worker who had been kidnapped in southern Afghanistan where he was held until the company paid a $100,000 ransom. In another instance, a couple of workers we interviewed said they had chipped in a week's salary each to secure a friend's release, but we couldn't track down the man himself.[1] Remarkably, they said, once the contractor was released, he went home, only to return to Afghanistan two months later. Part of the reason he did this, they said, was so he would have money to pay back some of what his friends had spent on his release.

Contracting in Afghanistan pulls you in and holds onto you.

There were a few stories in the media about these incidents, but most went unreported, whether it was due to a lack of media interest or attempts by both sides to keep the details quiet in order not to upset negotiations. One of the few things Nepalis seemed to have going for them in many of these cases was the argument that they too came from a poor country and their families could not meet ransom demands, and then hope that whoever was holding them would eventually decide they were not worth the risk or the cost of keeping.

In Hom's case, it took a long time for his captors to come to that decision.

Hom and his friend Pariwes met Dawa and me in a below-the-street café with low walls separating the tables from each other. Despite the sunshine outside, it was cool in the café, and the dividers gave the conversation a sense of intimacy. The waiter seemed to forget about us, and we were mostly left alone for the three hours that Hom spoke.

Hom looked older than many of the other Nepali guards and laborers we had already interviewed, and it was difficult to determine whether this was due to his age or the stress of some of what he had survived. His eyes were sunken, and he looked weary. His story was also fresher than many of the other accounts Dawa and I had collected, since he had returned from Afghanistan only three months or so before our conversation.

At first, Pariwes was the more talkative of the two. He led the conversation as he described how the two had originally decided to find work abroad, with Hom adding just the occasional detail. After ten minutes or so, Hom clearly warmed to the conversation, and Pariwes contributed less. As Hom's story descended deeper into a tangle of deceptions and lies, Pariwes, Dawa, and I became quiet, not even needing to ask questions to keep the narrative going. His story had a momentum of its own.

Hom had difficulty from the start of his journey. He came from the small town of Baglung Bazaar, northwest of Pokhara, where he grew up farming. Pariwes had heard about a local broker who could arrange jobs in Germany for a 50,000 rupee fee each (around $500). Liking the sound of the opportunity, they borrowed the money from various friends and neighbors, and he and Pariwes set off for Delhi with the local broker in spring 2015.

Once in Delhi, their broker handed them over to another broker, Karuna, a Nepali woman who lived in India and had arranged for them to stay in a hotel in Delhi's chaotic Paharganj neighborhood. After waiting there for twenty-eight days, Karuna came and told them that the posting was canceled, and if they returned to Nepal, the original broker would refund them their money. Pariwes returned, but Hom was already in debt from the hotel in Delhi and, he said, there was nothing back in Nepal for him. Karuna told him about several other Nepali workers she had found jobs for in Afghanistan. She told him their names and even showed him their photos. For an additional payment, Karuna suggested, she could arrange him a job there as well.

Hom took her up on the offer.

Karuna shifted him to another hotel with three other Nepalis looking for work in Afghanistan, and he paid her another 50,000 rupees that his family had wired to him for his Afghan visa, plus some additional funds for his accommodation. She took his passport and reassured him his visa would soon be ready. One of the issues was that by 2015, the process for securing a work visa in Afghanistan had become more difficult, and the embassy was stricter about issuing visas; six weeks later, the broker had been able to secure him only a visit visa that was valid for thirty days over the next three months.

As he described this, he pulled out his passport from his back pocket and gently thumbed through the mostly blank pages. He finally reached the page with the Afghan visa, explaining that while Karuna reassured him that he could eventually get a work visa in Kabul, he was nervous about the entire process and already exhausted from having spent two months in

Delhi. I noticed as he pointed to the Afghan visa in the passport that there was an entry stamp from earlier that spring but no exit stamp. It struck me as odd, but I didn't want to interrupt his story to ask about it.

With the visa secured, Karuna arranged him a ticket and flew to Kabul.

Things did not get easier after he arrived. The two young Afghan men who were supposed to pick him up arrived an hour and a half late. After leaving the airport, they drove through the confusing streets of Kabul, and the one in the passenger's seat turned and demanded $200 in advance payment for his lodging. He wasn't sure what to do. He had a cell phone, so he called Karuna back in Delhi, who sounded exasperated and told him not to give them a penny. Alone with them in the car, however, he was scared what they might do to him. Besides, if they abandoned him, he was in a strange and dangerous city where he didn't speak the language. What was he to do? He decided to compromise and give them $100 instead of the full payment.

They dropped him off at a compound that was run by an older Afghan man. The compound had a small courtyard in it surrounded by a house with a few rooms and three other smaller buildings. Those staying there referred to it as a camp, but really it was a typical small Kabul family compound in which sixteen other Nepalis and Indians crowded, all hoping to find work.

The compound itself was not luxurious, Hom continued, and they slept eight to a room, but it did not feel too bad, since it was spring and they could sit out in the courtyard. At least it was more comfortable than his grimy hotel in Delhi. On occasion the workers would go outside the camp to wander around, but there was not much to see on the dusty streets and the Afghans he met there were not particularly friendly. He was somewhat concerned about getting lost and had heard of Nepalis kidnapped, killed by "the Taliban," or harassed by the police. Since he was unable to speak the language, everyone outside appeared menacing. Most of the time he played cards and chatted with the other workers.

The compound included several of the Nepalis Karuna had told him she had secured employment for. None of them had jobs, and he didn't know if this meant Karuna had been lying or if the Afghan running the camp had lied to her. Either way, it was worrying. His optimism continued to fade as more Nepalis arrived, but no one seemed to be finding any work. Hom said that this was the first time he heard discussions of U.S. troops leaving the country. In Delhi, all they had heard about were the high salaries; now people were saying that going forward, there would be fewer and fewer jobs.

One day, at about 2:00 p.m., the police raided the compound without warning. They herded everyone outside into the courtyard and made them show their IDs. He could not understand what the police were saying to the owner of the compound, but during all this, he was only somewhat worried since he had heard the police became "demons" only if you lost your passport or overstayed your visa. The Nepalis and Indians who did not have valid visas had quickly hustled out the back door of the compound and down an alley, but because Hom's visa was still valid, he stayed. The police studied his passport and then left him alone.

In the aftermath, the owner of the camp sent all of the workers to different compounds. He took him to stay at a new guesthouse owned by another Afghan that he only knew as "Haji." This, Hom said, was where "the nightmare" really began.

FROM BAD TO WORSE

Haji's house was smaller and dingier than the previous compound. It had only four rooms in it, and three other Nepalis were staying there as well. One of them was a worker who had begun doing some work as a broker too. His name was Geet, and he became friendly with Hom.

By this point, Hom had been in Afghanistan for over three weeks and now had only two days left on his visa. "Not to worry," Haji said. "Give me your passport and two photos, and I'll get you a six-month extension." So Hom did. As the days passed, however, Haji did not return with the passport.

After a couple of days, it also became clear that Haji had other businesses on the side. He brought prostitutes to the house. Many of the male clients who visited were police. Haji threw parties for the policemen with alcohol, which is illegal in Afghanistan. The policemen lounged around on the cushions, their guns leaning against the wall, as they got drunk. Hom said it was disconcerting to see all those police when he didn't have a passport, but he figured that as long as the police were doing something illegal, they were unlikely to bother him.

At one of these parties, Haji accused Geet of owing him $22,000 for supplying him and sixteen of his clients with visas. Haji was particularly mad because he was so short on funds that recently he had to sell his car. Geet left the room where they were all drinking, but as he was leaving, Haji loudly asked the police chief sitting next to him to kill Geet for him. Hom said it was difficult to tell at that moment whether they were joking. An hour later, the police chief left the party without doing anything, and Hom thought Haji might just be trying to scare them. Nevertheless, he said, none of them slept that night.

The next morning as Geet was walking out the door to try to arrange some new business, the police seized him, taking all of his documents. Geet apparently told the police at the station that Haji was operating a brothel and keeping illegal workers, and so the police returned to search the house. As they were arriving, Haji pushed Hom out into the street and told him not to come back until that evening. Hom thought about trying to find some sort of help, but he didn't have his passport or any money, and he didn't know his way around the city at all. He thought it best to walk a couple streets down and simply sit on the corner and wait for time to pass, hoping the police would not spot him and that no one would try to rob him. It was one of the longest days of his life, he said.

When he returned that night, Haji immediately took his phone from him, and Hom realized that he was now the only Nepali left at

Haji's house and had no way to contact the outside world. Haji came to him and said his visa application had been rejected. He then returned his cell phone and pointed a pistol at him. He told him to call Karuna, his broker in India, and to tell her everything was fine and that Haji had found him a good job paying him $600 a month. Continuing to point the gun at him, Haji handed him a uniform. He ordered Hom to dress up in a chef's uniform, and he posed in Haji's kitchen holding a ladle as if he was working happily. Haji sent these photos to Karuna, and she later posted them on Facebook. His daughter and other family saw them back home in Nepal and assumed that he was doing well.

After this, Haji left for India, taking Hom's phone with him, to meet with Karuna to arrange for more money for securing jobs for other workers. Haji said he would be gone for three days but didn't return for sixteen. During this time, there was little for Hom to do. Only a gate-keeper and two Afghans who did odd jobs for Haji were staying in the compound. They kept the gate bolted, and it would have been difficult for Hom to slip out, but even if he could, where could he go? The world outside seemed filled with drunken, corrupt police and brokers who might be even worse than Haji. At least at Haji's, he was being fed. So he sat and waited for the time to pass, imprisoned in Haji's compound.

When Haji returned, it was clear that his meeting with Karuna had not gone well. Haji again gave Hom his phone, but this time had him call Karuna to tell her that he had lost his job, was homeless, and that she should send Haji money to look after him. Haji spoke Urdu, which is similar to Nepali, and made Hom speak in Urdu on the phone. On a couple of occasions, he pretended not to know a word in Urdu and slipped into Nepali, trying to let Karuna know that he was being held prisoner.

Eventually Karuna seemed to get a sense of what was going on through Hom's vague references. But it did not seem that Karuna was really interested in getting Hom released. It would have been bad for her business if he got back to Delhi or Kathmandu and started talking about the difficult time he had had. Better for her to convince Haji to find Hom work in Afghanistan. At the same time, Karuna was bothered enough to get Haji to give Hom his phone back. Somewhat placated, or at least getting tired of all these dealings, Haji relented and gave Hom his phone. That night Hom was able to call his wife back in Nepal for the first time in weeks and he explained everything that happened. Desperate, she got in touch with Pariwes, who had originally gone to Delhi with Hom and they started to devise a plan to rescue Hom.

Pariwes's scheming finally turned the tables on Haji and Karuna, but it was not easy. He used another Nepali he knew, Lekh, who had a legitimate job in Afghanistan, to help them. They constructed an elaborate story, and Lekh called Karuna to tell her that he had a job at his company that was available and that they were looking for a Nepali to fill it. He said that he knew Hom from back home and thought that he might be well suited for it. He said that they could charge Hom $3,000 and split the money. All she had to do was transfer $1,500 to his father in Nepal; he would set it all up. Karuna was sure she could get Hom to pay at least $3,000 for a job and a legitimate visa, so this sounded like a good plan. She called Haji and told him that Lekh was going to come by and that he should allow him to pick up Hom. Then once Hom started his job, he would pay Karuna, and she would then transfer money to Haji to cover what he felt he was owed for dealing with Hom for the past months.

Perhaps sensing the ruse, however, Haji was reluctant to let Hom go, hoping there was a way to still make more money off his precarious situation and perhaps nervous that Karuna would not come through with the money she had promised. As the time drew near for him to go with Lekh, Haji and Hom argued again and Haji again pulled out his pistol, pointing it at Hom and telling him he wouldn't let him leave.

At this point, Hom was so fed up he decided he couldn't let this continue: he pushed the pistol aside and walked out the door.

GETTING HOME

After almost six weeks with Haji, Hom was finally out of his compound, but his visa was long expired and there was no reason to trust the Afghan police. Hom might have freed himself from Haji's grasp, but there was no easy way out of the country.

At this point, an informal network of Nepalis, including the Nepali journalist Subel, who had previously helped Teer Magar get released from Afghan prison, began to mobilize. His first four nights away from Haji, Hom stayed at a security company compound, similar to the Pinnacle that Lekh knew of. They couldn't keep him there long, though, Hom said, since the Nepalis were afraid of losing their jobs for having an unauthorized guest, so one of the Nepalis working there got in contact with Subel.

Arriving at Subel's, he told him his entire story, tears falling down his cheeks. Subel started calling his contacts in the Afghan government and alerted the Nepali embassy in Islamabad. The embassy in Islamabad said that there was nothing they could do—Hom had technically broken the law by overstaying his visa—and the Afghan visa office said that the fine for overstaying his visa would be at least $3,000 if he wanted to leave the country legally. They were also afraid that once the authorities knew of his case, some might try to extract additional bribes or he might simply disappear again. Now that he was free, Hom had little interest in handing himself over to another potentially hostile set of captors.

Instead, Subel contacted another Nepali he knew at a private security firm, Nabin. At that point, Nabin, whom I later interviewed in Kathmandu, was a supervisor at a well-connected security firm in Kabul. This gave him the ability to pull some strings that those in lower positions could not. While Hom was not sure how Nabin pulled off all the details of the escape, a few days later, he found himself driven onto the International Security Assistance Force airbase in Kabul. Nabin told him to put on a badge and follow some of the other Nepali guards who were in the car with him. "Look like you know what you are doing," he was told. Silently he followed them through the first checkpoint.

From these bases there were international flights like those used by Supreme to get workers into the country without visas. After a cursory check of his ID, he was allowed to pass through and joined the other Nepalis waiting to board a private flight to Kandahar. Finally Hom was out of the country.

On his last day in Kabul, Subel had taken him to a party with a group of other Nepalis. It was a surreal event after his ordeal. Subel handed him $100 in case he needed it along the way and a bag of nuts for his children. He didn't want him to have nothing to give his kids when he returned home, and he said he could pay his father the $100 back once Hom was in Kathmandu. When Hom later called Subel's father in Kathmandu, he refused to take the money.

Hom's story took almost three hours to tell, and we left exhausted. Hours later, I could still hear the sheer despair in his voice as he described twist after twist and turn after turn. It was difficult to imagine a situation so helpless. The three-and-a-half-month period that Hom was in Afghanistan was short compared with the years that Teer Magar had spent in Afghan prison. Still, his experience was perhaps more scarring. Teer thought that much of the reason he ended up in jail was a simple misunderstanding that had long-lasting consequences. Hom was cheated and lied to at almost every turn by broker after broker. It was only the compassion of a few of his fellow Nepalis that eventually got him released.

Later that evening, Dawa and I sat at a café near the hotel we were staying at, looking out over Phewa Lake. On a clear day, the snowy peaks of the Himalayans reflected on its calm surface, but that day was hazy. The European, Chinese, and Japanese tourists who wandered along the promenade had to content themselves with watching the peaks poke out from the clouds to the north. It was difficult to reconcile the tranquil happy scene with the dark tale that Hom had told.

As I conducted more interviews, I thought about the extent to which Arda and Hom were being traded as pieces in much wider political and economic dealings. Arda was a representation of the American occupiers for the insurgents, albeit a slightly confounding one, since he was also a Muslim. Hom became a pawn in a different struggle between the various brokers he tried to use to get him a job. Held as collateral, he had little choice but to flee.

Contracting helped make war into business, but a desperate business, where individuals were forced to take risks and navigate the dangerous world of brokers, who seemed to hold all the power. Traveling to India in spring 2016, I turned my research toward the networks of brokers who trafficked these workers across Asia and into these war zones.

CHAPTER 18

THE BOREDOM OF BEING TRAFFICKED

ON TO INDIA

Arriving in Delhi was overwhelming. Plenty of writers have waxed poetic about Delhi's madness, but since it was the fourth country I was conducting research on over the past two months, my head was truly spinning by the time I settled into my room at the Guest House in Delhi University that Professor Jagdish Chander, who taught in the Political Science Department at the elite Hindu College, had arranged for me. In exchange for giving a public lecture, I had access to the library and was able to stroll the grounds of Delhi University, some of the most peaceful gardens in the city, in between the interviews I was conducting around the country.

The campus is located on the ridge where the British had retreated to during the 1857 conflict, called the First War of Independence or the Indian Mutiny, depending on what side you favored. The ridge saw some brutal fighting for both sides, and the British eventually pushed back into Delhi, in large part due to the support of Nepali troops. In return for the fealty of the Nepali troops, Britain returned some of the territory lost during the Anglo-Nepal war to the king of Nepal, which undoubtedly helped him preserve independence while other smaller principalities lost their self-governing status in the wake of Indian independence. It also complicated India's relationship with the smaller country, which led to numerous confrontations, small and large, such as the fuel blockade at the Nepal border that fall.

Not far from my guesthouse was a hospital where the main fortifications were the ones that the Nepalis occupied. Wandering around the grounds, I read British accounts of the role of Nepalis in defending the hill. In one, a British officer praised "the poor little Gurkhas [who] . . . are always jolly and cheerful as ever, and . . . anxious to go to the front when there is an attack."[1] The ridge was now overrun by monkeys and retired Delhites power-walking in tracksuits. It was a pleasant place to walk, particularly in the early morning or evening, as long as you kept your eyes out for the monkeys, which could be vicious. "Whatever you do, don't smile at them," one young woman in the park told me.

In the neighborhoods just below the ridge that the university is perched on, the bustling markets and narrow alleyways hold the thousands of small businesses that attract villagers from all over South and Central Asia to Delhi with the promise of jobs and the comforts of an urban lifestyle. The result in these neighborhoods where Old Delhi meets New was a tangle of cars, carts, and people. Among them were many of the young Nepalis I had met in Kathmandu and other towns, hoping to get jobs in Afghanistan or elsewhere. Most of these recruits, upon arrival, head to the backpacking hub of Paharganj, a kind of purgatory where bored recruits wait for visas from shady brokers who claim to be the only ones able to navigate the impossibly complex bureaucracy of embassies and contracts.

Paharganj is not one of the world's most pleasant tourist districts. In Delhi, however, where there are plenty of luxury hotels, sky-high rents mean that it's difficult to find a hotel cheaper than $200 a night but more than $8 a night. For those looking for the $8 and below option, Paharganj is the place to go. It's situated right across from the Delhi train station and jammed with rickshaws, and I dodged the Russian and Israeli tourists who were intermingling with local Delhi shoppers and the occasional American hippie who appeared as if he had been tripping down that same street since 1968.

After a little exploring behind the cluttered little shops selling trinkets, T-shirts, plastic clogs, and cheap fabrics, however, it became clear that the budget accommodations attracted another group of visitors: young men looking for work abroad. I convinced Dawa to come down from Nepal in March to prowl around Paharganj with me. I had been doing some initial interviews with Indians who had been contractors in Afghanistan but had also started exploring some of neighborhoods where young Nepalis hung out. Dawa was good at chatting with these younger job seekers, and I had missed having him to bounce ideas off of and talk to during my research in Georgia and Turkey.

Paharganj was far from the only neighborhood with Nepalis in Delhi. Several areas had enclaves of thousands of Nepalis, some of whom had been living in India for decades. You could find *momos*, Nepali dumplings, on many street corners, often cooked by Nepali immigrants. The Nepalis who settled in India had for the most carved out lives for themselves despite discrimination and the fact that many were confined to lower-paying jobs as cooks or guards. These settled Nepalis kept their distance from the young men who were coming through seeking work abroad. On occasion, a job seeker might stay with relatives in various suburbs, but it was Paharganj where the recruits and the brokers trying to secure them employment tended to congregate.

One such place was the Kathmandu Palace restaurant. Down a side street, toward the center of the market, the restaurant had a grill in the front of the storefront next to an annoyed-looking woman sitting at the cash box. It looked like hundreds of other similar cheap restaurants in the neighborhood. The menu listed several Nepali dishes but also had Indian items, Chinese dishes, and other foods catering more to a Western palate. Inside, two Japanese tourists were eating an early lunch.

As I continued deeper into the restaurant, a different world opened up. Behind a curtain and up a dirty concrete stairwell with no railing was a larger, darker room next to the kitchen. The room had once been painted red and white but now was covered in grease and other stains. There was a series of small tables with plastic chairs. The only luxury was two small speakers attached to the walls, with wires dangling loosely between them. This was the Nepali section.

Everything about the place was Nepali. The men chatting in the corner were Nepali, the food was Nepali, the music was Nepali, the waiter was Nepali, and even the busboy was Nepali. The waiter carried a plate of momos past us. Dawa and I ordered some dal bhat, which I had not eaten since leaving Kathmandu, and we tucked into filling plates of rice, lentils, and stewed vegetables. Eating it made most of the meals I had had in recent weeks in Delhi feel oily. A near-bottomless plate cost 100 Indian rupees ($1.50 or so), and it was easy to see why young Nepalis looking for work congregated here.

WAITING FOR VISAS

As I began to interview some of the young men there, I was also aware that this was a new phase in my research. Most of the Nepalis and Turks I had already interviewed had returned from Afghanistan, and talking with me did not have serious consequences for them. For these young men, however, who were actively working to get jobs abroad in Afghanistan, the stakes were higher. A few I had met had already been to Afghanistan, but most had not.

Their stories were just beginning.

The young men were understandably nervous about the process of securing work abroad. A Nepali journalist had told me previously that he received death threats from brokers after asking too many questions about the process. So after I explained my project to the first three young recruits I interviewed, I thought it would be better to talk away from the watchful Nepalis in Paharganj. Instead, we met in a pizza restaurant in Connaught Place, the busy Times Square of Delhi. It was only a fifteen-minute walk south of Paharganj, but with the trendy shops and restaurants with dress codes attracting the upper classes, it was a different world from Paharganj, and the first restaurant we stopped at turned us away since one of the young men was wearing sandals.

I thought that perhaps Connaught Place was close enough to Paharganj that they would have been there before—it was an easy stroll—but it became clear that the Nepali recruits lived in fairly confined worlds once they arrived in Delhi. The brokers tended to take them straight from the airport to Paharganj, explained the recruits I interviewed. Unless they had to go to an embassy for a visa (and usually a broker could do this for them),[2] most did not leave the neighborhood at all until they got jobs and were taken back to the airport. They had little interest in seeing the sights.

In this new setting, the young recruits were shy and didn't speak much at first as we sat around a couple of pizzas in a crowded restaurant filled with upper-middle-class Delhites. Yuba, the eldest of the three, could not have been more than twenty-five, and the other two were closer to twenty. I asked them how they liked Delhi, and they responded with a grimace. It was noisy and crowded, they said, and most Indians looked down on them as Nepalis from the countryside. All three appeared to be putting on brave faces when describing their current situations, though as our talk went on, their complaints grew louder. Two of them were working with the same broker; they said they trusted him but then laughed, saying that when he had no news, he was impossible to get a hold of. "I gave him 147 missed calls today!" one said.[3]

At least the three men had some experience abroad. One had worked in Malaysia for two years, which wasn't bad, he said, and another had worked in a shop in Dubai. The problem was that wages in these places weren't great for young Nepalis who didn't have any particular skills. The pay was low, and living costs, especially in Dubai, were high. It was difficult to save money, and many ended up in debt after paying for travel costs.

So the three had decided to head to Afghanistan, promised jobs in a company doing catering work by a broker. None of them knew anything about what life in Afghanistan was like. When I asked them about the danger, they said that they assumed that they would be on a large base, far from the fighting, but really, they were unsure. As the days passed and the jobs did not materialize, they were growing more nervous, but they knew their positions were not as bad as those of some others.

The night before, they said, there had been three Nepalis at another Nepali restaurant in Paharganj who were drinking and crying while they told their story to anyone who would listen. They had been promised jobs in Russia, but were told they had to fly through Sri Lanka (it wasn't clear at all why they were taking this route, but the more I heard stories about migration attempts westward, the more I realized that the most direct path was almost never the ideal one and, just before, I had heard about a group of Afghans who were stopped above the Arctic Circle trying to cross the short border between Russia and Norway). In Sri Lanka, the group had been arrested at the airport. To get out, they had to pay a fine, or more likely a bribe, of $8,000. They contacted their broker, who instead of helping, threatened them, telling them if they reported the incident, he would kill them and their bodies would never be found. For their families, having their loved one go abroad, never to be heard from again, was a fate worse than death. The young men I interviewed had been shaken listening to these ordeals.

Other Nepali migrants in Paharganj I interviewed were more hopeful, though some were paying much more than the three young men I initially interviewed; one group of a dozen Nepalis claimed to be on their way to Hong Kong, having paid $10,000 each. Some of the Nepalis ended up staying in Delhi more permanently. Two young Nepali women sat discretely at the Nepali café, their profession becoming more obvious when one of them on their cell phone arranged an appointment to give the caller "a massage." After we left the café one afternoon, Dawa commented on how sorry he felt for all the Nepalis there who were so far from home and with few prospects. It was frustrating seeing how confident they were that it would all work out while knowing that their the brokers were probably ripping them off, he concluded.

In some ways, considering how naive the young men seemed, I was surprised their brokers were not actually doing more to take advantage

of them. I expected the brokers to have connections with the hotel own-
ers and try to get their recruits to have to pay for meals, like a tour with
none of the extras included. In Nepal, I had heard of long waits in
India for visas and contracts, and I assumed the brokers must have been
making some money off these delays.

While during interviews I found a few cases of such constant exploitation, most, including the three young men I interviewed in Connaught Place, seemed fairly satisfied with the deal that they had worked
out with their brokers. For two of them, their broker was paying their
hotel fee and all their meals while they were in Delhi. The third was
receiving 1,000 Indian rupees a day from his broker (about $16), which
was more than enough for a decent room and three meals a day in the
neighborhood. In another case, one man I interviewed told me that a
broker had even paid for a prostitute to visit a fellow recruit who had
been there for a while and was thinking about returning to Nepal. In
several cases, I heard about brokers returning the original fee the workers had paid if they were not successful in finding them a job. While the
Nepali brokers in Delhi were certainly taking advantage of the young
men on one level, they were aware that their business hinged in part on
their own reputations. If the young recruits returned to Nepal empty
handed and taken advantage of, it would be more difficult for them to
get future clients. So the brokers seemed to work to keep the young men
just happy enough that they would continue to pay the broker's fee and
wait, albeit with slowly diminishing hope, for a job.

Perhaps as a result of this, the recruits I interviewed seemed bored,
stuck in a strange limbo. They seemed convinced that they could get
their money back from the brokers if they decided to return to Nepal.
If they did this, however, they would have to start the entire process
of trying to find work abroad again, and their friends and families
would want to know why they failed. So all of the recruits I interviewed
seemed willing to wait in Delhi in their rundown hotel rooms simply to
see what happened.

Hearing them talk longingly about securing work in Afghanistan, I found it difficult to resist giving advice. They all claimed to know that there were risks of getting injured or killed there, but they also pressed me to give them more details about what life was like there and whether it really was dangerous in Kabul. More than being concerned about safety from bombs and other military hazards, I was worried they knew so little about the companies they were working for and whether they were actually getting legitimate contracts and visas. I tried to warn them of the cases like Teer Magar and Hom Bahadur.

I asked the three I interviewed in Connaught Place who their brokers were arranging their jobs with, and they said they were not sure. One thought it was Reed, a private security company in Afghanistan and elsewhere, but all three were applying for catering positions and I hadn't heard in the past that the company was doing catering. They had not seen an actual advertisement for the position; it was all done through the broker. I tried to search for the positions afterward and couldn't find any details either. I told them that most of the Nepalis I knew who had gone to Afghanistan had a written contract before leaving, and they should ask their brokers if they were going to get one of these. "At least," I advised them, "make sure you know the names of the companies that you might be working for so that you can ask about them."

After discussing the paperwork some more, the youngest of the three took out his phone and showed me a photo of the Afghan visa his broker had sent to him, since his broker was currently "holding" his passport, which I found worrisome. The visa was for three months, he said, but when I looked at it, I saw that it was really only a thirty-day visit visa that could be used anytime in the next three months—the same kind Hom Bahadur had originally traveled on. I explained to him that this meant that if his broker did not find him a job and he went to Afghanistan, he would have thirty days at the most to find a job and get his employer to find him a longer-term work visa. This was possible, but thinking back to Hom's story, I told him that this was a risky strategy. I advised them to ask their broker about whether they would be getting a new visa after arriving in Kabul.

After dinner, the boys shook hands with us and seemed rather glad to have talked. They invited me out to lunch in Paharganj the next day with some other young Nepali recruits, unsurprisingly, at the Kathmandu Palace.

NO OBJECTION

After seeing the photo of the recruit's Afghan visa, I visited the Afghan embassy to get a better sense of how the visa process worked for these young Nepalis.

Only two miles from Paharganj is the wide grassy boulevard of Shantipath, home of the Afghan embassy as well as many other diplomatic compounds. It felt like a different world from Paharganj. Instead of the narrow alleys of Old Delhi, the wide street was filled with chauffeured Mercedes and BMWs. With razor wire, cameras, and neither parking nor sidewalks, it is not a neighborhood that invites the casual visitor.

The Afghan embassy is in a grand building sitting between the Dutch and the Russian embassies. Outside the building, some painters who were working in the nearby Canadian embassy lounged with about ten or so Afghans, Indians, and Nepalis who were waiting for papers from the Afghan embassy. The guards at the front gate shooed away a Nepali worker who was ahead of me, trying to ask a question. When I pulled out my own passport, they waved me through, and I was reminded how my American passport opened doors shut to others.

The man inside the embassy gave me the forms I would need to apply for a visa but would not answer questions about other nationalities. I had to put together pieces from these conversations with what some of the Nepali applicants told me in interviews in order to understand the hurdles for young workers. The process for both Americans and Indians applying for visas was fairly straightforward. The documentation was no different from what I had filled out for the Afghan embassy in Washington twenty times over the past eleven years. The form was even identical. Interestingly, regulations for Indians and Nepalis looking for visas were different. In particular, Nepalis looking to get a visa also needed to bring what was referred to as a "no-objection certificate" from a Nepali embassy.[4] This seemed a little odd, because it essentially meant that the Afghan government would give a Nepali citizen permission to enter the country only if he or she had permission from Nepal.[5]

At the Nepali embassy we tried to figure out what the experience of a typical young Nepali might be, so Dawa went in as if he were applying to work in Afghanistan and asked what he needed to do to get a no-objection certificate. The woman there was vague and simply gave him a blank piece of paper and told him to write a handwritten letter requesting one. He asked about whether he needed a contract from a company working in Afghanistan, and again, she did not give him a direct answer. Instead, she suggested that if he knew anyone in the Nepali government in Kathmandu, he should contact them, because the embassy in Delhi would then be sure to provide the certificate.

We were somewhat flummoxed by the vagueness around the entire process, and it clearly would be much more difficult for an illiterate young man from rural Nepal. Still, some people were benefiting from the ambiguity. Perhaps unsurprisingly, on the way back toward Paharganj, several well-dressed middle-aged Nepali men were gathered near the metro station close to the embassy. One approached Dawa, asking if he had just been at the embassy. When Dawa said yes, the man suggested that he could help him arrange the necessary papers to get a no-objection certificate. In fact, he said, he could arrange him a job too if he wanted; it would require only a reasonable payment. Dawa told me later that after glancing around him, the man said that the important thing was that he didn't tell anyone else about this.

We found that the brokers worked to reinforce the secrecy around the application process. The recruits in Paharganj I met said that the brokers they were working with kept reminding them that they needed to make sure they did not speak to anyone else and not to go near either the Nepali or Afghan embassy. It was for their own good, the brokers had said, and it was possible that they could be arrested if they tried to do so. With each interview, however, I was less and less convinced that this was true. Even if what the young men were doing was illegal under Nepali law (or maybe later under Afghan law), they were not doing anything wrong under Indian law. Even if they were doing something illegal, what interest would the Indian authorities have with three Nepalis in their twenties trying to find work abroad? Clearly the brokers didn't want the recruits to speak with anyone else since that might lead them to work with someone else. Ignorance was the trafficker's friend.

The young men had little solid information to work with and no help, only a daunting bureaucracy at the embassies, so back at the Kathmandu Palace, they sipped beers and traded rumors with the other Nepalis who came in and out of the room. Everyone had a story of the extreme lengths necessary to make it abroad. Two young men said that they were headed to Japan, where their broker had arranged work for them. Their broker said that on arrival in Japan, they would be put into a refugee camp, where they would stay for a month, but then they would be free to work. The broker was actually trying to arrange asylum status for the two young men. However, like almost all the other recruits I interviewed in Paharganj, the men were on the verge of spending much of their life savings often to go to dangerous places with almost no sense of how the migration process worked.

A few Nepalis in Paharganj looking for work were not as green. One day at lunch at the Kathmandu Palace, two men in their late twenties at the table next to us said they were looking for security work. With muscular, tattooed arms and tight T-shirts, they looked like much more promising job seekers than most of the recruits I had interviewed previously in Paharganj. The two men had both worked at Bagram Airbase. The company they had been with, like so many others, had lost its contract, and so they had headed back to Nepal. Now they had come back to India in the hope of finding work in Afghanistan again.

Perhaps since they were savvier than the new recruits, they refused to pay their broker his entire fee upfront. They said they were holding on to $1,500 of the $3,000 he was charging. Once he finalized the paperwork, they said, they would pay him the rest. In the meantime, since they had not paid the entire amount, the broker was making them pay their room and board in Paharganj, and they were not happy about the delay, which was cutting into their savings. The younger men asked them about work in Afghanistan, and for the next half-hour, the more experienced men regaled them with stories about rocket attacks at Bagram. One of the younger men rather breathlessly asked one about a scar on his arm, but he laughed and said it was something from his youth.

After lunch, I stayed behind as the three young men I had met initially went out to meet one of their brokers. I didn't want to upset any of the arrangements that they had made, and it seemed best to stay back.

The young men were not gone long and were not happy when they returned: the broker had not made any progress with their jobs and wanted them to wait another week. Also, they had asked the broker for the specific name of the companies that they were talking to. This apparently upset the broker, who began arguing angrily with them, telling them that they were ungrateful for his help.

Four months later, I heard from them again through Facebook. All three of them had ended up staying almost two months in Delhi before giving up and heading back to Kathmandu. The brokers in Delhi said that their original agents in Nepal would refund the fees that they had paid. The three young men all had signed documents showing what they paid, so they were optimistic that they could at least get their money back. The problem, of course, was that the brokers were nowhere to be found and months after their return, all three were fuming, trying to figure out what to do next.

As the weeks went by, spring came to Delhi and the cold morning fog gave way to pleasanter temperatures. Dawa headed back to Kathmandu to work on a project looking at migrant workers in the health care industry, but I continued to seek out Nepalis who were transiting through Delhi and others who had, for one reason or another, settled in India after working abroad. In India's bustling capital, with its populist president looking to expand India's influence on the world stage, particularly through business, I also started asking what Indians had gone to Afghanistan. Did their experiences differ from the Nepalis I had been speaking with? What did this say about India's complex situation in South Asia and how the war in Afghanistan shaped it?

CHAPTER 19

ACCOUNTANTS AT WAR

WHITE-COLLAR WORK

In Delhi, I sought out some of the Indians I knew who had worked as contractors in Afghanistan, and I met with Jai in the courtyard of a busy coffee house across from the Hanuman Temple, just south of Connaught Place. Jai was a geographic information system specialist who had worked in the United States for seven years doing various mapping projects before joining a French consultancy firm. After a month in Paris, his firm sent him to Kabul to analyze geographic data for them. He knew little about Afghanistan, but had read *The Kite Runner* and liked the idea of an adventurous posting. Once in Kabul, in part because of his affable nature, his network of friends, Indians and others, expanded quickly.

This social scene was one of the key differences between the Nepalis who tended to struggle to travel and make contacts and the richer, more mobile Indians. This network provided opportunities for them and a social and political safety net that contractors from poorer countries did not have. This was in part due to the cultural and historical links between the two countries and in part due to India's continued political interest in the country. At the same time, their higher status meant that Indian contractors often had insight into corruption and other issues that were less visible to those both above and below them, which made the Indian experience of contracting unique.

Jai got together on the weekends with a group of Indian friends to hang out and cook Indian food, he said. He tended to have so much work that he couldn't take the time off entirely, but instead he brought his laptop and worked on the sofa while the others chatted around him.

His account fit a pattern that the Turkish engineers and construction firm managers had also emphasized: there were links between nationality and positioning in the contracting food chain. These hierarchies extended beyond the tendency of Nepalis to be security contractors. In many Turkish firms, their upper management was all Turkish, their security was Nepali, and those in catering and housing were often Filipino. Companies relied on Indians for a series of jobs, particularly in administrative support positions. Indians in technical fields, I had been told in Turkey by Turkish contractors who were managing construction firms, were well educated, hard working, and inexpensive. When I asked Jai about this, he said that this was true and that Westerners in Afghanistan tended to treat Indians as "code monkeys," willing to put in long hours on the computer, and I did interview several Indians who were in a variety of IT-related jobs. More, however, worked in administration.

Most people argued that this trend was for linguistic reasons. Many Afghans understood basic Urdu, closely related to Hindi, which allowed Indians to communicate with their Afghan counterparts, particularly Dari speakers from the north of the country. Another reason there were so many Indians in these positions, Jai explained, was that they tended to blend in with Afghans better than others I had interviewed, and Jai went jogging many mornings through the Kabul streets, something that few foreigners (or Afghans, for that matter) tended to do. Almost all the Indians I interviewed were far more likely than the Nepalis, Georgians, or Turks to go out shopping or spend time just out on the streets. Afghans also like Indians, Jai said.

Part of this was cultural similarity, but there was also a significant amount of goodwill generated by India's foreign development aid, which contrasted with what most Afghans perceived as the political meddling of Pakistan. Indians, Jai explained, don't want to be confused for a Pakistani. Early during his time there, while out walking dressed in a traditional Indian shirt, a police officer had forced him to the ground and searched him. Once he found his Indian passport, he quickly apologized and shook his hand, telling him he had mistakenly thought he was from Pakistan. After that Jai tended to go out only in Western dress.

While being Indian made everyday security better for the Indians living in Afghanistan, those who lived in Kabul did think that it made them more susceptible to attacks by the Pakistani intelligence agency ISI, and the groups it was rumored to support, particularly the criminal syndicate, the Haqqani network, which had been involved in Arda's kidnapping. Jai said that one day when he was walking to work, a Toyota pickup truck trailed him for several blocks. He tried to cut down a side alley to lose it, but the truck picked up speed and pulled in front of him, blocking his path. The driver leaned out the window and asked him where he was going in Hindi. At this point, Jai ran back to the main street, though the pickup continued to follow him at a distance to his office. He reported the incident to the Indian embassy in Kabul.

Jai didn't feel safe going home that day, and for the next three weeks, he spent the night at various friends' houses. Eventually he moved into the Indian embassy with some diplomat friends and spent his remaining months in Kabul there. He certainly made the Indian embassy sound more welcoming than most of the Turkish contractors made the Turkish embassy sound. It was not as fun as having his own place, he concluded, but it was certainly safer and allowed him to maintain his relationships with the Indian community he had become a part of in Kabul. Unlike the Nepalis and Turks in Afghanistan, the Indian community there had longer, more continuous ties.

ACROSS THE BLACK WATER

While Nepal, Turkey, and Georgia had moments when their histories overlapped with Afghanistan's, it is difficult to even begin to separate the history of Afghanistan from that of northern India. For centuries, Afghan rulers raided India to keep their coffers full. At various moments, Delhi was part of empires based in Afghanistan, and Kabul was part of empires based in India.

More recently, the return to power of the Bharatiya Janata Party (BJP), with its Hindu nationalistic rhetoric, deemphasized India's multicultural history and ties with neighboring countries while focusing on India's role as a global power.[1] The current populist head of the BJP, Prime Minister Narendra Modi, gained fame by bringing international business investment to the Indian state of Gujurat. Now many hoped he could do the same to internationalize India's business ties more generally through a neoliberal, trade-focused approach to government and economics. However, Delhi's Mughal monuments remained reminders of the long ties between the two countries.

Historically, the commercial links between communities in Afghanistan and India meant a regular flow of travelers across what are fairly recent (and not at all agreed-on) national borders. India's Mughal rulers had originated in Central Asia, moving south to Kabul before eventually setting their sights on the riches of Delhi. Merchants and itinerant craftspeople moved back and forth between the cities of what would become Afghanistan, Pakistan, and India. Goods, particularly fruit, were grown in Central Asia and Kabul and shipped south to Delhi.[2]

Populations intermingled, though often in ways that were not immediately visible. Large groups of Afghans have lived in India for years. When I was living in Kabul, a few of my neighbors were members of the dwindling Sikh minority that had remained in Kabul during the Soviet period. In my interviews, Indians described how Hindus, both living in Kabul and there for work, gathered at temples scattered across the city.

Relations between Afghans and Indians had largely frozen with the partition of Pakistan and India in 1947. The violence of the partition saw many Hindus living in Pakistan forced to move to India, while Muslims in India were forced to move to Pakistan. Delhi, previously the capital of the one of the great Muslim empires, was now a city of Muslim monuments largely inhabited by Hindus. Grievances from that period, particularly over the mountainous state of Kashmir, continued to define relations between the two states. Pakistan actively opposed any moves by India that attempted to influence Afghanistan, which they saw as their backyard. This split was made worse by thirty-five years of war in Afghanistan. Still, the social and economic ties between India and Afghanistan remained, and these shaped the flow of contractors to Afghanistan from India.

One of the places where Afghanistan and India remained linked locally was in the market for certain goods. On a Tuesday morning, I headed to the bustling backstreet bazaar of Old Delhi to a northern section of the maze of backstreets called Khari Baoli. Here the spice and dried fruit sellers are clumped, along with some jewelers and fabric sellers. In general, the Old Delhi bazaar is a little cleaner and more electrified than its cousin in Kabul, but with similar products and amiable bartering among merchants, it felt similar to the Kabul market, making it easier to imagine goods moving back and forth between these similar locales.

In the first shop I approached on the outer ring of the market area, the owner was sitting between sacks of apricots, walnuts, pistachios, and figs. Pointing to various containers, he showed me the products he imported from Afghanistan. They were primarily figs, walnuts, and a few other dried fruits, and constituted about 10 to 20 percent of his stock, he said. He didn't travel to Afghanistan himself, but an Afghan merchant with a large truck came by his shop every two months or so. He would pay for the goods through the *hawala* system, a traditional money transfer system where he would give money to one of the moneychangers in the Delhi bazaar and the supplier could then pick up the money from a related merchant in Jalalabad, Kandahar, or Kabul.

Of the shopkeepers I briefly surveyed, most bought some of their product from Afghanistan, though none actually ever went there. One even ordered his goods online through an Emirati company, which then had an Afghan supplier deliver the goods. The logistics firms that transported these goods were essentially the southern, and more rugged, cousins of the truckers who transported Özgen's goods through the more stable north of the country. Those in the Delhi bazaar agreed that while there were still security concerns and occasional interruptions, the market for Afghan products was much improved. It had been more difficult, particularly before the Taliban period, since the border was not predictable and goods could end up stuck there, rotting.

Down a narrow alley that a shopkeeper had pointed me toward, the wholesaler section of the market was a series of slightly different shops with similar products, all off a row of long internal hallways. Here men sat at bare desks, armed with multiple cell phones, with a sack or two of their wares beside them as samples, and purchasers could come and buy 50 or 100 kilograms or even a truck full of figs, apricots, or whatever else the shop was selling. Money would exchange hands, phone calls made, and deliveries arranged.

Unlike in the shops outside, most of these shops specialized in one or two products and imported from one area in particular. When I asked about Afghanistan, several people along the hallway pointed me toward a slightly larger shop near the end of one of the halls. Sitting outside was a middle-aged man with a calculator who seemed to be doing most of the selling. Behind him in a smaller office with a plastic door sat a stout older man, with a nameplate that read K. K. Nanda.

K.K. went to Afghanistan first in 1961, he told me. He had picnicked on several occasions in Istalif, the town where I had conducted my dissertation research, and we chatted about its scenic river. Things were better in Afghanistan then, he said, but even at that time, Kabul was not a particularly desirable posting for an aspiring young merchant. It was considered something of a backwater. Hindus also believed crossing the sea, or the *kala pani* (literally, black waters), was polluting and could result in a loss of status.[3] Usually this was applied to the Indian taboo on sea voyages, but it was sometimes applied to Hindus in India to travel overland to Afghanistan as well.

K.K.'s father and a business partner had founded the dried fruit importing company, but his father had died when K.K. was young, and by the time he was old enough to join the company, there was no room for him in the Delhi office. He was sent off, to prove his worth, by managing the Kabul side of the business across the kala pani. Living in Kabul for two to three months of the year, he would roam the bazaar, chatting with Afghan merchants and farmers, placing orders. It wasn't easy, he said, because Afghans were tough and could be "slippery," but he seemed to mean this in a complimentary manner. Relationships between Indian and Afghan merchants developed and persisted in ways that they certainly had not for their Nepali and Turkish counterparts. K.K. spent part of each year in Kabul for over twenty years, and it was only in 1992, amid the worst of the fighting of the Afghan civil war, that his company closed its office there.

After the American invasion, he had visited the country again in 2003 to see what business opportunities there were. He spent a few weeks there and was able to reinitiate some of his old contacts and establish a few new ones. Most Afghan merchants were now willing to make deals on cell phones and then drive their stocks to his warehouse. In this way, he said, he could take advantage of his Afghan contacts without ever needing to go to Afghanistan.

Twenty percent of his stock came from the far reaches of Afghanistan's various corners: Kandahar to Kunduz and Herat to Kabul. Afghan imports to India had boomed in recent years and prices had dropped. Part of this, K.K. reasoned, had been the rise of solar pumps, which provided water more regularly, so that many more rural areas in Afghanistan were finally stable enough to see some investment in agriculture. The biggest issue was that through a strategic partnership agreement between the two countries, the tariff on select goods imported to India from Afghanistan had been lowered to 0 percent.[4] All of this was good for business, he said, waving his hand out toward the corridor where several workers struggled with sacks.

K.K. said as we were concluding that the businessmen in India have a tendency to look forward more than back. These days, Afghans did not need visas to visit India and Indian businessmen were afraid of kidnapping and crime in Afghanistan, so he thought it was unlikely that Indian merchants would return to Afghanistan. The markets were all global now, and it was more important that India emphasize its geostrategic positioning as a regional power. India's ability to flex its muscle in the region was paramount, whether it was on the cricket pitch or it was the fuel blockade in Nepal aimed at demonstrating its displeasure at the new constitution.

This was particularly true when the subject of Pakistan came up, which happened in most of my interviews in India.

PAKISTAN, NOT AFGHANISTAN

On the same day I visited the dried fruit market, I met with a well-connected senior journalist at the India International Club, a private club for diplomats, reporters, and policy wonks. The club's carefully manicured lawns and large fountain were where many of India's diplomatic elite gathered. The guard looked at me a little askance as I hopped out of a rickety taxi and walked up the drive; everyone else seemed to be arriving in BMWs or Mercedes that were then valet-parked.

In the tearoom, our conversation was regularly interrupted as various people came in and out, everyone stopping to greet each other. Praveed introduced each one and give a brief bio and pointed out the foreign minister's mother who was sitting out on the patio. Praveed had a PhD from Columbia, and much of his family was now in the United States. He had been tempted on several occasions to join them, he said, but he had trouble extracting himself from his role in Indian politics.

Praveed had started going to Afghanistan as a young journalist and eventually became one of India's primary commentators on the country, developing close ties with leading Afghan politicians. The dark days of the Soviet occupation and the following civil war had been difficult to bear, he said, and he hoped that the American invasion would have done more to transform the country. His connections meant that he did not just report on the news, but made it happen, in part by connecting diplomats and key political leaders to each other. He was close to both Karzai, the former president, and Abdullah Abdullah, the current chief executive of the country. He had also visited the former communist president, Babrak Kamal, after he fled to Russia and was a long-time friend and supporter of the exiled king in Italy. He clearly cared about the country and believed that India had a place to play an important role in all this, but the problem he kept returning to was India's focus on Pakistan.

While the governments of Nepal and Georgia had minimal interest in the long-term politics of Afghanistan, this was certainly not true for India, and this had an impact on the lives of contractors. The Indian government, Praveed said, historically hoped for a stable, pro-Indian government in Afghanistan. This was a possibility that terrified Pakistan. A stable Afghanistan would put pressure on Pakistan on a number of issues, including Kashmir, and force the Pakistani military to shift some of their attention away from the Indian border. Many in the government saw this as a good thing and an opportunity to use soft power to undermine Pakistan from behind.

Praveed had been more optimistic than some of his colleagues with the election of Modi as prime minister that some of this stalemate would end. He hoped that this could be the beginning of something of a rapprochement between the two countries as India looked to improve its standing internationally. Modi, he said, however, proved to be "illiterate with foreign policy." His bumbling and the continued tension with Pakistan seemed unlikely to do anything other than freeze India's position in the region, Praveed concluded. Most of those in the club seemed to agree with him. This was the old guard establishment of Indian foreign policymakers who tended to be from the Indian National Congress, the party of Gandhi and Jawaharlal Nehru, India's first prime minister. Few were happy with the current government led by Narendra Modi and his increasingly market-based approach to foreign policy.[5]

What all this meant, Praveed explained, was that for many in the Indian government, Afghanistan was no more than a tool in India's attempt to expand its influence while undermining Pakistan. Mostly the current situation was Pakistan's fault (almost everyone in India agreed on this), but the government was missing a clear opportunity to both expand its influence and encourage some important things, like democracy and development.

To help expand their soft power in Afghanistan, India funded some key, and highly publicized, infrastructure projects, which meant more job opportunities. This included a dam in the west of the country, which provided energy to Herat, and the new parliamentary building, which had been inaugurated only a few months before my arrival in Delhi. India had a sizable and visible diplomatic presence in Kabul to support these projects, and relations between the two governments appeared to be strengthening. President Karzai, who had been educated in India, in the British summer capital of Shimla, was considered particularly close to the Indian government.

Clearly, Praveed concluded, this growing cooperation made some in Pakistan nervous. This was what made Afghanistan dangerous for Indians. ISI, Pakistan's intelligence service, was long said to provide support to insurgent groups that targeted Indian interests. Almost all Indian contractors I interviewed could recite a litany of attacks that targeted Indians in particular: The embassy was attacked by suicide bombers in 2008 and 2009. In 2013 the Indian consulate in Jalalabad was attacked, and in 2014 it was the consulate in Herat. Indian civilians were also targeted, and a large, coordinated attack in February 2010 was aimed particularly at a group of Indian doctors in Kabul.

It was not surprising that most of the Indians I spoke with were far more concerned about targeted killings and kidnappings, as opposed to more occasional rocket attacks that Nepali contractors described. The historical ties between Indians and Afghans and their cultural similarities allowed them to move between the worlds of Afghans and internationals in ways that others could not. But it also meant they were more exposed to some of the underlying tensions in the war between Afghanistan, Pakistan, and India. All this gave these Indian contractors a rather unique view on a variety of issues.

KICKBACKS, MONSTER ENERGY DRINKS, AND OTHER PERKS

As my interviews continued, I kept track of the various ways in which the war seemed to affect the Indians that were different from what I had heard from Nepalis and Turks. What made the Indian position unique?

Over the course of six years Sajit had worked for four international contractors in Afghanistan as a procurement officer, which essentially meant he was in charge of buying or arranging the contracts for anything that the company needed. This involved working closely with both the American and Europeans above him who oversaw his budget and made various requests, as well as the local Afghans who ran the supply companies that Sajit made many of his purchases from. Balancing these relationships was complicated, though Sajit's rather jolly demeanor and his tendency to constantly offer cigarettes to those around him probably helped. Still, as he lit another of the many cigarettes he smoked during our interview, it was clear that the tension wore on him.

Indian accountants and managers like Sajit tended to have a wider view of how the pieces of the international intervention came together in Afghanistan than many others I interviewed. Even the American and European managers who ran these companies were limited by the fact that they mostly interacted with a small group of other international managers and, occasionally, the most senior Afghans in the organization. They were unlikely to speak much to the guards or the laborers doing the actual work or to interact with the Afghans around them. Those like Sajit who did much of the accounting and logistics for contractors often had a better sense of the experience of people at all levels of the company. It also put him in a better position to talk about some of the practices that went on among contractors at almost every level in Afghanistan. As Sajit told me, contractors quickly learned that "it was all about the money."

One of the things I had been struck by while living in Afghanistan was how often the international community complained about corruption in Afghanistan.[6] I was not immune to this and would fume at the airport when the policeman suggested a little contribution might speed up the process of getting through immigration. Still, it seemed, whenever anything went wrong, the U.S. government and others in the international community tended to blame corruption. This was convenient, since it was not the design or implementation of development projects but "corruption." This was such a simple excuse that few looked any deeper at what was driving the corruption or how it even worked.

As more and more efforts were made by the international military to weed out corruption, including a much-publicized joint task force starting in 2010, Afghans I knew, particularly government officials, complained irately that the international community was deeply involved in these practices as well.[7] Why, they wanted to know, were the Americans not focusing on the internationals who were involved in this corruption? Moreover, how could there be all of this Afghan corruption if the internationals who were providing the money were not involved somehow?

In researching for my previous book, I decided they had a point. I found numerous soldiers and contractors who had been caught in complex cons and scams, and it certainly seemed likely many, many more went undiscovered. I found stories about contractors or soldiers caught sending boxes of cash by express delivery back to the States, but these were embarrassing to both the military and contractors, and so, when possible, they were hushed up.[8] While the international cases were often described as outliers and isolated cases, Afghan corruption was almost always described as systemic.

Sajit argued that more often than not, corruption was linked, even if simply by the fact that contractors created an environment that allowed these practices on all levels to flourish. He lit another cigarette and explained several of the deceptions that he had encountered as a procurement officer for Supreme. Procurement offices might have been the best place for the various types of scams, he said, since the Afghan and international firms were desperate to try to secure contracts with these larger companies. The easiest tricks involved simply slipping a procurement officer some cash to help make sure that the contract went to a specific bidder.

As monitoring for this grew more rigorous, subcontractors had to resort to other means. They were also willing to use violence to try to scare other firms away. In one case, Sajit had given a contract to an Afghan company that later called to say that they had "made a mistake" in the bidding estimate and were withdrawing, so the contract should go to the second-place firm. He found out that someone connected to the second firm had threatened to kill those working for the first firm, convincing them it was in their best interest to withdraw. In another case, Sajit dealt with a firm that lost a contract and set up a roadblock of armed men outside the office, not allowing anyone from his team to leave until the contract was renegotiated. If such an incident had occurred with American or European administrators, it would have likely become a political incident involving embassies and the Afghan government, but since the low-level procurement officers were "just Indians," such incidents tended to escape notice. Violence and attempts at bribing low-level officers, however, could get only so much money, Sajit said. As the years went by and monitoring systems improved, the methods for extracting bribes and other funds became increasingly creative and seemed to increasingly rely on an international partner.

Even the simpler scams seemed to have some international tie-in. Sajit described an operation at a large international contractor that had started rather small but grew rapidly. When he arrived at a camp with about a hundred contractors working in it, he was flooded with numerous complaints about how bad the food in the dining hall was. As he investigated, he heard rumors about several rackets that were being run out of the kitchen. Perhaps it was because he seemed more honest or perhaps as the highest-ranking non-American or non-European working in the compound, other workers began coming to him to complain about the facilities manager who was letting conditions slide. The workers, he said, had a point. The rice, in particular, was terrible. And while South Asians were often more critical of the quality of rice than others, it had gotten so bad that some of the Indians and Nepalis on the base were buying food from the Afghans who came to the compound each day and cooking quietly in their rooms—something that was against regulation. How had the quality gotten this bad?

Sajit complained to the British manager of the dining hall that something was amiss. The manager agreed that the rice that day had been below par, but told him it simply hadn't been washed properly and he would fix it tomorrow. Later, when he was sampling some of the rice from the local market, he came across sacks of rice that were supposed to be delivered to the base. Instead, it turned out, the sacks were being pulled aside at the last minute and exchanged for cheap local rice, leaving the manager to pocket the difference. And this was just the beginning of the problems in the kitchen.

Sajit later discovered the kitchen staff had taken to throwing out entire unopened boxes of food, especially chicken and other meats. The trash guys would then collect the garbage and pull out the unopened boxes once they had left the compound but before disposing of the trash. This practice eventually become such an industry that at one point, he had seen a flat-screen television and exercise equipment being loaded carefully into the garbage trucks to be "thrown out." Such theft could not have happened without the notice of the British manager who was overseeing the kitchen, but the Indians and other workers were afraid to complain. The company, as demanded in its contracts from the U.S. military, had an ethics and complaints office, but those working in the compliance office were friendly with the upper-level management, so there was no way they were going to take the side of these workers. Better to keep quiet and keep your job, the workers told hm.

Sajit and others I interviewed marveled at the complex array of scams the Afghans and their international partners were managing. From his rather privileged position, Sajit knew there was far more going on than he would ever understand. At one of the fuel depots he oversaw, trucks were supposed to drive out of the gate and head to the nearby U.S. base. When he stood on one of the towers, he said, he could watch the trucks leaving the depot. The base was to the left, yet about two out of every ten trucks instead turned right, heading away from the base and toward the border with Pakistan.

In his years of overseeing the depot, no one at the base ever complained to them about the missing trucks, he said. But there was no way that those on the base who were supposed to be receiving the fuel couldn't have noticed the missing fuel. Someone must have been paid off to sign for higher numbers, someone must have been adjusting the amounts recorded that were leaving the depot, and someone must have been paying off the drivers to divert the fuel, but he did not know how that was happening. Sajit had little to gain by investigating. He was still being paid on time, and he was nervous that if the scam involved anyone important on the base, it could get him fired.

In other cases, particularly when powerful Afghans were involved, there was little that anyone could do. In another case Sajit described, Supreme got a contract from the Canadian government to supply generators and fuel to an Afghan National Army (ANA) base that the Canadians were supporting. They bought two generators for the camp and supplied it weekly with fuel. Soon the police commander started calling and saying that they had run out of fuel for the generator and needed more. Sajit was puzzled: they had just delivered 10,000 liters of fuel and the generator company said that there was no way the generators could burn through fuel that fast. So Sajit called the Afghan driver who delivered the fuel. The driver admitted that the commander had instructed him not to deliver the fuel to the generators, but to unload it in an additional fuel tank in the back of the police station. What could he do? There was no way the driver would defy the commander. Sajit said his bosses didn't want to hear about this; it made them sound incompetent and, moreover, they were getting paid for how much fuel was supposedly delivered. Stopping the practice would lose them money as well.

Other Indians I interviewed pointed out that many assumed that Indians were more corrupt than other internationals there because they could speak Urdu and communicate with the Afghans. In the case of one small fueling base that three people I interviewed worked on, the Indians there were a small part of the workforce that included Zimbabweans, Kenyans, Sri Lankans, and Filipinos as well, but the Indians were also the only ones who could communicate easily with the Afghan drivers who delivered fuel to the base. When they chatted with the drivers, their bosses and the other workers often assumed they were engaging in some sort of scam or kickback scheme.

At this base, the line of fuel trucks parked along the side of the road waiting to be checked at times had more than five hundred trucks in it. The Afghan drivers sometimes would try to buy their way to the front of the line by giving the guards small gifts from the market. Monster energy drinks were apparently one of the more desirable gifts, and they could sometimes help the drivers get ahead of others waiting to be searched. On other occasions, since the Indians had to stay on the base, and therefore had little access to local markets, they would pay the drivers to bring them goods from the market. There was nothing illegal about that practice, but their bosses often saw little difference between these two activities, and anyone seen with a soda or other food not supplied by the contractor was automatically assumed to be evidence of corruption.

In his office overseeing procurement, Sajit received offers of bribes, he said, usually cash, but all sorts of other perks too, even a car at one point. Much higher up than his Indian colleagues overseeing fueling,

Sajit was earning enough to make a bribe not worth the risk of losing his job. But the sizable salary managers made also meant they were more likely to keep their eyes open for new, better-paying job opportunities.

Over the course of his five years in Afghanistan, Sajit worked for four companies, and the Indians I interviewed were generally far more likely than their Nepali counterparts to move between companies. Perhaps this was because the web of Indian managers meant that Indians were more likely to hear of new openings than their more isolated Nepali counterparts. Or perhaps it was that many were better at both English and Urdu, allowing them to network more easily.

This also meant that while almost all Nepalis seemed to take the well-trodden road to Afghanistan through Delhi and Paharganj, it became increasingly apparent during my interviews that Indians had other options for finding jobs and getting to Afghanistan. Still, despite these options, I struggled to find cases of contractors who were getting ahead due to their work in Afghanistan, so I began to cast a wider net.

CHAPTER 20

CLASSES AND GENDERS AT WAR

FORKLIFT DRIVERS

As the weeks passed, I began traveling outside Delhi, visiting other parts of India where I had contacts among international contractors. Some of these areas were unique in the way that they contributed contractors to the war in Afghanistan. For example, the small state of Kerala contributed a disproportionate number of contractors to support America's wars, despite the fact the state had only 3 percent of India's population.[1] This was largely due to an international network of Keralite workers across the globe.

Located on the southern coast, facing the Arabian Sea, the region was connected to more distant parts of the world far sooner than most of the rest of the subcontinent as a center for trade halfway between the Middle East and China. Instead of traveling through brokers in Delhi to arrange their jobs, many headed to the Gulf, relying on local brokers and networks they had that linked Kerala to the United Arab Emirates, Kuwait, and other wealthy Arab states.

The Keralites I interviewed came from a range of backgrounds and worked for international contractors I had become familiar with, like DynCorp, Supreme, and Ecolog and also for a variety of Indian and Afghan firms. I spent ten days there, interviewing fuel operators, in particular, who dominated the offices of several subcontracting companies in Afghanistan. There were also many Keralites in the telecom industry, which boomed as an estimate $2.6 billion of U.S. government money entered the strategic sector.[2]

In other ways, however, the experiences of Indian contractors even from those from very different regions were more similar than they were different. One of the overriding patterns was the way that migration to dangerous areas continued to be driven by the stories that young men told each other about the possibilities of life-changing wealth that one could secure abroad. The more I traveled to different areas, however, the more I questioned these tales and realized how little contracting in conflict zones did to alter the social status of those working abroad.

From Kerala, I headed back north to Kurukshetra, the site of the epic battles of the Mahabharata in the state of Hayrana. More recently, the city was home to a camp for 300,000 Hindu refugees from West Punjab following partition in 1947. In the years that followed, 7,000 officials worked to redistribute 1.9 million hectares that had been left behind by those fleeing to Pakistan.[3] Hayrana saw another influx of Hindus in the mid-1980s who were fleeing from the neighboring state of the Punjab as the result of religious violence between Sikhs and Hindus there. In more recent years, Hindu voters in the state have strongly backed the Bharatiya Janata Party (BJP).

Kurukshetra and the rest of the state of Haryana are connected to Delhi by a busy superhighway. Compared to the bustle of Delhi, the city felt quiet as I drove around the downtown area with three former contractors, who had insisted they give me a tour of the area while chatting about their time in Afghanistan. Raghbir, who was driving, was the most vocal of the three. He had grown up in a nearby village, the son of a poor farmer, so he had no real opportunity for good work in India. When he was just eighteen, he got a job working for a Turkish construction company in Iraq that was building bases on a KBR contract. The money was not great, only about $600 a month, but that was more than he could earn in India, so he took it and stayed there for four and a half years, during which time he was not able to return to India once.

Tired and homesick, he returned home and quickly married. He spent a year with his new wife but couldn't find a job in India. When a friend from Iraq found work in Afghanistan and offered him a job, he took it. Afghanistan was better than Iraq, he said. The pay was higher and he was able to return home twice a year. With the money he saved, he bought the car we were driving in and set up a small clothing shop in his hometown. He now had a daughter and was still struggling financially.

In contrast with Raghbir, earlier that day I had met with Kanwar, who was an acquaintance of the other three. Kanwar was a mountain of a man, at least six feet four inches and well built. He had worked as a forklift operator at several U.S. bases for DynCorp. His pay was better than Raghbir's, $1,250 a month, he said, but now his father needed him back in India. His father was a local politician with the BJP, and in recent years, the BJP had won resounding victories in Haryana. Now, as Kanwar put it, he and his father "had sixty or seventy villages," meaning they were loyal to the BJP and under his father's political influence.

After coffee, we agreed to drop him off, and he directed us to a small storefront in a nearby town. Kanwar got out of the car and headed into the shop, while we started to turn the car around. But then Kanwar came back out, leaned his large frame down to talk into the car window, and insisted that we go inside with him for a little while. The late request seemed somewhat odd and Raghbir wanted to show me another town to the north, but we decided we could delay a bit.

Inside there were eight other young men lounging around a spartan room with no decorations, which I later found out was the area political headquarters for the BJP. Kanwar introduced me to the various people around the room, but didn't really explain what any of them were doing there. He sat me in a chair in the middle of the room and offered me a cup of Mountain Dew.

I sat with the three other contractors I knew better, sipping our cups awkwardly, with Raghbir and his friends looking at me intently but saying little. Trying to make some conversation, I asked Kanwar if he worked in the office, and he shrugged his shoulders and said, "Sometimes." His other job, he said, was working as a security supervisor at a government power station, and he pulled out his phone and showed me some photos of him leaning against a gate at the station looking severe, with the power plant's name behind him. The other seven men listened, and then we all sat there and looked at each other self-consciously. Finally, Raghbir looked at the door, and I made our excuses, saying something about how we were heading north. Kanwar seemed fine with us going.

As soon as we got back in the car, I asked Raghbir and the others why Kanwar had suddenly wanted us to come in after we dropped him off. The entire visit to the office felt odd. Raghbir speculated that Kanwar hadn't realized the other men would be there, and since they were, he wanted them to see that he had an American friend. Now that his father was a big member of the BJP for the area, it was important that his family had contacts and that they showed them off, another of the contractors in the car suggested. Kanwar's time in Afghanistan clearly helped his father's rise in the BJP, and the office was a point where global and local networks came together, Raghbir explained; that's also why Kanwar got a good government job at the power plant. He didn't need to work hard there or even show up more than occasionally, but the paycheck always came. "I wish I could get a job like that," he concluded wistfully.

From there we drove north, stopping at the university's heritage museum, which was filled with old tools and traditional instruments from the region, and then stopped by the science museum, which was mostly filled with loud students banging on the various exhibits. To get out of the heat and away from the kids, we went to a dim bar at the edge of town. Here, we picked up on the topic of Raghbir's return to India. It was difficult, he said, having a fairly good job in Iraq, coming home and finding no work, then having an even better job in Afghanistan, coming home and again finding no work. But here he was again, with a small shop but nothing too promising, and he was considering going abroad again.

The problem, he said, was that businesses in India were conservative. They did not look at anyone's skills or experiences. It was a question of just "showing your diploma" and having connections. Really, he said, the two were related. To get a good job, you needed to go to a good school, and to go to a good school, you had to have connections with someone in one of the major political parties. "Look at Kanwar," he said. "His job in Afghanistan was driving a forklift, but now he has a respected job here because of the connections that his father has and because people assume he has connections abroad. When I came back, the only thing that I could do was set up a store and work on my own," he concluded.

Raghbir was not the only Indian contractor I interviewed who had such complaints. It was as prevalent, if not more so, among those working in more skilled positions. Vishal, who worked on public relations for a telecommunications company, finally decided that he had had enough of Kabul and so started looking for work elsewhere. No one in India would hire him at a comparable salary, and none of them seemed to give him credit for the years he had spent in Afghanistan, he said. He had a deal with his boss that he worked for three weeks in the company's Delhi office and then spent one week in Afghanistan each month. Since the Delhi-Kabul flight was short, it was not a bad setup, he said, but his wife and parents still worried about him.[4]

The contrast with Nepal was striking. The few skilled workers and administrators I found in Nepal who had been in Afghanistan previously, particularly hydroelectric engineers and even some government officials, all had benefited greatly from their experience in Afghanistan.[5] They returned to promotions and better jobs, with their employers in Nepal viewing their time in Afghanistan as exposure to the international community and a valuable asset.

The opposite was true in India. Working in Afghanistan is like a scar on your résumé, one contractor told me. Indian firms like to hire internally; experience in Afghanistan was not beneficial, and in fact could actually hurt. Most Nepalis I interviewed were intrigued by the notion of working in Afghanistan. For the Indians I interviewed, perhaps it was because they were closer and their histories were more intertwined, Afghanistan was seen as something of a dead end. It was a place you went to, I was told by several people, only when you had no other career options. Any other place was better.

When I asked Raghbir if he would consider going back, he replied, "Certainly, as long as the money is good." However, he was hoping that something might turn up someplace else. He stayed in touch with several of his colleagues from his time in Afghanistan and they emailed occasionally about job openings in the South Sudan or the Democratic Republic of the Congo where they now worked. Even better, though, would be America.

He had even thought about marrying the sister of one of his American friends, whom he called his best bud. Apparently she was willing to do it for $6,000 so that he could get his green card. Ultimately he decided this was probably not a good idea. She had two children from a previous marriage and that seemed complicated. Still, he asked if I knew anything about what life in Wyoming was like.

IN A MAN'S WORLD

Work in Afghanistan, particularly for those in India, did not seem to translate into social or economic mobility as much as it created the desire (and sometimes the need for those who fell into debt) to travel farther and farther in the search of opportunities that always seemed just over the horizon. Contrary to narratives about migration and opportunities of free markets, work in Afghanistan seemed to me to be solidifying social categories and divides, not breaking them down.[6] This was also true of gender.

While I had interviewed women in each of the countries I had visited before India, it was clear that from these countries, travel to Afghanistan was something that primarily men did. It was not that women couldn't go, just surprising when they did. The assumption, particularly among private security contractors, was that this was men's work.[7]

On military bases where many of the Nepali security contractors worked, there were women who worked in the dining halls or in low-level administrative positions, but they tended to come from Southeast Asia, the Philippines, and other less patriarchal societies. In Kabul, there were female sex workers, and multiple brothels in Kabul had been shut down by the Afghan government in raids that appeared to be about public relations more than anything else. But in Afghanistan itself, sex trafficking did not seem to be the same issue that it was at U.S. bases in other countries.[8] While several of the men I interviewed were open about having visited brothels, there were not many of them, they reported and they were not close to the bases. Perhaps this was because the strict norms regarding sexual segregation in Afghanistan made them easier to police, or perhaps this was simply due to the remote nature of most bases.

With some of these tales circulating, combining with Nepal and India's largely patriarchal families, there was an informal taboo on women from these countries working in lower-level contracting jobs. Instead, the Nepali and Indian women I interviewed tended to have mid- and upper-level administrative jobs. Diya, an Indian woman in her mid-forties, had been incredibly successful and in just a few years had worked in Afghanistan for Chemonics, Louis Berger, and International Relief and Development (IRD) in offices in provinces across the country. She had given a paper on using geographic information systems (GIS) to map crops at a conference in Delhi and an American who was setting up an office for Chemonics in Afghanistan was impressed. He offered her a job on the spot, and the next thing she knew, she was getting on a plane.

The office was just being set up as she arrived, and getting in on the ground floor like this, she said, allowed her to shape the project, which she really liked. In an Indian company, there was often "one guy who controlled everything," and all that happened was that you received your orders and did your work. You might spend one year drawing lines and not know why, she said. In contrast, in Afghanistan, because she had GIS skills, she was treated like a king, she said. She designed the project, oversaw its setup, and monitored its process. Plus, she said, they bought her great software.

Still, being a woman made other things difficult.

When women did go to Afghanistan, they were often treated as unique creatures. Another female contractor told me that IRD had a house for Indians, a house for Filipinos, a house for Nepalis, and then a house for women, where it didn't seem to matter where anyone was from. One of the Georgian contractors characterized international women as a third gender. Their Afghan colleagues, in particular, did not treat them like Afghan women; their Afghan colleagues listened to them in meetings, ate with them, and socialized with them in ways that they would never do with an Afghan woman. Yet they were treated cautiously, in ways the Afghans would not treat international men, with continual, subtle discrimination; they were given additional escorts, faced stricter security restrictions, and were not permitted to travel alone.

Sani, a Nepali woman I interviewed who had worked in management for a large international relief organization, said that the only way she could get her job done was by accepting a certain amount of gender discrimination. Trying to fight all aspects of discrimination would get her nowhere, she said. This meant that although she wanted to sit in the front seat of the car, she knew that her colleagues would never let her, so why argue about it? Unfortunately, it also meant keeping quiet even when she didn't want to.

In one instance, a case was brought to her office of an Afghan woman who had eloped with her boyfriend. When the taxi driver realized they were eloping, Sani said, he robbed them and raped the woman. Despite these crimes, the woman had been arrested for adultery and her family had threatened to kill her for the shame she had brought them. Sani wanted her organization to do something about all this, but her colleagues did not think this was a good idea. They felt that the woman had technically broken Afghan law by eloping, and the community was against her. If they tried to intervene, it could be bad for all of the other important projects they were working on. Sani thought the men in the organization were dismissing her concerns since she was a woman and considered to be too emotional about the situation.

Sani suggested that as fellow South Asians, Indian women were at least more aware of the issues that gender raised than their Western counterparts were. She described a few occasions early in her time working there when Afghan guards or drivers at her compound would brush by her when walking past or reach over her to lock a car door. Such physical contact might seem benign to a Westerner, but in Afghanistan's highly segregated world, such touching was illicit and most probably not inadvertent. Sani said that by taking a firm line with the drivers, they began to treat her more appropriately. The Western women who were not even aware of these underlining tensions, she said, essentially allowed the stereotypes about their loose morals to continue to flourish.

Female contractors also stressed that the presence of women in war, a place where masculinity was emphasized, served to highlight the divide between genders and social classes. When I asked Sani if she had interacted with any other Nepalis in Afghanistan, she replied, "Almost none." There were a handful at other organizations, but her group did not have any on their team and their security guards were all Afghan. Even if the guards had been Nepali, she suggested, socializing with them would have seemed inappropriate for her as a woman. This was in contrast with some of the Nepali officials I spoke with who worked in administrative positions and reminisced about celebrating Nepali festivals with their fellow countrymen who worked as security contractors. The social and gender distance between these security contractors and the female administrators was bigger and harder to cross.

The one time Sani ran into a group of Nepali men was at the airport in Kabul for a flight back to Delhi. When the men realized that she was Nepali, they quickly befriended her, calling her "sister" and sharing some snacks they had brought. They were telling stories about their time working in Afghanistan, but it seemed that not all had the correct paperwork and were worried about transiting through Delhi. Sani thought they must have assumed she was perhaps kitchen staff on one of the bases, since they gave her careful instructions about how to pass through immigration, how to respond to various questions, and where to hide her money.

She didn't have the heart to tell them that she had passed through immigration numerous times and wasn't going to Kathmandu but to a conference in Europe. When they arrived, they all walked together toward immigration, and Sani ducked into a bathroom to let the others get ahead. She didn't want them to see that up ahead a man was waiting with a placard to take her through the VIP lane, then put her in a Mercedes and drive her to a five-star hotel for the night, before flying on the next day. She did not want the men, who were poor workers, to see how easily she was whisked through security while they struggled. The men had assumed that as a Nepali woman abroad, she was both poor and relatively inexperienced. Their work in Afghanistan had done little to shake some of their assumptions about gender or class.

While many contractors described earning good wages, few Nepalis or Indians were able to use a few years' salary to significantly change their position on the social ladder once they returned home. After the travel costs and cuts for brokers, there was only so much left for those actually doing the work.

In some cases, workers back at home seemed fine with this. The lucky ones built extensions on their homes, redid their roof, perhaps opened a small shop, and continued on with their lives. Especially for some of the younger ones, however, going abroad, passing through Dubai and Delhi, hearing about how some of their international colleagues lived back home, they wanted more. They thought that risking their lives for years at a time in Afghanistan would be more transformative and open doors, but instead they found themselves living in similar neighborhoods, unable to break through class, caste, and gender divides.

Returning to India and Nepal, young men like Raghbir became more and more desperate to find a way out of their countries and began discussing sham marriages and other schemes that tended to leave them even more vulnerable.

BEYOND AFGHANISTAN

A few months after I interviewed Ashim at the mall in Kathmandu about his time in Afghanistan working for Supreme as a baker, he sent me an email. This was not unusual: I stayed in touch with many of the people I interviewed over the year, checking to see how job hunts were going, asking follow-up questions, and in one case attending a birthday party. I ended up helping more than a few with a cover letter and proofreading résumés. Ashim had a similar request. He had received an email offering him a job and wanted to know what I thought about it.

I was always somewhat suspicious of emails like this. Many of them were a close cousin of the emails you might have received from a Nigerian princess who needed help transferring her recent inheritance to your bank account. For those looking to leave India or Nepal, their desperation was so great that it was tempting to start believing in fairy tales with Nigerian princesses.

As I scrolled down and read the email he had forwarded, however, I found fewer red flags than I had been expecting. The email was well written and clear. A hotel in New York City was recruiting bellhops and other workers, the email said. Ashim had been recommended to them, though the email was vague on who did this recommending. They wanted him to go to Delhi to collect his visa from an Indian lawyer. The lawyer's contact information was below. Even while I was suspicious, the email had links to the hotel's website, as well as Yelp reviews. I looked up the hotel on Google Maps and was promptly taken to a touristy neighborhood in Manhattan where they said the hotel was located. It seemed too good to be true.

And of course it was.

It took some more digging before I could find the cracks in the offer, but they began to appear. First, while there were reviews of the hotel, there were only a handful of them, all recent. Some of the latest reviews seemed to describe a hotel, but the earlier ones made it sound like more of an Airbnb homestay. I also had problems on Google Street View seeing anything that resembled a hotel and asked some friends who worked nearby if they had heard of it. They had not. It seemed that the writer of the email had gone so far as to build a website, set up an Airbnb from his apartment (or the apartment of a friend) in New York, while also having a contact working as a lawyer in Delhi. If Ashim went to the lawyer in Delhi, I grew increasingly sure that he would be asked to pay a deposit for the visa, either to be told to wait in limbo, like those other Nepali workers in Delhi, or perhaps they would try and send him someplace else. It was difficult to guess how deep the scam went. Regardless, it did not seem like a good path forward for him.

I bumped into more evidence of such hoaxes when researching Contrack, the company that Teer Magar had been working for before he was arrested. I came across a job offer on a website that was supposedly signed by the CEO of Contrack. The letter was impressive because of the details it gave. It was a call for applications for a position at Contrack's headquarters, not in Afghanistan, but in the Washington, DC, area, offering a salary of $10,000 a month. The letter claimed to be from the CEO of the company, and the address of the company's headquarters and many of the details of the job were accurate. It was only the language of the last few paragraphs of the letter and the fact that the letter writer was not asking the recipient to contact someone in the United States that made the letter suspicious.

The letter concluded, in language that was much less fluid than the rest of the writing, "Please be very important that you humble yourself before the agent Barrister Elvis Good Luck, respect him and follow his instruction for him to get your visa in due-time to ensure confirmation with our record for immediate delivery of your Air-Ticket together with your Job Offer Letter hard copy direct to the embassy in New Delhi India via DHL Courier service."[9]

The rest of the language in the letter was cleaner, and I could only imagine that whoever was running the scam had manage to procure a genuine contract letter from Contrack and then inserted a few paragraphs about applying for a visa and paying fees to the Indian lawyer at the end of an authentic letter.

For Nepali and Indian contractors who had gotten a taste of the experience working abroad and met Americans and Europeans who had access to what seemed to be infinite opportunities, it was possible to see how such a letter might have appeal. Workers in Afghanistan and other war zones had been the first step; now they were anxious to take more. And there were brokers, con artists, traffickers, and others who were willing to take advantage of them and their frustration with the fact that no matter how hard they worked, they never seemed to be able to advance as far as they would have liked.

While most of these scams resulted in nothing more than lost money and wasted time, there were Nepalis and Indians who had worked in Afghanistan who ended up settling abroad. I had heard particularly of British Gurkhas who had earned British citizenship, but there were others too. I wanted to understand to what extent their experiences abroad matched their expectations and how working in Afghanistan might have prepared them (or not) for work in the West.

So with the monsoon season looming, I packed my bags and turned in my key at the Delhi University Guesthouse to head to Britain to see what the ripples of contracting in Afghanistan might look like there.

CHAPTER 21

RETURNING ABROAD

ALDERSHOT

Aldershot is a little less than an hour's train ride southwest of London. My train passed through the green fields of the English countryside, punctuated regularly by the expanding towns that surround London. At first glance, the Aldershot station looked no different from the eight or so other stations before it. Past the ticket office and coffee shop, several buses waited outside to pick up passengers.

On my walk toward the town center, it took only a few blocks to see evidence of some of the ways in which Aldershot is distinct from its neighbors. As I passed the Gurkha Kitchen, the Momo Station, and Gurkha Villa, I realized there were more Nepali restaurants than fish and chips shops. There were jewelry shops with Nepali names, such as Laxmi Jewellers and Gurkha Bullion. Travel agencies advertised cheap flights to Kathmandu, and a taxi drove by with "Gurkha Cab" painted on the side.

Nepalis made up only around 10 percent of the population in Aldershot, but as I strolled through one of the parks in town, it seemed even higher than this.[1] In the weekly market, women in traditional Nepali dress lined up to purchase fresh produce, and Nepali teenagers walked to school in British uniforms. Older retired Gurkhas and their wives took evening strolls and sat chatting on park benches. When I greeted several with a "Namaskar" and "Sanche cha?" typical Nepali greetings, a few raised their eyebrows, but others stopped and chatted with me, perhaps assuming I was a British officer who worked with the Gurkhas.

It was as if I had landed in a strange merging of a British town and Nepali village.

Aldershot is not the only town with a retired Gurkha population. Other areas, particularly in Hampshire and Kent, which have military bases, have sizable Nepali populations, but perhaps because Aldershot was home to the camp that the Gurkhas first moved to in 1997 when their base was moved from Hong Kong, the population around the town felt more established. Home to 10 percent of the entire Nepali population of the United Kingdom, it had become emblematic of the Nepali presence in Europe. The Dalai Lama had even visited twice since 2012.[2] It was the best place for me to look for both Nepali soldiers and contractors who had been to Afghanistan but were now living in the United Kingdom.

For many of the contractors I had interviewed in Nepal, the intervention in Afghanistan was the story of going abroad to fight in a war and then, in a rather abrupt fashion, attempting to come home and resume life as it was before, but better off financially, they hoped. Most of those who were in the United Kingdom already had come from a different economic class and a different view of the world from those I had spoken with in Nepal.

I interviewed Manish at the Himalayan Java Coffee House, adjacent to one of the town's rundown central shopping malls. (I had previously conducted multiple interviews at branches of the Himalayan Java Coffee House in Kathmandu, so this seemed like a fitting place to meet.) We could chat only for about an hour, he said, because that afternoon he was taking a flight back to Kabul, but he was happy in that time to tell me about some of his experiences.

Manish was originally from a small farming town between Pokhara and the town of Gorkha. He had joined the British Army at eighteen and served for twenty-two years before retiring. Most of the time he had been stationed in the United Kingdom, but he had also done three tours in Afghanistan, in 2007, 2009, and 2011. He served in some of the worst areas during the worst times: a stint in Kandahar, a stint at Camp Bastion, and a stint at a more remote outpost in Helmand.

Even before he had retired, a friend contacted him about the possibility of doing contracting work, and he moved almost directly from the British Army into working for FSI, a British firm that was supplying security personnel on a subcontract to DynCorp at that time. The difference between working for FSI in Kabul and being in the military in Afghanistan, he said, was extreme. He had been used to living in a tent and going on foot patrols, but now he lived in a solid compound near the airport and had a fairly light rotation of three months in Afghanistan and then one month leave at home.

He was working with six other Nepalis who were all former British Gurkhas. They were responsible for supervising, training, and generally overseeing another 197 guards, most of whom had served in the Indian Army. They were almost entirely of Nepali descent, but many were from Deradun or Darjeeling in northern and northeastern India, and so technically they were Indian citizens. FSI referred to itself as "a specialist supplier of ex Gurkha personnel," though Manish was clear on the fact that he and the other six who had served in the British Army were the "real" Gurkhas, not those who had been in the Indian military.[3]

Working with these former Indian soldiers, he said, could be challenging. For the most part, they did have some weapons training (particularly compared to those in the Nepali Army, he said), but the Indian Army did not like to expend ammunition, so a lot of soldiers had rarely fired their weapons. Another issue was age. Manish was in his early forties while a lot of the guards from the Indian Army were ten years older. This made supervising them awkward in a culture where younger men were expected to listen to their elders. There was also something of an imbalance since Manish was considered better off and with more opportunities given his status as a British citizen.

Language separated the two groups. Most of those who had been in the Indian Army did not speak English very well, which meant that Manish and the others who had been in the British Army were liaisons in most of their interactions with other internationals. The upside was that most of the guards from India spoke Hindi, which Manish was not fluent in, and so they could often communicate better with the Afghans they worked with.

This model that FSI and DynCorp were relying on had spread slowly to other contractors as well as they streamlined their structures and maximized their profits over several years of the war. Security companies discovered they could charge a premium for contractors they labeled as Gurkhas, but those like Manish who had retired from the British Army were expensive and reluctant to travel to places like Afghanistan and Iraq. Manish estimated that only 2 percent of the British Army Gurkhas he knew had retired around the same time as he did took jobs contracting abroad. Most preferred looking for work in the United Kingdom after retirement, and many moved into Military Provost Guard Security (MPGS), which guarded bases in the United Kingdom.[4]

Companies discovered that it was far cheaper to provide either former members of the Indian Army or the Nepali Army or Singapore Police Force, or in some cases, even Nepali civilians and using the ambiguity around the term and calling them "Gurkhas" in their advertising. The problem, then, was then how to manage these "Gurkhas." Most didn't speak English, and their training was uneven, so companies hired a few retired British Gurkhas like Manish; they were expensive, but could speak both Nepali and English and oversee training. Hiring a few expensive British Gurkha managers then allowed the companies to hire cheaper, less qualified guards for the rest of the positions.

Manish was clearly proud of his Gurkha background, but was more pragmatic about his current situation than some of the other retired Gurkhas I interviewed in Nepal. His father had also been a Gurkha in the British Army, and his grandfather had fought for the British in Malaysia. Both of his brothers had also served in the British Army. Manish had a son, but he was not sure whether the tradition of fighting for the British would continue with him. For one thing, he was already a British citizen, so he would not need to go through the onerous trials of Gurkha selection if he wanted to just enlist as a British citizen, a far simpler process. Moreover, Manish said he liked computers and wasn't sure that the military life was for him. He would have to see.

As we finished our coffee, I asked him how long he thought he would keep returning to Kabul, and he said that he wasn't sure. As long as the security conditions remained more or less what they were, he would keep going back for the time being since the pay balanced out the risk. Eventually, he said, he would settle down in the United Kingdom.

As I interviewed other British Gurkhas who had recently retired or were soon to retire in and around Aldershot, it became clear that not only were retired British less likely to take contracting jobs, they were also less likely to return to Nepal. Manish said that he liked to go back every year or so, but it had been three years since he had had an opportunity to make the trip. His wife and kids were here, and so was his mother, so there was less reason to go back on one of his few leaves. Everyone was settling in the United Kingdom these days, he concluded, and the only ones he knew who were going back to Nepal regularly had political aspirations or business interests.

All of this reinforced the divides between Nepalis who managed to secure British citizenship and earned higher wages and those who primarily supplied the labor at much lower rates in places like Afghanistan. Contracting companies then could hire people who could speak Nepali and manage their workforce without paying most of them the salaries that Nepalis who were now British citizens demanded. The Gurkha system allowed the companies to reinforce this divide, but as I spoke with more people in Aldershot, it became apparent it was also changing the way that Nepali soldiers related to the United Kingdom and racial and economic tensions more broadly.

GURKHAS NO MORE

The next night, I walked down the street from my hotel to the Gurkha Villa restaurant. The restaurant seemed to be a cross between a Nepali restaurant and a typical British pub. At the bar, a group of men were sharing a large plate of steaming momos. Two younger couples were sitting at tables on what appeared to be dates, and two other tables of middle-aged men were drinking Nepali beer.

I was meeting a unit of ten British soldiers at Gurkha Villa for dinner. They were in town for a shooting competition that was taking place over the next two weeks. Nine of them were from Nepal and one was British from Sandhurst, the prestigious royal military academy, a twenty-minute drive north of Aldershot. The British officer had been temporarily attached to the unit and was, other than me, the only person in the bar who was not of Nepali descent.

The pub next to my hotel where I had eaten the night before looked very different. It appeared to have been filled entirely with professional white Brits, having a drink on their way home from work; there were no Nepalis. It seemed this racial segregation was maintained throughout the town. The only real exception to this was in cheaper establishments, fast food restaurants like McDonald's and discount grocery stores, which seemed to be accommodating to both poor Nepalis and poor whites.

Vinod, the team leader, ordered a round of momos and other Nepali appetizers, as well as Nepali beer, after we sat down at the table. I had interviewed his father, who had served in the Indian military, in Kathmandu about six months before, and he told me that if I was ever in Aldershot, I should look up his son. As the food arrived, we chatted about the hometowns of the different soldiers, some of which I had visited in Nepal. They talked about how these places compared with their postings with the military in England and abroad. About half of the unit had already done a tour in Afghanistan, and more Gurkha units were training to deploy again.

Over the course of dinner, those who had been in Afghanistan joked with those who had not. At one point, one of the men who had been in the British Army for several years but had not yet been deployed there was asked where he had been "hiding" those years. In fact, he had a fairly good chance of being deployed there. Although the British military was technically not engaged in Afghanistan at that point, a protection force of approximately 450 remained in the country. I contacted the headquarters of the Brigade of Gurkhas to find out how many of those were Nepalis, but the figures were not publicly available. I found from my interviews that Gurkhas, despite composing less than 3 percent of the British military, certainly made up a much higher percentage of the soldiers who were still going to Afghanistan (I interviewed ten in a fairly short time who had been deployed there or were about to be), though there were no official statements on whether this approach of sending a disproportionate percentage of Nepali soldiers to Afghanistan was an official strategy.

Vinod worked in communications in the Signal Corps and had done two tours in Afghanistan. During the second one, he was part of a Gurkha squadron, but in the first tour, he had been on a team with no other Gurkhas on it. This type of integration of Gurkhas into other British units was rare in the past, he said, but becoming more common. Part of the reason for the change was the more effective distribution of skills. He had computer training that allowed him to manage a variety of communications systems simultaneously, and so when the British needed a team to oversee their communications in Helmand, he was a logical choice.

Vinod's background and training in computer engineering meant that he could easily get a good job in the tech industry if he decided to leave the British Army. He said he had already been approached by a couple of defense contractors who had jobs available. In the army for thirteen years already, he had had his British citizenship for years, so there was less holding him to the military. This was different from the experience of his father and other Gurkhas from previous generations, who would not have thought of leaving before the mandatory retirement age.

But, he said, the British Army was changing, and so too was the place of Gurkhas in it. Far from the image in the British press of the Gurkha recruit from some remote Nepali village, whose only chance at a better life was to join the British military, Vinod had grown up following his father's postings and attended some of the highly reputable military schools where he had started his computer training.

In addition, the British Army was having trouble recruiting and maintaining British soldiers in its ranks.[5] So while the number of Nepalis the British Army was recruiting had dropped precipitously after the end of World War II, more recently, the annual numbers had increased slightly for the first time in fifty years. Gurkhas, I was told by an army official, were now being perceived as more than just loyal and fierce; they were also a good economic investment since their retention numbers were better than for soldiers recruited in the United Kingdom. Even here, however, the numbers were shifting. Soldiers like Vinod had more options than their fathers had. They could stay in the United Kingdom and were increasingly being lured into the private sector by defense contractors and the growing community of Nepalis who owned businesses in the United Kingdom. The salaries were better and deployments to places like Afghanistan when working for a private contractor were voluntary and far more lucrative.

The changes that Vinod outlined were not simply with military structures. Nepalis in the United Kingdom were changing culturally as well. He pointed to the fact that he had not gotten married until he was twenty-nine. "Before my generation," he said, "everyone was married by twenty, but now young people are waiting and not always listening to their parents." Besides, his parents were in Nepal most of the time, so it was easier to dodge the pressure when it was being applied only by Skype and email.[6]

The other problem, he said, "is that my wife is a nurse and she gets paid well. But my postings change every two years, and so we have to pick up and move and she has to start a new job somewhere else. She's good at what she does and she wants to take a more administrative position, but that's impossible when she has to keep starting over because of my work." For now, it seemed, he would continue trying to rise in the ranks, but who knows what might happen in the longer term.

His wife's father had also been a Gurkha in the British Army and had retired as a lieutenant colonel, a rank very few Nepalis had reached before him. He was living in the area and was active in the Nepali community. Vinod said that while he was considering leaving for the private sector where he could earn more money, he realized not making a similar rank as his father-in-law would cause family tension. This was part of a growing divide between younger and older generations.

Walking around a place like Aldershot, he said, you could see a lot of poor Nepalis, who were mostly recent immigrants, but there were also numerous Nepalis, some retired Gurkhas and some the sons of Gurkhas who had received citizenship through their fathers and had been successful in business outside the military. The jewelry shops were one example, he continued, but the one that fascinated him most was the old Aldershot Theatre, which had been bought and renovated recently by Gurkha Event Planner. Its extravagant reopening, complete with medals for the mayor and Nepali traditional dance, had been covered in the local media.[7] He had spoken to one of the Nepali businessmen who was behind the deal and had come away impressed with their plan. They had leased the building for fifteen years and 500,000 British pounds. While that might seem like a lot, he said, the building had become something of a Nepali social hub. It was now by far the best place in the area for Nepali weddings, engagement parties, retirements, and other celebrations. It was booked every Friday, Saturday, and Sunday night and at a cost of 1,000 pounds per event, so even when the costs are factored in, the company was going to be generating a profit in no time.

When I was at the British camp in Pokhara, I had heard both Brits and Nepalis talk about the way the Gurkha tradition lived on after two hundred years. Yet Vinod saw such business opportunities in the United Kingdom as an increasingly appealing way forward. That night in the

Nepali restaurant in Aldershot, drinking beer with the team of Nepalis, each one was aware of these changes and the fact that what it meant to be a Gurkha was very different from what it had been in the past.

TWO NEPALS

While Nepalis in Aldershot like Manish and Vinod were becoming more integrated with British society (and, in the case of Manish, more and more removed from some of the Nepali security guards that he was overseeing in Afghanistan), it was also apparent that there was not one Nepal in Aldershot: a divide had arisen through a series of British policy decisions and shifting trends in Nepali migration.

Gurkhas became a more substantial presence in the United Kingdom in a series of phases. First, in 1997, the Gurkha headquarters moved from Hong Kong to Britain, bringing more active Gurkhas to British soil, even though smaller numbers rotated through at different posts in the past. The Gurkhas I interviewed who had retired during this period were far more likely to have retired to Nepal, unless their sons or other relatives were living in the United Kingdom. In 2004, Gurkhas serving in the army at that point for four years or more won the right to settle in the United Kingdom with their families, as well as securing pensions equal to those of their British counterparts. In 2009 this was expanded, and all Gurkhas serving four or more years in the British Army were granted the right to settle in the United Kingdom, but, notably, they were not granted equal pensions.[8] This meant a large number of retired Gurkhas who served the British in Malaysia and Borneo, many of whom had never been to Britain before, could now settle there, but were supplied only with pensions meant to sustain them by Nepali living standards.[9]

Not only were those who served eligible, but for Gurkhas who died, their widows were also eligible. This meant that as the Nepali population grew in the United Kingdom, its demography shifted significantly, with the population getting both younger and older, as retirees and their widows were added to the number of young soldiers and their wives and children already in the country. As a result, the Nepali population also had a decreasing percentage of Gurkhas actively serving in the British military.[10]

This changed the way that the Nepalis were perceived by the local population in part because Gurkhas in the military tended to work on bases and have military support structures and now many Nepalis did not have these. Both the teenaged and older Nepalis were more visible, and were more visibly accessing British social services.[11] Beyond the use of the National Health Service and other services available to all British citizens, the town offered a series of resources aimed specifically at retired Gurkhas and their dependents. A branch of the Gurkha Welfare Scheme in Aldershot on High Street linked recently arrived Nepalis with the various social services available to them. Beyond this, several of the churches in town advertised free English lessons, and when I stopped by the Aldershot Library, the perplexed woman at the front desk pointed me toward a corner in the back when I asked if there was a section of books in Nepali.

In recent years, building on wider anti-immigration sentiments in the United Kingdom, Nepalis were specifically targeted by both conservative newspapers and politicians. A white supremacist newsfeed headline stated: "White People Replaced by Gurkhas for No Reason." More mainstream conservative papers, like the *Daily Mail*, spoke of "Joanna Lumley's Legacy of Misery," while showing photos of elderly Nepalis at a bookmaker's or looking solemn at a remembrance service.

One of the photos in the *Daily Mail* article was captioned, "Many Gurkhas Pass the Time Away by Aimlessly Walking Around Aldershot and Browsing Shop Windows," as if this were a particularly deviant practice. Certainly High Street had some casinos with slot machines blinking in the dark windowless rooms, and outside, a group of volunteers collected funds for homeless veterans, but none of the articles linked any economic changes in town with the actual presence of Nepalis. Instead, it seemed, like much of England outside London, the town was trying to navigate its way through wider economic changes. These were places that were struggling, like the half-empty shopping mail downtown, but also areas of growth, with neighborhoods of new condominiums rising just outside the center, close to the base. The area was not more or less depressed than most other towns in the area with no Nepali communities.

The change, instead, was primarily demographic. The local government district of Rushmoor, of which Aldershot is a part, had gotten significantly more diverse over the past decade according to official numbers. The percentage of the population that was white dropped from 92 percent in 2001 to 80 percent in 2011. Still, Rushmoor remained only marginally more shaped by immigration than the rest of the United Kingdom with 78.8 percent of the population born in the United Kingdom compared with 80 percent nationally.[12]

Instead, it seemed the visibility of the Nepali presence was most helpful as fodder for anti-immigration politicians. Some of this appears generated by concerns over available resources; local council member Charles Chaudhary complained that Nepali immigration was "putting a lot of pressure on social housing and the integration of the community."[13] Other attacks tended to display little knowledge about the Nepalis in the community or suggestions of real policy solutions, as opposed to simply expressing racial discomfort with diversity.

Gerald Howarth, the local Conservative member of Parliament, criticized Gurkha immigration on multiple occasions, calling it a matter of "grave concern."[14] He complained on BBC that "I was walking around in Aldershot on Saturday and everywhere I went there were Nepalese just basically sitting out in the open, sitting out on the park benches."[15] His constituents, he concluded, were concerned about finding space for themselves on park benches in the future. While his comments generated some backlash from other politicians, particularly those concerned with veteran's rights issues, such anti-Nepali comments clearly were seen as effective in persuading certain voters. Independent Party candidate Bill Walker later referred to Nepalis in Aldershot as "parasites" and tweeted that "we should be not be taking any more in."[16]

For the Nepalis themselves in Aldershot, many of those I spoke with seemed to have problems not that dissimilar from those that I had interviewed in Nepal and India, despite the growing distance between the groups. For example, it was the eldest who often had the most trouble adapting to life in the United Kingdom both socially and financially. Neha Choudary, who did an ethnographic study of Gurkha wives and widows around Aldershot, found that those who arrived in Aldershot late in life were most likely to be dissatisfied with the move and to have trouble integrating themselves with either British society or the younger generation of Nepalis already living in Aldershot. Even here, the global network of brokers and agents had become part of the system, and some widows had been pressured by their families to take out loans in order to pay for their travel expenses, along with the costs of setting up a new life in the United Kingdom. This was often based on the faulty assumption that they could bring their adult children with them eventually. Many of these widows ended up with little in the United Kingdom beyond their former husband's meager pension, a bus pass, health care, and the barest social services, all while dealing with growing racial resentment.[17]

The day I was set to leave, the United Kingdom was voting on a referendum on whether it should leave the European Union, and the issue dominated conversations. Both "Leave" and "Stay" posters hung in shop windows and on the lawns of some homes. Most of the Nepalis I had interviewed did not seem to feel strongly one way or the other. More were probably in the "stay" camp, since EU membership was a clear expression of the increasingly international world that the Nepalis and British were living in, but others, particularly retired soldiers, were more nationalistic, more loyal to the crown than some transnational organization like the EU. But even these Nepalis were concerned and angered by the growing anti-immigration and racial rhetoric. Here was a group of people, many of whom had risked their lives for Britain and others who were now actively contributing to small business growth, yet politicians saw them as a prime target for nationalist attacks. The sense was that it was fine for Nepalis to serve the United Kingdom in its military, but the United Kingdom didn't want them to stay there when they were done.

The same day that I was leaving the United Kingdom, presidential nominee Donald Trump was flying to visit one of his golf courses in Scotland, where he applauded voters in favor of the so-called Brexit. His election victory four months later would be fueled by similar anti-immigration nationalism. Having looked closely at the struggles that these Nepalis, Gurkhas, and contractors had finding a place for themselves in the remnants of the British Empire, I turned to compare the experience of Gurkhas with international contractors in the American empire amid Trump's nativist rhetoric and anti-immigration policy. What will happen as the United States increasingly relies on global labor flows to fight its wars, but simultaneously makes its own borders less crossable for those who have fought for it as contractors?

CHAPTER 22

WHEN YOU CAN'T GO HOME

BROKEN CELL PHONES AND
BROKEN BUREAUCRACIES

As you drive into the United States from Mexico south of San Diego, the first exit you pass has a K-Mart, a Marshalls, and an outlet mall with an assortment of stores catering to those who have just driven into the country. Inside the mall is a small kiosk that repairs cell phones. It is a multicultural business: most of the repairs are done by a Chinese American immigrant who was a whiz with cell phones, the Mexican American sales assistant deals primarily with the Spanish-speaking customers, and the co-owners are two recent Afghan immigrants. While the kiosk may be small, it represents an incredible amount of perseverance and luck by Waheed and Ahmed, the co-owners.

For most of the contractors I spoke with, their fates and futures were distant from American soil and received little media coverage in the United States, but Waheed and Ahmed were in the small category that did receive some attention: were both former translators for the U.S. military and had come to the United States on the so-called Special Immigrant Visa (SIV) program.

The U.S. State Department, after receiving significant pressure from American veterans in particular, began a visa program meant for Iraqi and Afghans who had worked as translators for the military and were in danger as a result of their assistance in the war effort. The program, which began for Iraqis in 2006, was expanded to include Afghans who worked for a year or more for any part of the U.S. government or on any U.S.-funded program.[1] This essentially meant that anyone who had worked on a U.S. primary contract had a route to a U.S. visa. The expansion of the program opened the door for tens of thousands of Afghan contractors to apply.[2]

This led to a sharp increase in applications, and soon many of my former Afghan colleagues who had worked with me at the United States Institute of Peace were asking me for letters attesting to their employment there. The Congressional Research Service reported that 9,107 visas were issued in 2014.[3] The program, however, was not implemented smoothly. Initially both applicants and the veterans supporting them criticized the lack of transparency around the application process and the time it took for applications to be processed. Waheed's took about a year and a half and Ahmed's a little under a year. Compared to others they knew, that was fairly quick.

After completing my research in the United Kingdom, I returned to Vermont, but only had time to repack my bags before I was back on a plane, first to the Bay Area and then to Southern California, to interview Afghan contractors I had reached out to through contacts in Kabul and some of the Nepalis I had interviewed. Many of the themes in my conversations about visas, brokers, and the immigration process remained the same. As we sat in Waheed's apartment overlooking a busy San Diego highway about 30 minutes north of their kiosk, both Waheed and Ahmed discussed the process with a sense of relief, as if they were still, two years later, surprised that they had somehow made it through the bureaucracy. The thing that was most annoying about the process, Ahmed said, was that there was no real information given about your status. You submitted your application and heard nothing for months, then suddenly you'd get a call to go to the embassy to take a lie detector test, then you'd hear nothing again for months, and then all of a sudden you had a visa. It was nerve-wracking. Applicants did not know whether they should be packing to leave the country or whether they were going to remain in Afghanistan, constantly in danger due to the work that they performed as contractors for the United States. Both Waheed and Ahmed knew applicants who were in grave danger but had been rejected with no significant explanation, and they also knew several who had applications come through in half the time it took them to get theirs. None of it made sense to them.

It was not simply securing the visa that was frustrating. Once granted a visa, the potential immigrant was then offered assistance in resettling, which the U.S. government had outsourced to the International Organization for Migration (IOM). IOM would purchase plane tickets and then arrange resettlers' initial lodgings in the United States, as well as provide a small initial cash payment to help them get settled. The concern was that not only would the immigrant later have to pay the price of the ticket back (something that several said was not made clear to them beforehand), but ultimately the Afghans I interviewed said they felt little control over the entire process. In multiple accounts, I had heard of representatives from IOM or partner organizations who arranged for apartments for recent arrivals, paying for the first three months, only to have the immigrants later discover that they were trapped in year-long leases in overpriced, substandard apartments in dangerous neighborhoods while the representatives allegedly received kickbacks from their new landlords. Listening to these stories, I began to note the ways in which these resettlement organizations were distant cousins of the brokers selling jobs and visas in Nepal and India.

Others complained that the slow IOM bureaucracy was both taxing and potentially put them at risk. One former translator showed me a string of emails he received from IOM following the approval of his visa. IOM had sent him an initial email when his visa was approved and he promptly responded saying that he was interested in receiving assistance. Two months later, however, he had not received a response despite sending three follow-up emails. Nervous that his visa might expire before IOM arranged his travel, he went ahead and did this himself, sending IOM an email about this change. This time, IOM responded almost immediately, stating that by setting up travel himself, he was no longer eligible for any assistance. It seemed, he concluded, that IOM really would respond only when it meant less work for them.

For those who were being directly threatened by the Taliban or criminal groups who targeted SIV recipients, such delays could be deadly. For those who already had family in the United States or a good amount saved already, Ahmed said, it was better to skip the IOM assistance than to accept it and get potentially slowed in the bureaucracy.

After interviewing other Afghan immigrants to the United States over the next couple of weeks in San Diego, Los Angeles, Seattle, and Tampa Bay, I contacted several nongovernmental organization advocates who had been working on behalf of the SIV recipients and the State Department as well. While the information from the State Department clearly outlined what was involved, the fourteen-step process was complicated. After receiving criticism from several groups, included a vocal group of bipartisan senators with John McCain and Jeanne Shaheen, the Department of State had begun releasing statistics about process time, which as of July 2016 averaged 374 days for processing.[4] The step with the longest wait time was number 13, which was labeled "administrative processing" and took on average 185 days. The Department of State report concluded, almost apologetically, "Although step 13 is lengthy, process enhancements have resulted in improved efficiency."[5]

Much of the process didn't seem logical to me even as I gathered more information. For example, in the second quarter of 2016, 1,501 applicants who had previously been denied visas filed appeals. Of the appeals that were adjudicated, 54 percent were eventually given visas.[6] If their cases were strong enough to be given visas, what was being done during the initial review that more than 50 percent were essentially being misjudged?

This was even more frustrating for those involved in the process. Even while the Afghans I interviewed expressed immense gratitude for the opportunity to come to America, they also found the morass created by the program and some of its technical quirks frustrating. One distraught father I met the day after interviewing Waheed and Ahmed said that the Afghan SIV program permitted those receiving visas to bring only their spouse and any children who were minors, while the Iraqi program, for no known reason, allows parents and all children to accompany the applicant. This father had a nineteen-year-old son who had been left behind in Kabul, and now all they could think about was getting him out. The Taliban, he said, certainly did not differentiate between children who were minors and those who were not. The son was considering either crossing Iran illegally in an attempt to get to Europe or taking the risky sea voyage to Australia in the hopes of reuniting their family eventually.

In some ways, the father said, applying for U.S. visas could actually put a family in even more danger. Before applying, perhaps only the Taliban wanted to kill you. But as soon as your neighbors heard that you or someone in your family had a U.S. visa, they assumed you had money. This made you an appealing target to the criminal gangs in Kabul who many felt were actually far more effective than the insurgents with whom they often partnered.

Views of the program from those still in Afghanistan were even more negative. Timor Sharan, an Afghan colleague of mine who worked at the Asia Foundation, conducted interviews of a group of Afghan contractors and translators in Afghanistan who were in the middle of applying for the Special Immigrant Visa and another group who had had their applications rejected.[7] As he gathered accounts in Afghanistan, I interviewed Afghans in the United States, and we compared results.

In Kabul, rumors ran wild among applicants, many of whom had worked together and were heavily invested in the success or failure of their colleagues during the application process. Once one person working for a contractor was successful in securing a visa, others quickly followed with applications. When someone failed to receive a visa, there was widespread speculation about what had caused the application to be rejected. Some of those who were well educated and worked in high-level positions at the U.S. embassy were more confident about the process, but for those with less education—the gardeners, cooks and others—the process was confusing and daunting. Rumors spread about specific tricks one could use for getting accepted and who should be asked to write the applicant letters of recommendation.

As was the case in Nepal, there was a group of Afghan brokers who were more than happy to take advantage of the ignorance of applicants and the rumors about the process. A growing industry of "travel agents" popped up in Kabul to help applicants fill in their application and, particularly, craft their all-important "threat letter," a statement that demonstrated the danger they were in. When Timor visited some of these agencies, they were charging between $80 and $100 per step. All told, a complete application (successful or not) could cost almost $1,000.[8] One of the travel agencies had 200 applicants currently at various stages of the process during Timor's visit. Like their counterparts in Nepal, these brokers advertised unrealistic success rates, and when the applications were rejected by the U.S. State Department, the agents kept the fees that they collected. The immigration industry, wherever it sprung up, seemed to enrich only the unregulated brokers and agents who helped spread rumors about the complex and mysterious process of migration to the West.

AL CAJON

As I spoke with Waheed and Ahmed, we ate Kabuli palau, an Afghan rice dish cooked with mutton, and mantu, a dumpling that is the Central Asian version of the Nepali momo, cooked by Ahmed's wife. I asked where they got the ingredients, since it was amazing how close the food tasted to the real version I had eaten for years in Afghanistan. Ahmed explained that with the growing Afghan community in San Diego, it was possible to find stores with authentic ingredients or, perhaps in a pinch, you could substitute something you picked up at an Indian grocery store.

Both Waheed and Ahmed considered themselves lucky. The transition to American life had not been easy, but they were still success stories. Ahmed arrived in Los Angeles first, and not finding a job, headed up to the Bay Area with his wife and four children. He worked some odd jobs, clerking at a gas station for a while, earning minimum wage—not enough to support a family of six. For a while he was commuting between Sacramento, where rent was cheap, and the Bay Area, where he had found a better-paying job, but this never seemed like a long-term solution and there was no way he could afford to live in the Bay Area. Working for other people as a recent immigrant, he said, it seemed you could never really get ahead. Bosses were always trying to take a cut out of your paycheck, the tax system was often mystifying, and it wasn't clear where you could get help if someone was trying to cheat you. Better to strike out on your own, he finally decided.

He met up with Waheed, and they discussed opening their own business. They found this to be difficult since neither had established credit histories. They had social security numbers, but no history to back them up and no one would give them a loan. Finally, another Afghan immigrant introduced them to a representative from a chain of malls. He could help them out, they were told, as long as they set up shop in one of the designated mall kiosks close to the border. They agreed, signed a lease for the kiosk, and moved down to Southern California, joining the growing Afghan population in San Diego.

Both Waheed and Ahmed moved to the cheaper suburbs that climb into the hills east of San Diego, away from the beautiful beaches. One of the last towns in the suburban sprawl east of San Diego, El Cajon is referred to by many in the area as Al Cajon, due to its large and growing Arab population from Iraq, Syria, and other parts of the Middle East. El Cajon gained national attention later in the summer of 2016 when a police officer shot a Ugandan immigrant who was holding an e-cigarette that he thought was a gun. Protests followed, and tensions between immigrant groups and local authorities and between immigrant groups themselves increased. The Afghan population here is smaller than the Arab one, but has increased rapidly, with most of the recent arrivals being supported by the SIV program.

The businesses in El Cajon are those that tend to spring up in areas where there are a lot of new arrivals: payday loan centers, pawnshops, ethnic grocery stories, cheap furniture stories, and an Enterprise Rent-a-Car sales lot. Banners advertised no-money-down loans for cars, kitchen sets, and sofas. Particularly for new arrivals with no credit, it was easy to see how quickly someone could descend into a quagmire of debt.

The only resource that many former Afghan contractors had, Waheed and Ahmed explained, were the various settlement agencies that were assigned to them. All of these agencies were themselves independent of the government and were essentially contractors themselves, getting a government payout for each immigrant case that they managed. The support these agencies provided varied widely. The settlement agency responsible for Afghan arrivals to the San Diego area was the Alliance for African Assistance. It specialized in resettling refugees from various parts of Africa but had recently taken on Afghan cases as well. This gave it additional funding to cover its overhead while maintaining a focus on the populations it was originally set up to serve. Such uses of funding were common among settlement agencies.

Several Afghans I interviewed who used them said the Alliance was good at paperwork and getting your kids into school, but beyond this, it didn't provide much assistance. Their programs, it seemed, were aimed at families, and one young Afghan man who immigrated alone said that they put him in a rundown boardinghouse with drug addicts. It was so bad that he ended up sleeping on the sofa at the apartment of a fellow immigrant. Finally, after speaking to an Afghan American lawyer, he sent an email to the Alliance's headquarters. The people in the headquarters were concerned about bad press if he spoke with a journalist—this had occurred in a few other cases—and they said they would send someone out to help from the main office to look at this case. Before this happened, however, someone from the local office came to his boardinghouse and told him they had found him an apartment and implored him not to say anything negative about the local office.

Even with the help of these agencies, arrival in the United States was often rocky. One translator I interviewed had flown over with another Afghan refugee, arriving at Los Angeles International Airport at two in the morning. They were picked up by a sullen Iranian refugee who worked for the local resettlement agency and seemed none too happy to greet these new arrivals from a country that he saw as a motley group of backward tribesmen. He took them to a run-down hotel near the airport, amid a sea of concrete, and left them, hungry and tired from their long flight. Someone would stop by later, he said, but they weren't told who or when.

The next morning, the translator awoke and found his companion staring out the window at an abandoned lot next door with broken glass and dangling electrical wires, typical of some of the rundown neighborhoods around the airport. The man turned to his awakened companion and said, "Are you sure they brought us to America?"

IMMIGRANT LABOR

As I traveled to the various poor, urban neighborhoods where most of these recent immigrants lived, each person had a different story of the hardships they endured. What came up most often was their desire to find work and how difficult it was to find good employment as a recent immigrant.

The settlement agencies were legally required to help them with this, but what this assistance looked like could vary. One SIV recipient said that after he had been in the country for a couple of weeks, a man from his local agency had picked him up and driven him to a job fair, where he was left at the curb. This, he was told, was the extent of the "job assistance" the agency was required to provide in order to continue to receive government funding.

It seemed that the problem in part was that the American immigration system is set up to help only the most destitute or the very rich. For those coming from conditions of extreme poverty, working for minimum wage to support a family may seem viable. The very rich have money, personal connections, and degrees that can translate into good professional jobs in America. Most of the Afghans on the SIV program were somewhere between these extremes.

Mohsin, for example, was a lawyer I had known while living in Kabul. He had been one of the U.S.-sponsored advisors at the Ministry of Justice and was considered one of the keenest legal minds in his department. As we chatted on the floor of his living room in San Diego, which had been outfitted with Afghan carpets, he described how difficult it was earning enough to support his family of six.

He had studied law in Iran for over a decade, receiving the equivalent of a PhD, but none of that translated in the United States and most people did not even give him credit for the BA which he also held. "How can I translate my deep knowledge of sharia into a job in this country?" he wondered aloud. The settlement agency had at least helped him find a job as a teacher's aide at a local middle school, but the pay was low and it certainly wasn't a good use of his skills. This is what bothered him the most when looking at his friends: all the wasted potential. The SIV program had essentially just given them visas and sent them to far-flung corners of the country. Why hadn't anyone thought about how to help them secure employment using the skills they actually had?

After conducting interviews in the Bay Area and Southern California, I flew east and spent a week in the Washington, DC, and northern Virginia interviewing SIV recipients. I hoped to find some of the immigrants there to be more optimistic than those on the Pacific Coast. For one thing, there was a long-established community of Afghans in northern Virginia, but there were also the headquarters of many of the contracting firms, like Chemonics and Creative Associates, that the Afghans I interviewed had worked in, as well as offices for contractors like DynCorp and KBR.

Washington, DC, however, was not particularly friendlier to recent Afghan arrivals than California was. None of the contractors I interviewed found employment at any of the large contractors in the area.

The only one I could find who was doing anything remotely related to what he had done in Afghanistan was working at the headquarters of the United States Institute of Peace. One contracting firm, PAE, that had run the State Department's large Justice Support Project, had been rumored to be holding one position for an SIV recipient, and numerous new arrivals quickly sent in résumés. Nothing came of it, said one of the former PAE contractors I interviewed. Whereas the U.S. government might have felt it owed a debt to these Afghans and was thus supplying them with visas, it was clear that the contracting firms felt that their obligation had ended when the contracts of these Afghans expired.

Mohsin was now worrying about his son, a senior at the local high school. He was doing extremely well in math and physics and dreamed about working for NASA. Mohsin was not sure how they could pay for college. Even the local community college was beyond their means. His son had heard an army recruiter speak at his school; perhaps he should join the Army and use the GI bill to pay for college later. Having just escaped from Afghanistan, however, it felt strange sending his son back to the war.

While some of the Nepalis in Aldershot might have been struggling financially, they had a support network that Afghans in the United States lacked. First, all the Nepalis who served in the British Army were receiving some sort of pension, something no company gave to Nepali or Afghan employees. Further, the British military offered jobs such as the Military Provost Guards who guarded military base gates, which were sort of a phased partial retirement. For those who did not still work for the military, there was a network of Nepalis who had started private security firms, driving schools, and dozens of other businesses that tended to hire Nepali employees. For Afghans in the United States, there was not this type of support and Afghan contractors who immigrated here were far more likely to struggle.

When I asked Waheed and Ahmed why they felt that they had been more successful than some of the other SIV recipients in setting themselves up, they both said they were younger and more flexible than some of the other immigrants they knew who had come on the SIV program. The older immigrants who were less willing to move between jobs had a harder time, they said. Still, their business was in the tenuous early phase. Not enough foot traffic, and they could easily fall behind on the rent.

Waheed told me that a friend of his who had worked as a guard on a U.S.-funded project just received his visa and had called him asking for advice. He was planning on leaving for the United States soon with his wife and eight children and wanted to know whether San Diego was a good place to come. Waheed was at a loss. He couldn't tell him not to come, but at the same time, he didn't think that his friend understood the challenges he was sure to face when he arrived. His friend didn't know how to drive. How would they get around San Diego, a city of highways? In Afghanistan, a well-paid guard could support a family of ten, but in San Diego? That was going to be far more difficult than he could imagine.

WHERE THEY GO FROM HERE

Most of the SIV recipients I interviewed were unsure what the future held. Some were considering moving to more rural parts of America where they heard that their government benefits could stretch further—maybe Texas, some said. It was a state that had a lot of other immigrants in it, and its conservative social values appealed to many of the Afghans I spoke with.

Perhaps because most SIV holders tend to settle in larger cities, they had not attracted the same media attention that the Nepalis of Aldershot have. Yet those articles that did make their way into the American press tended to focus on these Afghans as "a young generation estranged from its parents" and suggested that police fear of "pockets of extremism [that] may take hold undetected in this cocoon of a community."[9] Several newly arrived Afghans discussed how they planned on trying to ensure they continued to live in primarily immigrant communities where diversity and difference were more normal. Here people did not look at them strangely when their wives went out in headscarves or when they attempted to set up a mosque. Perhaps since most of the SIV recipients I interviewed continued to live in poor, minority filled neighborhoods, economic concerns and worries about employment were clearly more important than some of the cultural differences they encountered. Still, a couple of them brought up Trump's campaign for the presidency and how it left them feeling conflicted. Many Afghan immigrants who tend to be more socially conservative than the average American are naturally inclined to lean toward the Republican Party. Given Trump's statements about Muslim immigrants, Ahmed asked me, "Can I really vote for him?"[10] This all added to the feeling among many that America was a land of opportunity, but it was also a country of racial, social, and economic divides, and most opportunities were guarded and only accessible to a select few.

I did find several cases of Afghan SIV recipients who returned with their families to Afghanistan after finding life in the United States too challenging. Everyone I interviewed had heard of at least one other person who returned despite the potential danger there. In most cases, this was driven by the inability to find enough work to support their families, though some also had trouble with the culture and that women were expected to be able to leave the house and work just like men.

In Washington, I spoke to several impassioned defenders of the SIV program. These were primarily American military veterans who had worked closely with translators in Afghanistan and had then worked to get them visas through the program. As one veteran explained to me, a U.S. senator putting pressure on the State Department could speed up the process significantly, but this meant that each veteran was in a position to get one Afghan a visa and that was it.

While I understand the emotions driving this argument, the problem for me was that the numbers simply didn't add up. In 2010, the Department of Defense reported 97,276 people who were "Foreign and Host Country National Contractors" working for the U.S. military.[11] While many of those may have been from places like Nepal, India, and Turkey, as the war went on and the military emphasized the "Afghanization" of their projects, focusing on hiring more and more Afghans, at least half of the contractors hired were Afghan. Considering the turnover of fifteen years of hiring and war, tens of thousands, if not multiple hundreds of thousands of Afghans, had been employed on U.S.- funded projects. Surely the State Department was not likely to grant visas to all of these Afghans who were potentially just as likely to be targeted as the ones who had already received visas.

And what about local elders, parliamentarians and villagers who in some way or another visibly supported the American presence? If the United States was committed to being responsible for those Afghans that it had put in harm's way by pulling out of the country before a meaningful peace had been reached, wouldn't it be easier to work to support the Afghan government and police so that these Afghans could support themselves?

Instead, the SIV program created an exit that tens of thousands of Afghans were hoping to slip through, though since only a fraction were successful, this meant they began looking for alternative routes as well. All of this was contributing to a massive exodus from Afghanistan. By 2015, Afghans made up the second highest number of refugees trying to get into Europe, trailing only refugees of the war in Syria.[12] Immigration clampdowns meant that these immigrants were using more and more dangerous routes, relying on brokers, and living in extremely difficult conditions while en route.[13]

In the meantime, by removing those Afghans who had worked closely with the United States or at U.S.-funded programs and NGOs, the SIV program was actually removing the Afghans who were most qualified to lead their country in the future. Civic groups, think tanks, and other organizations that were instrumental to creating peace in Afghanistan were suddenly having trouble finding qualified people to fill their ranks.

After fifteen years of war, the SIV program signaled to many that the United States was essentially leaving Afghanistan to the warlords. Beyond this, as these Afghan contractors relocated to San Diego, Omaha, and refugee camps in Turkey, the repercussions of America's war in Afghanistan continued to ripple out into other conflicts and wars, whether it was Europe's immigration crisis, tensions around Muslims immigration to the United States, or simply the gross disparity between rich and poor in the West. Even if all the U.S. troops left Afghanistan, the ripples of war are sure to continue on.

CHAPTER 23

WHERE THE WAR WENT

WORKING FOR THE MOB

It's difficult to image a place farther from Afghanistan than Khabarovsk, Russia. Located on the Chinese border about 500 miles north of Vladivostok and a full seven time zones ahead of Moscow, Khabarovsk is far from just about everywhere, except for the endless tundra of Siberia. Yet for a Russian businessman with interests in Hong Kong and Singapore, looking for a place to lay low while being accused of tax evasion, it is a fairly convenient spot. And particularly for a businessman nervous about rivals infiltrating his inner circle, Nepal was a good place to look for bodyguards.

A few months after leaving their jobs in Afghanistan where they worked for DynCorp, Bhek and Prawjol found themselves working as two of a five-member team of all-Nepali bodyguards protecting a young billionaire who had gotten his start in supermarkets. Bhek and Prawjol were from Matta Tirtha and were friends with Yogendra, though they were a few years younger than he was. So when all of their contractors in Afghanistan finished up a few years before, they went off looking for new opportunities. Most of the jobs they had applied for were in Iraq, but they were casting a wide net and were open to almost anything abroad that paid close to what they had earned in Afghanistan.

They both heard of the job through a retired military friend, and since they knew the other three Nepalis who had been recruited to work for the businessman, it seemed like a fairly safe new opportunity. The pay was approximately the same, and it was nice to no longer be in a war zone. None had been to Russia before or had much of an idea of what Siberia was like, but it was hard to imagine that conditions would have been worse than what they experienced in Afghanistan.

The differences between working for an established international company in Afghanistan and for an individual businessman, however, became apparent quickly. Instead of having their pay direct-deposited to their accounts as it had been in Afghanistan, the Russian businessman would go to the ATM, take out fistfuls of cash, and hand it to them. They would then go to Western Union to wire it home. The process for issuing weapons was similarly informal, and the businessman had weapons all over the house. After they arrived, Bhek said, the businessman handed him pistol after pistol until he couldn't hold any more.

The businessman also used some creative methods to get around visa regulations. First, he got the Nepalis to apply for tourist visas from the Russian embassy in Kathmandu. Once they arrived, he enrolled them in a local college taking language courses, ostensibly so they could learn Russian. It soon became apparent that he was really using these courses to get them student visas, even though the men were all in their late thirties and seasoned security contractors who did not look anything like typical students (nor was Khabarovsk a place that would attract Nepali students).

The businessman had a paranoid streak and would stay up at night, plotting business moves and fretting over cameras that he said the police were setting up. One minute he was trying to convince Bhek to go into business with him and set up a wine exporting firm, and then the next minute he was accusing them of spying on him and searching his house for bugs. He regularly descended into long profanity-laced diatribes that shocked the Nepalis despite the fact that they had all spent years with American contractors who were also fond of colorful language.

Once after doing tequila shots with his sister-in-law, the businessman decided to go out to a club despite already being very intoxicated. When the bouncer stopped him, the businessman pulled out a knife and started waving it in the bouncer's face. Bhek stepped in, but diffusing the situation was tricky; Bhek's English was good, but his Russian was not despite the language classes. Even after he got the knife away from him, the businessman, a former boxer, decided to use his fists instead. The night did not end well. They eventually got the drunk man home, but everyone was bruised, and Bhek worried about what he might think in the morning and that the Nepalis would somehow be blamed for what happened.

After a few of these incidents, Bhek and the other Nepali bodyguards decided that they had had enough. Bhek had already spent seven years in a war zone; he had some money saved up, and this job didn't seem like it was worth the risk anymore.

But leaving was not as easy as coming. They tried to check in at the nearby hotel, but the man at the desk perhaps knew the Nepalis worked for the influential businessman, and he told them the only rooms available were $1,000 a night—clearly a message that their business was not welcome. Since they didn't know anyone else in town, they called the businessman's ex-wife, who had also treated them nicely on a few previous occasions. Sympathetic to the mood swings of her former husband, she let them stay at her place, until they could arrange flights to Hong Kong, then Delhi, and, at last, back to Kathmandu.

Retelling the story in the house he had purchased with his salary from Afghanistan while his wife served us tea, Bhek chuckled at the absurdity of it all. He laughed, particularly, at how he had convinced a Russian traffic cop that his Nepali driver's license was an international driver's license. Working for DynCorp in Afghanistan had been fine, he said, but nothing made sense in Russia, he said. It wasn't clear whether he was supposed to be protecting the businessman or protecting other people from the man's temper, and there was certainly no one there who would have helped them if they had gotten into more serious trouble. Violence always seemed to be just around the corner.

Still, he said, while he had started a business in Kathmandu, if someone offered him a decent security job abroad working for a reputable company, he'd probably take it. The pay and the lifestyle were too appealing to turn down an offer. Compared with some others, however, the experience in Russia had clearly taught him that not any old job was really worth it.

A NEW OLD WAR

The war in Afghanistan was a distinctly modern twenty-first century one, fought using satellites and drones. It was also distinctly modern in the way in which it relied on global labor flows to outsource the vast majority of the labor of war. It lured workers from across the world who, taking advantage of porous borders and globalized business, came from the migratory flows out of poorer countries that supply the world's labor for rich destinations like the United States, Saudi Arabia, Israel, Australia, and Europe. At the same time, the war in Afghanistan built on colonial pathways and processes that have been in place for centuries. Ports like Kerala, which ship laborers to Afghanistan, have been connected for hundreds of years to trade, commerce, and conflict, while in the bazaars of Delhi, the prices of dried fruit have moved up and down in reflection of the way that the war has made it more or less difficult for individuals and goods to cross the border of Pakistan and Afghanistan.

Today, the United States is continuing British colonial practices, using Nepali private security contractors to exert its influence globally in a manner that it is politically and economically incapable of doing by relying solely on U.S. soldiers. While there are real questions over whether outsourcing to private security contractors actually saves money,[1] it seems difficult to imagine that the U.S. government would have been able to bear the political costs at home of fighting the longest war in American history in Afghanistan if the country had supplied its own labor to run the war.[2]

While the number of contractors on U.S. contracts has dropped as American troops have withdrawn from Afghanistan and Iraq, the lure of relying on contractors to do much of the real labor of war for the United States remains and the apparatus of military firms has not stood by idly, but instead has strongly advocated for even more reliance on private contractors. In an op-ed for the *Wall Street Journal* Erik Prince, former head of Blackwater, called for the installation of an American "viceroy" in Kabul, suggesting the government look to the East India Trade Company as a more effective model.[3] While the Trump administration has thus far resisted such a turn, the lure of no longer dealing with the political consequences of relying on American troops may still prove too strong to resist.

Eight years after Obama's surge, this suggests the contracting model is likely to remain a fundamental part of the U.S. military strategy unless a serious reversal of course is made. This will have repercussions for countries like Nepal that supply the labor as well as the United States. Ultimately, this form of contracting is likely to cost both more Nepali lives and more American dollars, particularly if contracting firms remain more concerned with securing a contract extension than winning wars. More generally, it will contribute to the perception of America as an empire of hypocrisy, whose rhetoric about democracy and liberty is not supported by its disposable approach to those actually fighting their wars.[4]

It might be tempting to say that in Afghanistan, this is somehow a passing of the colonial baton from the British Empire to the American empire (with perhaps a Soviet period in between), but this ignores some of the different ways in which the United States relates to its so-called colonial subjects in contrast with the British relationship with the Gurkhas. For the Nepalis who served in the British Army, as well as the Indian Army, there are pensions, health care, and even retirement homes. For those who are injured while defending Britain, Britain compensates them. There are even fairly clear paths to citizenship now, even if the reality of living in Britain is more difficult and racially charged than many admit. By relying on contractors, the U.S. government avoids taking responsibility for the hundreds of thousands of contractors, subcontractors, sub-subcontractors, and so on who have worked for the U.S. effort in Afghanistan. There are no pensions or retirement homes for them. In a tiny fraction of the best cases, there is potentially a visa to support immigration into America's poorest, most racially segregated, and economically marginalized communities in the United States.

While the British recruit Gurkhas each year using British military officers, U.S. contractors rely on shady manpower agencies that exploit the nontransparent ways in which contractors are assigned. The one attempt to help Afghan contractors in danger of retributive killing by the Taliban, the SIV program, has stalled amid bureaucratic red tape and a lack of support for recipients who wonder whether destitution in the United States is worth leaving Afghanistan for.

All of this encourages potential contractors to take incredible risks and traffic themselves across borders for jobs that often pay only marginally more than they could make at home. Even those who do get better-paying jobs doing things like defending U.S. bases, if they are injured or killed while defending American soldiers, it is up to the company they work for to make sure that they are compensated.

And this is where, for Nepali and other migrant war workers, the real problem is.

There are certain regulations and stipulations to protect these workers, the most notable of which is the Defense Base Act. To date, the only cases that have been successful are when, usually as a result of media attention, laborers can find a U.S. lawyer to represent them and they can then sue their company in a U.S. court. This would have been unimaginable for almost every contractor I interviewed in Nepal.

Currently the United States has moved to a model where the main business of government is outsourcing contracts, yet most government employees have little training in this.[5] The government demands that companies adhere to regulations like the Defense Base Act, but then asks companies to self-report. In the meantime, the oversight mechanisms that have been successful, like the Special Inspector General for Afghan Reconstruction, have focused on the ways in which contractors have wasted taxpayer dollars through fraud and ineptitude, while paying far less attention to the human costs and the ways in which contractors have failed to protect human rights while taking advantage of global inequalities.

More rigorous oversight measures are necessary, but it's not clear that some of the approaches being considered would help.

Following the killing of thirteen Nepali contractors working in Kabul for the Canadian embassy in June 2016 while I was in the middle of conducting this research, the Nepali government announced it was considering a ban on Nepalis working in Afghanistan.[6] This pronouncement seemed to be at least a recognition that if Canada, the United States, and other funders were not protecting Nepali laborers and if the Nepali government was not able to protect them, perhaps making the practice illegal was the simplest route. The problem is that the ban is not only unenforceable but could actually make things much worse.

The Nepali government lacks the ability to protect its citizens diplomatically, but it also lacks the means to stop them from traveling to conflict zones. When the Indian government stopped allowing Nepalis to fly from Delhi directly to Kabul, they found other routes, passing through Dubai or Istanbul. By making the process illegal, the government would only be driving migration further underground. Migrants would rely more and more on dangerous paths and exploitative brokers. With remittances making up 30 percent of Nepal's GDP, cutting off these funds would also further impoverish one of the world's poorest countries.

The other option is to push contracting companies to be more self-regulating by incentivizing certain practices and protections for their employees. Private security contractors, in particular, are aware that bad press over scandals in Afghanistan and Iraq has damaged their ability to secure contracts. Partially as a result of this pressure, government officials from many key countries supplying contracts gathered with industry representatives, including G4S and Aegis, in Switzerland in 2008 to draft the Montreux Document.[7] This agreement outlined the obligations of private security firms and laid out a code of ethics in an attempt to try to create some business norms. Despite this, critics pointed out there was no real way for signatories to enforce or even monitor the guidelines outlined in the document, with many feeling that such attempts were little more than industries' attempt to pacify its critics.[8] How can you expect companies in an industry widely considered unethical to hold themselves to a code of ethics with no enforcement? As both contracting and conflicts expand, none of this is likely to get simpler.

Smaller-scale attempts have included partnerships between security firms and international nongovernmental organizations (NGOs). In Kathmandu, I met with two representatives from the private security firm FSI, which was piloting a project with the International Organization for Migration (IOM) that seemed promising. It was a rare case of a contracting company partnering with a civil society group. FSI was convinced that by working to create an ethical code with IOM's advice, it could ensure that all its employees were "unexploited," something that it could then advertise to potential clients, allowing it to charge more and offsetting the costs of avoiding brokers and trafficking. For these initiatives, the United Nations honored Tristen Forester, CEO of FSI for showing leadership in "in finding creative and sustainable solutions to combat modern slavery across the globe."[9]

While the FSI plan was the most robust attempt by private security firms to police themselves that I could find, meeting with the representatives was still frustrating. One, a retired British Gurkha, asked me for suggestions I might have about their policy. I said that I thought the plan could do some good, but first they would need to get other firms to sign on. Only if they had a critical mass in the industry would such ethical labeling, like "organic" farms, really catch on. He didn't think much of this advice, and such cooperation was clearly not in his plans. It is against the nature of these firms to assist each other, and in the cutthroat world of contracting, I found it difficult to image this changing anytime soon.

The competitive, flexible nature of contracting made any sort of co-operation or self-regulation unlikely. Just as Blackwater became Xe and then Academi, private security companies tend to go out of business and recreate themselves with different names. They rely on little infra-structure, since they are basically mobile military units, which means headquarters can be almost anywhere in the world. This makes them difficult to regulate without pressure from the countries actually supply-ing the funds.

What is necessary instead is an attempt at coordination, most likely led by the United States and European countries, through which they more thoroughly vet companies and oversee the work being done by contractors taking their funds, while simultaneously working with countries like Nepal that supply this labor to register and certify work-ers. Without getting donor countries, labor-supplying countries, and contracting companies together to work on the issue, little is likely to change.

The problem with this is that the more reliant the United States has become on contractors, the more difficulty it is to regulate them. This has allowed a growing set of perverse incentives to dictate the outcome of war, with profit trumping peace, as the key objective.

During the war in Afghanistan, there were significantly more con-tractors than soldiers involved. This means there were more people in-volved in a conflict in order to earn a paycheck than there were who were fighting to end the war. That is not to say that contractors actively tried to perpetuate the conflict, but in an increasingly neoliberal world, where the market is meant to dictate economic and political realities, if the United States continues to outsource its wars without significantly more oversight, war is going to become increasingly market driven, kill-ing those who are willing to work for the lowest wages while ensuring that those at the top continue to profit.

If this happens, war will become an even bleaker affair. Particularly if the United States continues to rely on a model of warfare that does not appear to be effective, while simultaneously putting tens of thousands of workers in danger, with only the blind hope that companies will self-regulate and not violate human rights, this will be bad not just for the United States but will have potentially global repercussions.

THE FUTURE OF WAR

Eighteen months after my last visit to Bagram I returned to Kabul. It was early in the fall of 2016, and the city was on edge from a spate of recent bombings. Little had changed politically in the last year and a half—with the Taliban unwilling to negotiate seriously, the Afghan government corrupt and hampered by infighting, and the United States trying to extricate itself from the conflict without seeing the government it invested so heavily in fall. Still, as I drove through the familiar streets, past my old house, it seemed to me that as the conflict had gone on, the walls of the various compounds had only grown higher and the atmosphere more hostile. The massive influx of funds over the past fifteen years and the constant presence of international troops had changed the place.

But as I had seen over the past seventeen years, it was not just Afghanistan that had been changed by the war. In the seven countries I had visited, people's lives had been shaped by their experience of the war. Bhek, Arda, Teer, Hom, and countless other men continued to deal with the consequences of war.

Some had earned enough to build homes, while others had been injured, even disabled for life. Many of the effects of the war in Afghanistan are still ongoing. Young men in Delhi wait to see if they can get visas to work in Afghanistan and, if not there, perhaps in South Sudan or Yemen; some of those injured in Afghanistan continue to work with Matt Hadley and others to see if they can get some compensation for their injuries; and Afghan contractors, unable to return home, attempt to adjust to lives of poverty in the United States.

I was left with the feeling that it would be easier if contracting and this migration of underpaid laborers toward conflict zones was more clearly a purely good or bad thing. Instead, it was clear that it is primarily a messy thing, helping certain people get ahead (usually those already in positions of strength), while greatly exploiting others (usually the poor). Nevertheless, this messiness is a reason for optimism, since there are also ways that it can be done better. However, as private security contracting grows, there is also the possibility that it will grow so large that the United States and other large countries lose control of it, and it becomes impossible to regulate.

I called Pratik, who I had interviewed in Kathmandu almost a year before. We had emailed back and forth since, and he was still managing security at a large housing compound just down the Jalalabad Road from Camp Pinnacle, where so many Nepalis had been injured. To get inside his compound, we had to pass through three gates. At the front one, a group of Afghan guards checked the vehicle, but each of the subsequent ones was manned by Nepali contractors, who, I greeted.

Once inside, we settled into comfortable, overstuffed sofas in the recreation hall and drank coffee we had purchased at the café staffed by Filipino women who also ran the attached spa, restaurant, and bar. Pratik said that while the numbers were decreasing, they still had a hundred Nepalis on his staff. The recent killing of Nepali guards working for the Canadian embassy had made it more difficult for some of them to get their paperwork organized, but it had not done much to slow the flow of young men looking for work in war zones, just made them look elsewhere.

Pratik was thinking about settling down in Kathmandu and maybe setting up his own business, but then backtracked and admitted that if a good job came along working abroad, he'd probably take it. The problem, he said, was that Nepali youth think only about working abroad. There's no investment in the country. It was remarkable, as I thought back on my interviews; I could remember only a handful who were eager to settle down in Nepal. Instead, once a young man went abroad, he was likely to continuing to look abroad. This was a process that was speeding up, not slowing down.

Contractors I had interviewed who had been in Afghanistan were now in Iraq, Turkey, the Ukraine, Somalia, the Democratic Republic of the Congo, Sudan, South Sudan, China, Russia, and the United States. Unfortunately, Pratik concluded, for those heading to new zones of conflict, the only way to get there was by relying on crooked brokers, who at best took their money and at worst could get them killed. He cursed a few of the brokers he had met in Afghanistan, who were doing nothing more than exploiting Nepalis: "Make sure you write that in your book," he said.

I drove back to my guesthouse from Pratik's compound just as the sun was going down. What my grandmother, who had been raised in Istanbul, would have called a Turkish moon was setting in the pink light. This was always my favorite time of day in Kabul. The dusty air seemed soft. As I looked out at the city, the scars of the war were visible, from the near constant police presence and concertina wire to the students who were afraid to go to school following the recent Taliban attack on the American University that had killed thirteen. I thought about the other scars on workers now in other cities that were less visible, but with potentially devastating long-term consequences.

My account, of course, looks only at one type of ripple of the war in Afghanistan: the effect that it has had on primarily Asian contractors who came to the country to work for the American intervention. Countless other repercussions are either unreported or are only beginning to be documented by scholars and journalists, ranging from posttraumatic stress disorder among veterans to communities in Pakistan terrorized by America's drone war to the environmental damage caused by American bases supporting the war.[10] Still, the consequences of hiring so many contractors are something that we are only beginning to understand, and its impact is likely to continue changing in surprising ways.

It is also a ripple that could grow into a wave as other countries and nonstate bodies increasingly mimic U.S. methods of outsourcing war.

One of the effects is that tens of thousands of the private security contractors who worked in the country are now unemployed. Many of these contractors are Nepalis, but as security companies looked to find cheaper labor, they expanded much of their recruiting across the globe from Fiji to East Africa. Some like Bhek are now looking for other forms of security work in growing economies (and in places with insecurity or high crime rates) like Russia and the Ukraine. Many of the Afghans who worked for international security companies in Afghanistan are now ideal employees for local commanders and warlords who are reasserting themselves as the international presence leaves.

The U.S.'s use of private security contractors in Afghanistan and Iraq over the past two decades has justified the practice internationally. In 2015, the United Arab Emirates had begun deploying Colombian private security contractors as a military force fighting in Yemen's civil war. According to the *New York Times*, the United Arab Emirates had a brigade of 1,800 soldiers from various Latin American countries who were ready to deploy globally.[11] It's possible that in the future, rich countries, particularly those with small populations in the Gulf, for example, will be able to fight wars without using any of their own citizens. They will have a supply of willing recruits from poor countries that are currently part of an unemployed standing army of private security contractors who are increasingly willing to work for whoever is willing to pay them, whether it's Vladimir Putin or an emerging strongman somewhere else.

In other countries, particularly in Latin America, private security guards now outnumber police. As Mark Duffield argues, the use of private security supplied by international firms by elites in poorer countries is likely to continue to expand as governments fail to provide protection for their citizens, leaving the rich to purchase their own security privately.[12] As this practice continues, states will lose their monopoly on violence and it will become increasingly unclear who is actually responsible for protecting citizens' rights.

Companies are going to continue to take advantage of instability and conflict and respond more quickly to conflict than diplomats do. Moving forward, this is likely to blend technology and violence in more difficult ways to track and monitor. Already on the Alibaba shopping site, the Amazon of China, listed under "Labour & Employment," shoppers can order "Nepali Workers and Ex-Gurkha Security Guard from Nepal" by "adding them to your cart." This online hiring of security contractors has a "10 pack" minimum order, with each "pack" costing $300 to $600, with "500 packs" currently available. "Packing Details" are listed as "By Air" with a delivery time of "2 months." Payment was possible through Alibaba or offline through Western Union.[13] I emailed for more information and received a prompt reply that the contractors were ready to head wherever I needed them.

With a credit card, you can now buy an army online.

As such new modes of creating and commodifying violence continue to enter the marketplace, states and international bodies like the United Nations are going to have difficulty regulating them. Governments are losing control of war.

What's more, we are struggling to simply understand the current consequences of conflict.

As I traveled from interview to interview, took bus rides across the plains of the Terai in Nepal, climbed into the lush mountains of Kerala, and sat by the Mediterranean in Turkey, person after person I met had been shaped by the war in Afghanistan. Each was part of the ripples that had sprung out of the conflict. Flows of laborers, the various policies and approaches of countries to conflict, and the increasingly globalized world we live in mean that these consequences of war are going to come in more surprising ways to even the remotest corners of the globe.

In the future, war is only going to become more complex, its consequences more difficult to trace. It is our responsibility to keep chasing the ripples of conflict and attempt to understand them so that we can limit the human cost of war.

A NOTE ON THE RESEARCH

The text names and some identifying characteristics of those interviewed have been changed except in the case of public figures and when interviewees explicitly asked that their name be used. With the interviewee's consent, I recorded most interviews. However, while writing this book, I often found that paraphrasing quotes from interviews was far more effective at conveying the intended meaning as I understood it. In the cases where precise phrasing is important, I have indicated direct quotations by using quotation marks.

I attempted to confirm various details of these experiences to the extent that I could. For example, in Teer Magar's case, most of the details of his trial were verified by a reporter whom I interviewed later. In this and most other cases, the contracting firm he worked for did not respond to my emails. In some cases, confirming specific details was often impossible because the contractor was alone or had lost contact with those he was with. Throughout the analysis, my emphasis is thus on perceptions of the war and the individual experience described by the contractors.

The sources I have used for various statistics are available in the Sources section. In general, however, there is a dearth of quantitative data on contracting (and, I would argue, these numbers often distract us from the more human aspects of contracting I explore here). This, I believe, is in part due to the chaos of the surge, when money was spent haphazardly and record keeping was not prioritized. More broadly, however, the suppression of data is a useful strategy by contracting companies to prevent the public from getting a real sense of the consequences of their practices. The U.S. government similarly benefits from the public's having little sense of the economic or human costs of these ongoing wars. Furthermore, by keeping the process muddled and nontransparent, companies, brokers, and corrupt officials benefit while the workers suffer. I would thus argue it is not coincidental that data on contractors are often difficult to find.

As a result, the only way for workers to make it through these convoluted processes is by relying on personal networks. This research was also possible due to the generous sharing of experience and personal networks by a variety of people.

In particular, this book has four godfathers: Lieutenant Bodh Bikram (ret.) in Kathmandu, General Levent Gözkaya (ret.) in Ankara, Archil Kikodze in Tbilisi, and Dr. Jagdish Chander in Delhi. Each provided me with instrumental introductions in their respective countries, advice on logistics, and general moral support during formative phases of this project.

Dawa Sherpa was my research companion (*assistant* is not a strong enough word) for most of my time in Nepal and for some of my work in Delhi as well. Talking with him as we attempted to track down our initial contacts was instrumental to the shaping of the project, and his sharp eye and empathetic nature make him a fabulous researcher. Our coauthored work (Coburn and Sherpa, 2018) served as a place where I tried out some of the initial ideas that appear in this book. In addition, Kriti Upadhyaya provided me with research assistance in Delhi, and Ben Simpson relentlessly searched archives back in the United States in an attempt to untangle where the American government's money in Afghanistan actually went. I am grateful to both.

In Nepal, my greatest debt is to Deepak Thapa and Bandita Sijapati at Social Science Baha and the Centre for the Study of Labour and Mobility (CESLAM). Some of the material in this book builds off a preliminary report I wrote for CESLAM, "Labouring Under Fire: Nepali

Security Contractors in Afghanistan." They graciously hosted my research in Nepal and provided valued context and guidance, particularly in getting the research started.

The Fulbright office in Kathmandu, headed by Laurie Vasily with the assistance of Yamal Rajbhandary, was immensely supportive, as was Manohari Upadhyaya at the Association for Nepal and Himalayan Studies. Also in Nepal, I am also grateful to Subel Bhandari, Bhojraj Bhat, Sara Schneiderman, Dinesh Regmi, Padam Prasad Upadhyay, Jacob Rinck, Neha Choudhary, Bhim Kala Limbu, and Amanda Snellinger, who provided me with valuable introductions and feedback on key aspects of the project. In Kathmandu, Sugat and Roshani Manandhar provided me a warm and welcoming place to come home to.

The following people all offered valuable introductions, as well as logistical, intellectual, and moral support for various aspects of the project: in Turkey, Z. Tuba Sungur, Navid Fozi, Sanem Guner, Asli Mutlu, and Christina Bache Fidan; in India, Taran Khan and Rajpal Yadav; in the United Kingdom, Neha Choudhary and Anna Larson; in the United States, Thomas Barfield, Michael Carrol, Ishmael Hakimi, Baheerullah Safi, Matt Zeller, Matt Handley, and Naeem Anis; and in Afghanistan, Shahmahmood Miakhel, Mohamad Hassan Wafaey, Muneer Salimzai, and Sediq Sediqqi.

I presented various material for this book in a variety of venues. I am grateful to David Gellner, Ina Zharkevich, and Bandita Sijapati for their conference, Circuits of Labour, Obligation and Debt: International Labour Migrants, Their Families, and the "Migration Industry" in Nepal," hosted by the University of Oxford, Institute of Social and Cultural Anthropology. I also presented material at Security, Society, and the State, a conference sponsored by the Gerda Henkel Foundation and organized by Wolfgang Seibel, and at a lecture arranged by Heather Hindman at the South Asia Institute at the University of Texas, Austin.

At Stanford University Press, Michelle Lipinski thoughtfully shepherded this manuscript through the editorial process. Dawa Sherpa,

Michael Hutt, Matt Handley, Ronald Neumann, Catherine Lutz, Deepak Thapa, Philipp Münch, and my students in Bennington's Social Science Senior Seminar read and provided feedback on early versions of various pieces of this book. I am particularly indebted to Beth Ruane, Laurie Rich Salerno, Dawa Sherpa, and Michelle Lipinski, who took the time to read and comment on more complete versions of the text. I am also grateful to the two anonymous reviewers at Stanford who provided insightful feedback.

The funding for the travel and the time it took to complete various aspects of this work came from a Fulbright Regional Scholar Fellowship, a grant from the Gerda Henkel Foundation, a fellowship from the American Institute of Afghanistan Studies, and a joint grant with Timor Sharan from the Hollings Center for International Dialogue.

Bennington College provided me with valuable time away to do most of the travel for this project and the flexibility with scheduling to complete the writing of it. It is a terrific place to teach and learn.

Beth Ruane-Coburn tolerated my peripatetic nature (during this research and in life more generally), and her companionship during this research including a willingness to take long bus rides and put up with my excitement at finding details of a particularly obscure construction contract, made the process more bearable and made this text more relatable on a human level. I will never stop being thankful that she decided to come along on this adventure.

Finally, my greatest debt remains to the over 250 contractors and others who agreed to sit down for one or more interviews with me. In addition to sharing their stories, both good and bad, many opened their homes to me, fed me, and even provided me a place to stay, often refusing to allow me to repay them. Their bravery and humanity are truly extraordinary. I hope that this book does some justice to their remarkable lives.

TIME LINE OF RELEVANT DATES

1816   Treaty of Sugauli begins the practice of Nepali recruitment first into the military of the East India Trade Company and later the British Imperial Army, the British Army, the Indian Army, the Singapore Police Force, and other foreign militaries

1857   India's First War of Independence, also known as the Indian Mutiny, is put down by the British with the assistance of Nepali troops

1947   Indian independence followed by its violent partition with Pakistan

1979   Soviet invasion of Afghanistan

1988   Soviet withdrawal from Afghanistan

1992   Collapse of the Communist regime in Kabul amid widespread civil war

1996   The Taliban take Kabul

1997   Gurkha Brigade headquarters is moved from Hong Kong to the United Kingdom

2001   Initial U.S.-led invasion of Afghanistan

2003   U.S. invasion of Iraq, which leads to an increase in contracting jobs in Iraq and less U.S. focus on Afghanistan

Rose Revolution in Georgia

2004   Hamid Karzai is elected president in the first post-2001 elections

Twelve Nepali drivers are kidnapped and later killed in Iraq, sparking protests in Nepal

Eleven Chinese workers in Afghanistan are killed, the first large-scale attack on contractors in Afghanistan

2006   Taliban insurgency gains strength in the east and south of the Afghanistan

Maoist insurgency in Nepal ends following peace treaty

2009   President Barack Obama initiates American surge, leading to a massive influx of troops, funds, and contractors

Hamid Karzai reelected president in disputed, fraud-marred elections

2011   Four Nepali guards are among those killed when protesters attack a U.N. compound in Mazar-e Sharif

Number of U.S. troops in Afghanistan peaks at just below 100,000

2012   Number of Department of Defense–funded contractors in Afghanistan peaks at just under 120,000

Drawdown of American surge troops in Afghanistan begins

2013   Attack on DynCorp contractors at Camp Pinnacle on July 2

2014   Contested election in Afghan leads to a split administration with Ashraf Ghani as president and Abdullah Abdullah as chief executive officer

2015   Two earthquakes cause devastation across Nepal

2016   Thirteen Nepalis working for the Canadian embassy in Kabul
       are killed, leading to a temporary ban on work in Afghanistan
       by the Nepali government

2017   The Trump administration authorizes increases in U.S. troops
       in Afghanistan, expanding the effort there again

NOTES

PROLOGUE

1. Peters, Schwartz, and Kapp 2015, 3–4.

2. President Obama declared an end to U.S. combat missions in Afghanistan on December 28, 2014, and by spring, just under 10,000 U.S. troops were to remain in the country primarily in a training and support capacity, drawing down by the end of the year (White House 2014). As the Taliban reasserted itself and the Islamic State began organizing more attacks in Afghanistan, debates over troop levels renewed, and the White House decided to maintain troop levels through 2016. The Trump administration then slightly increased troop levels again in 2017 and 2018, while relying on an intensified bombing campaign For ordinary Afghans, perhaps the most important fact was that by 2015, it was clear the numbers would not reach the 2010 levels, when there were almost 100,000 U.S. troops in the country, along with another 50,000 from other coalition countries (Rosenberg and Shearoct 2015).

3. There are numerous analyses of various aspects of this era and the effects of the massive influx of funds. Some of the more useful are Suhrke 2011, Wissing 2012, and Chandrasekaran 2012.

4. Much of this research can be found in Coburn 2011.

5. Many of the local Afghan firms providing security were little more than formalized militias made up of many of the men who had fought against the Taliban or during Afghanistan's civil war in the 1990s. See Sherman 2015, 106.

6. Belasco 2014, 6. This number does not include long-term costs like veterans' benefits. See a more thorough discussion of costs in Stiglitz and Bilmes 2012.

7. Joya, Nassif, Farahi, and Redaelli 2015, 17.

8. Names and some identifying characteristics of those interviewed have been changed in most cases. For more, see A Note on the Research.

9. See, in particular, Coburn 2016a.

10. U.S. Department of Defense n.d.

11. U.N. Assistance Mission in Afghanistan, and United Nations Office of the High Commissioner for Human Rights 2016, 1.

12. Crawford 2016.

13. Stiglitz and Bilmes 2012, 305; Crawford 2016, 18.

14. The destruction brought on by the war was far from solely isolated in Afghanistan. When the Department of Defense releases casualties for Operation Enduring Freedom (the operational name for the U.S. mission in Afghanistan), a footnote explains that these numbers include not just American deaths in Afghanistan but also fatalities in "Guantanamo Bay (Cuba), Djibouti, Eritrea, Ethiopia, Jordan, Kenya, Kyrgyzstan, Pakistan, Philippines, Seychelles, Sudan, Tajikistan, Turkey, Uzbekistan, and Yemen." U.S. Department of Defense n.d.

15. In personal communications on the subject, Ronald Neumann, former U.S. ambassador

to Afghanistan, suggested to me that the increasing "nationalization" of the press, with American journalists reporting only from areas where there was an American military presence and British journalists reporting only from areas where there was a British military, contributed to the narrow focus of many media reports.

16. Peters et al. 2015, 3. As if to prove the dollar theory and that troop levels have more to do with public relations than military need, by spring 2016 there were 9,800 American troops in Afghanistan but, notably, not 10,000.

17. For more on the increasing U.S. reliance on contracting, see Stanger 2009. Krahmann 2010 also reviews the implications of outsourcing on democratic oversight of the military and the relations of soldiers, citizens, and the state.

18. For an analysis of the Taliban during this period, see Giustozzi 2012. For more on the ways in which Western conceptions of the Taliban were constructed during this period, see Strick van Linschoten and Kuehn 2012.

19. See Coll 2005.

20. Looking at the experience of these contractors also provided an opportunity to examine how the war was experienced not just by America and Afghanistan, but through the lens of the other countries that had become involved in the conflict in numerous ways. Several analyses of the experiences of some of these countries in the country have been done on a country-by-country basis, such as Kaya 2013 and Çolakoğlu and Yegin2014 for Turkey, but few have looked comparatively at how the experiences of those countries that were not major troop contributors in the intervention.

21. Some of the best of these are Avant 2005, Singer 2007, Krahmann 2010, Dunigan 2011, and McFate 2015. I tend to avoid the term *TCN*, except when it was used by an informant or in literature I am quoting, since the phrase tends to be used to homogenize this group, particularly those from poorer countries. TCNs, like their American and Afghan counterparts, were almost all contractors, the term I primarily use in the text.

22. This figure comes from the Department of Defense's "Boots on the Ground" monthly report to Congress, compiled by the Congressional Research Service in Peters et al. 2015, 3.

23. I am not the only one to have encountered difficulty in determining these numbers. The Special Inspector General for Afghan Reconstruction (2013, 12), the lead U.S. oversight body for reconstruction in Afghanistan after citing Department of Defense numbers, complained: "We could not obtain similar data for the number of non-Afghan contractors supporting State and USAID." To demonstrate the way that contracting numbers can conflict with each other, even when they come from within the U.S. government, the Iraq Study Group estimated in December 2006 that there were 5,000 civilian contractors in Iraq, while Central Command released a review reporting that there were 100,000 government contractors in Iraq. The Associated Press, three months later, put the number at 120,000 (Isenberg 2009, 8).

24. For example, I noticed in the U.S. Central Command's monthly reports to Congress in three quarters in 2010 that the number of international contractors swung wildly from 88,000 to 49,000 back up to 68,000, changes in personnel that were logistically impossible to make in three months (Peters et al. 2015, 3). Sure enough, my research assistant, Ben Simpson, found that a deputy assistant secretary of defense had later issued a correction stating that "this reported decrease is not due to a large reduction in contractor personnel, but rather is due to counting errors (primarily in one specific reporting activity), that have been perpetuated throughout this Fiscal Year. These errors have been identified and subsequently corrected" (U.S. Department of Defense 2010). Six years later, Congress was still using the original erroneous numbers in their reports on the war.

25. The U.S. military tends to outsource more regularly than other countries, but it is far from alone in the practice. For more comparative analysis, see Krahmann 2010 and Dunigan and Petersohn 2015.

26. While sex workers are rarely counted, a bipartisan congressional committee did acknowledge the U.S. role in encouraging this migration: "Globalization of the world economy has spurred the movement of people across borders, legally and illegally, especially from poorer countries, to fill

low-skill jobs in support of the U.S. contingencies in Iraq and Afghanistan. Exploitation includes forced labor, slavery, and sexual exploitation" (Commission on Wartime Contracting in Iraq and Afghanistan 2011, 92).

27. U.S. Department of Labor n.d.a, n.d.b. Notably, these are also the source that the Department of Defense uses to count contractor fatalities (Commission on Wartime Contracting in Iraq and Afghanistan 2011a, 31).

28. Isenberg 2011.

29. Peters, Schwartz, and Kapp 2017, 5. There is a similar trend in the number of contractors employed by other countries in conflict zones as the conflict winds down. The United Kingdom, for example, had 2,000 contractors supporting 46,000 troops deployed in Iraq in 2005, but that ratio swung wildly by 2008, when the Ministry of Defence listed 2,200 contractors supporting just 4,100 troops (Krahmann 2010, 201–2022).

30. U.S. Department of Defense 2014, 10. These numbers do not indicate how blurry the concept of being a contractor is and how difficult it is to use fixed categories to explore the edges of the war in Afghanistan. For example, how different was the Republic of Georgia's sending 2,000 troops to Iraq in exchange for a lucrative military aid package from the Bush administration than the State Department's paying Blackwater for 2,000 private security contractors who were essentially being employed in the same manner? A large amount of policy literature and material generated by contractors attempts to differentiate categories such as military combatant companies, military consulting firms, and military support firms (e.g., Isenberg 2009, 25). I attempted to avoid becoming mired in such debates because many of the firms I looked at moved between these categories as their contracts changed and they filled different roles, particularly when the most U.S. money was flowing into the country. More important, these categories were almost never relevant to the individuals I interviewed. If you are guarding a compound and getting shot at, it does not matter to you or whoever was doing the shooting whether it was a combat company or a consulting firm you were employed by.

31. Some of the best accounts by academics and journalists include Rashid 2008, Suhrke 2011, Chandrasekaran 2012, and Gopal 2015.

32. Special Inspector General for Afghan Reconstruction 2015, 2016.

33. Towell and Else 2012.

34. Schwartz 2014, 18. Not just U.S. spending on security contractors has increased. Between 2009 and 2011, the United Nations almost tripled its spending on "security services" (Pingeot 2014, 6).

35. "Supreme Group Holdings Sarl in—Contracting Profile" n.d. While the size of the private military industry is debated and extremely difficult to measure, Singer in 2004 put the size of the global market at $100 billion (quoted in Hodge 2011, 195).

36. For other work that has various methodological similarities in approaching transnational migration and conflict, see Nordstrom 2004, Vine 2009, Lutz 2009, Bruslé 2012, Chandrasekaran 2012, Duffield 2014, and Gopal 2015.

CHAPTER 1

1. This assumption ignores some of Afghanistan's minorities, such as Hindus and Sikhs who became important later when I interviewed Hindu contractors in India. However, the notion of Afghanistan being a strictly Muslim country dominates political discourse in the country. (See Barfield 2010, chap. 1.) While many Nepalis I spoke with felt that the key difference between them and Afghans was the Afghans' strict devotion to Islam, I could still see similarities in how the two cultures practiced their religion. Much like Matta Tirtha, the Afghan town I originally conducted research in, Istalif, had a shrine that served as a focal point in town life. There, a holy man would sit and bless amulets or other trinkets, particularly for the women who visited it. Shrines like these and celebrated local saints were common throughout Afghanistan, illustrating Islam's local variations, even though most Afghans, and the international press, tended to describe the country's religion as more monolithic.

2. The British relied extensively on merchants from nomadic tribes on the frontier to provide them with transportation and supplies, forefathers of the Afghan construction and logistics subcontracting firms the United States would rely on following its invasion in 2001. For more on the role of these figures in nineteenth-century Afghan history, see Hanifi 2008, 67–76.

3. Particularly following the U.S. invasion of Iraq, there was an upswing in debates in the popular media and in academic literature over whether the United States should be considered an empire and if it was an empire, how it was different from those that preceded it. Was the United States a "liberal" empire (Fergusson 2004), a "lite" empire (Ignatieff 2003), an "incoherent" empire (Mann 2003), an "inadvertent" empire (Odom and Dujarric 2004), or an "empire of bases" (Vine 2015)? Various contractors I interviewed would have agreed with the implication of a hegemonic U.S. global presence, in either the benign or bellicose sense of the term, yet most would point to the United States as primarily dominating through its control of military and economic resources. In the eyes of the workers from Nepal and elsewhere, the United States empire was marked by its highly outsourced nature more than anything else.

4. DynCorp, based in Fort Worth, Texas, was one of three companies receiving funds through the so-called LOGCAP contract. This, the holy grail of defense contracts, provides logistical support for all aspects of the U.S. Army's activities. When the current iteration of the contract was issued in 2007, it had a base value of $5 billion per year, with a ceiling of $15 billion. Axelrod 2014, 297–300 and U.S. Army Sustainment Command Public Affairs 2007.

5. Some have argued that the rise in contracting in recent years is simply a return to a more normal state of affairs, with the national conscription armies of the twentieth century being the actual exceptions. For a history of how private security contractor use by the U.S. military has evolved, see McFate 2015. Krahmann 2010 provides an interesting comparative look at how the industry has developed differently in the United States, the United Kingdom, and Germany. Eichler 2015 provides a look at private security through the lens of gender.

6. Through the first two hundred years of American conflicts, Isenberg (2009, 4) points out, the ratio of contract to soldier actually remained fairly consistent. It was 1:6 in the American Revolution, 1:5 in the Civil War, and 1:7.5 in World War II. It dropped to 1:2.5 in Korean before returning to 1:6 during the Vietnam War.

7. Vine 2015, 218.

8. Commission on Wartime Contracting in Iraq and Afghanistan 2011a, 18.

9. This number includes only contractors receiving funds from the Department of Defense (McFate 2015, 19). While the focus of my research was on contractors who received funds on U.S. government contracts, I occasionally interviewed those who had worked on contracts from other NATO countries in Afghanistan. The number of these contractors was lower than those working for the United States, which is not surprising considering these countries committed fewer troops and less money. Nonetheless, the numbers were still substantial; for example, the United Kingdom had 2,000 defense contractors in Iraq and, simultaneously, 46,000 troops. While other NATO countries often favored companies based in their countries, many of the contracting companies were similar and the United Kingdom, for example, had a contract worth up to 50 million British pounds with KBR in Iraq to feed their troops there, despite that KBR worked primarily for the U.S. military (Krahmann 2010, 201–204).

10. There's no breakdown in numbers of contractors who were involved in private security and those doing other work in part because definitions are so fluid, though private security contractors have certainly gotten much more media attention than other contractors. However, spending on "guard services" in Iraq and Afghanistan between 2002 and 2011 was $3.8 billion, while "logistics support services" and "technical assistance" were $46.5 billion and $10.5 billion, respectively, suggesting on a dollar-to-dollar basis at least that private security was a smaller part of the overall contracting picture (Commission on Wartime Contracting in Iraq and Afghanistan 2011a, 23). This blending of soldiers, security contractors, and civilian laborers can be thought of as the extreme example of the militarization of civilian society, a process that Catherine Lutz (2001) traces in Fayetteville, North Carolina, home of Fort Bragg. Lutz's work triggered a wave of

anthropological literature on American bases and military labor; some of the best are Vine 2009, Lutz 2009, and Macleish 2013.

11. As an anthropologist concentrating on the lived experience of these workers, I focused on the bottom-up view that sees many of these categories as deeply related. In general, the tendency to insist on categorical differences is legal and, perhaps more important, financial, since companies can budget different amounts for various types of contractors. These categories can shape the experiences of those contractors, but this is often subtler than the notion that clear distinctions suggest. For the tension between contractors as criminals as opposed to victims, for example, see White 2016.

12. The use of the term *mercenary* is contested. See Singer 2007, Isenberg 2009, Krahmann 2010, Dunigan 2011, Axlerod 2014, McFate 2015, and others for debates on terminology. There is disagreement over the term *private security contractor* (PSC), who are almost always armed, and the more numerous private military contractors (PMCs), which includes those doing logistical support work. Some others refer to both groups collectively as private security providers (PSPs). As Isenberg 2009 (Preface), points out, in actuality, these groups intermingle freely in the field (and in accounting ledgers). The debate is largely academic, and none of those I interviewed brought up the distinction. The debate over the term *mercenary* is more important because according to the Geneva Convention, mercenaries are not granted the same rights as combatants or prisoners of war. For the most part the U.N.'s Mercenary Convention excludes many of the private security contractors in Afghanistan with its stipulation that a mercenary "is neither a national nor a resident of the State against which such an act is directed (Axlerod 2014, 188), though, notably, this would not exclude Nepali contractors. While most in the private security industry argue strenuously against the label, it is not clear at all that this is justified beyond attempts by these companies to disassociate themselves with the negative imagery associated with the term. I tend to avoid the term *mercenary* in this book, not to make any sort of political statement, but since those interviewed almost never referred to these international treaties and definitional debates when referring to the work that they did and as an anthropologist, my interest tends to be in how individuals classify themselves.

13. One of the reasons they resembled traditional military strike teams was that they were composed of retired Special Forces soldiers. For a case in Iraq, see Krahmann 2010, 221.

14. Military officials are, unsurprisingly, the most adamant that security contractors do not take part in military operations. As I interviewed contractors about the experiences, however, it seemed the military was defining "military operations" by the presence of soldiers (and the absence of contractors) rather than by any type of "kinetic" activity, for example, the presence of enemy combatants or the complexity of the operation. This is clearest in the case of supply convoys detailed below. The circular nature of these definitions creates an opening that security companies used to become more integrated into military operations.

15. The Commission on Wartime Contracting Report (2011b, 17–18), stated, "Contractors who perform movement security in Iraq and Afghanistan are likely to traverse hostile environments and enter into high-risk situations."

16. Hodge 2011, 101.

17. U.S. Department of Defense n.d. and U.S. Department of Labor n.d.b.

18. An analysis by Ann Hagedorn (2014) suggests that in 2010, contractor casualties in Afghanistan and Iraq were greater than troop casualties, though as with all other studies of contractors, it is difficult to determine the accuracy of these claims since records are poorly kept and often not publicly released. Several U.S. government sources suggest this is the case,. with the Commission on Wartime Contracting (2011, 30) reporting that "significant contractor deaths and injuries have largely remained uncounted and unpublicized by the U.S. government and the media" (Commission on Wartime Contracting in Iraq and Afghanistan August 2011a, 30).

19. Some have commented that these divisions between contractors don't seem to be racial as much as based on country of origin. This appears to be misleading: of the twenty plus senior management people at the security firms whom I interviewed or interacted with, all were white. None were Americans or Europeans of color. Similarly, white South Africans were in the upper

levels of several companies, whereas black South Africans never were. Some argue that this is the remnant of white officers' losing their jobs at the end of apartheid, but if so, with apartheid ending twenty-two years before, it must also be reinforced by racial biasing.

20. Chisholm (2015, 101) describes how language and training are used by the white men who dominate the upper ranks of security firms to create a world where the "desired skill set reinforces the white security contractor as the ideal and the positions the TCN as forever lacking."

21. There were exceptions to this, including some examples mentioned below, but this hierarchy was predictable and similarly structured in almost every compound I visited in Afghanistan.

22. I did preliminary research on the topic in both Afghanistan and Nepal in 2014, but the sustained field research for this project was completed in 2015, 2016, and 2017.

23. National Reconstruction Authority 2016, 1, and Food and Agriculture Organization of the United Nations 2015.

24. The majority of my interviews were in English, since most contractors who worked with internationals spoke English fairly well. It was not always the best educated who also spoke English. Those who were less educated but were in positions that interacted regularly with a variety of internationals tended to speak English the best. I relied on translation (more on this below) in cases when the contractor did not speak English or felt uncomfortable being interviewed in a different language.

25. This is based on a compilation of the various companies that employed Nepalis and the approximate number of Nepalis working at each company.

26. For an account of the assistance that the Pakistani government provided in the Taliban retreat and the acquiescence of the Bush White House in this process, see Rashid 2008, 91–92. For a particularly prescient analysis of the opportunities missed by the international community during this period, see Johnson and Leslie 2005.

27. For an account of some of the missed opportunities during this period when the Bush administration was focused primarily on Iraq, see Neumann 2009.

28. For an analysis of some of these figures and how they contributed to the fractioning of political power across the country, see Mukhopadhyay 2014.

29. See Neumann 2015.

30. See Partlow 2016 and Rashid 2008 among others.

31. Joya, Nassif, Farahi and Redaelli 2015, 17.

32. For a review of the debate within the Obama administration around this shift, see Woodward 2010.

33. "Supreme Group Holdings Sarl in—Contracting Profile" n.d. Companies like Supreme did not just contract for the American military. Supreme, for example, also had a contract to supply the German military, among other countries in Afghanistan, with catering services (Krahmann 2010, 213).

34. "DynCorp International—Contracting Profile" n.d. Much of this increase was due to the issuance of the fourth Logistics Civil Augmentation Program (LOGCAP IV), the contract that supplies much of the logistical support for U.S. Army bases, including those in Iraq and Afghanistan. LOGCAP III went solely to KBR (Kellogg, Brown and Root, which at that point was a Halliburton subsidiary), but after cost overruns, charges of corruption, and significant political pressure, when LOGCAP IV was issued, it was divided between KBR, DynCorp, and Flour (Hedgpeth 2007). As a result, while KBR did much of the initial building in Afghanistan and Iraq (including much of the building at the Guantanamo Bay facility; Axelrod 2014, 287), much of the later work was done by Flour and DynCorp. When I found contractors from the early years after the invasion, they were far more likely to work for KBR, whereas many in later years tended to work for DynCorp or Flour as KBR's U.S. contracts declined steadily from $7.5 billion in 2004 to $500 million in 2012 to $21 million in 2014 (KBR Inc. in—Contracting Profile n.d.).

35. The U.S. government has largely acknowledged the issues of their approach. Its primary finding in a 2014 report was that "the U.S.'s initial support of warlords, reliance on logistics contracting, and the deluge of military and aid spending which overwhelmed GIRoA's [Government

of the Islamic Republic of Afghanistan] absorptive capacity created an environment that fostered corruption and impeded later CAC [corruption/anti-corruption] efforts" (U.S. Department of State 2014, 9). For a more thorough review of the surge and related issues, see Coburn 2016a.

36. This was not always the case, and for those assigned to positions such as static security, in many instances they were cut off from the wider political and strategic issues of the conflict. Others, however, were very knowledgeable about the various nuances of the conflict.

CHAPTER 2

1. There are multiple spellings of the word *Gurkha* depending on which transliteration of Nepali is used and which type of soldier is being referred to. Following the independence of India in 1947, the spelling "Gorkha" was adopted to designate Nepalis serving in the Indian army, while "Gurkha" was still used for those in the British Army. The term may also be used to refer to Indian citizens who are of Nepali descent, but are not Nepali citizens, serving in either the British or Indian armies, complicating the issue further (The British historically recruited "Gurkhas" from both Nepali and Indian territory). In the British Army today, there are still Gurkha regiments, which are led by white British officers, and these white officers are also referred to as Gurkha officers even though they are not Nepali. Further confusing things, in the contracting world, "Gurkha" has now become a gloss for any Nepali contractor regardless of their experience (though Nepalis who have served in the British Army passionately reject this expansion of the terminology). In general, many of the contractors I interviewed referred to themselves as "Gurkhas," whether they had served in the British or Indian armies or not. More, however, referred to themselves simply as "contractors" and a few used "Lahure," a synonym of "Gurkha" that came from the city of Lahore where many used to originally go to enlist. Through the text I try to use the term as my interviewees used it. For a far more thorough account of the debates around the term, see Caplan 1995. For more on the image of the "brave Gurkha warrior," see Des Chenes 1991. Chisholm 2015 considers the issue from a gendered perspective.

2. In fact, in 2011 there were 70,000 non-U.S. citizens in the American military, though these figures are rarely mentioned in discussions around citizenship and the military in America, as opposed to the case of the United Kingdom where the Gurkha presence is more celebrated. McIntosh and Sayala 2011, 14).

3. This has not been helped by the tendency of retired members of the Gurkha Regiments, both British and Nepali, to write orientalist memoirs glamorizing the heroics of the men whom they fought with from Burma to the Falklands.

4. Most people in the retelling of this story actually assume that it was the British Imperial Army fighting in this war, not the East India Trade Company; the United States did not invent the outsourced empire, and long before both British and Russian expansion in Asia was driven largely by overly enthusiastic businessmen and merchants.

5. Before Indian independence, Britain also recruited a large number of soldiers of these Nepali ethnicities but who lived across the border in India, and these groups still make up a large percentage of the international security contractors who come from India. See Des Chenes 1991 for more on the establishment of some of these recruitment practices. The accounts of Gurkhas as brave and savage Nepali warriors who needed only the discipline and structure white British officers could provide continues to be retold today, yet in reality, the treaty was politically useful for both sides. The Nepali king granted the rights to recruit from the east and west of the country from the so-called hill tribes, such as the Magars and Gurungs from the west of Nepal and Rais and Limbus from the east. These groups were generally considered below the ruling class, but above the Dalit, or so-called untouchables, who make up the lower rungs of Nepali society; only recently did the British lift the ethnic restrictions on their recruiting. Shipping many of the young men from these castes and ethnicities out of the country helped prevent them from seriously challenging some of the higher castes and has deeply shaped the politics and cultures of these groups. MacFarlane in his ethnography, for example, suggests, "The Gurungs may be looked at, from one angle, as a tribe of migrant labourers, a strange off-shoot of the British Empire set amongst other less affluent and mobile peoples" (1974, 14).

6. Farwell 1984, 59.

7. There is not widespread agreement on these numbers, and Gurkha casualties are, in particular, thought to be potential higher. See Parker 1999, 214, and Rathaur 2001, 21–22.

8. Rathaur 2001, 22–23. Various sources disagree on this number, and Axlerod (2014, 70) puts the number of Nepalis killed or wounded during World War II at 43,000.

9. Cross 1986, 96. For more on this colonial image of the Gurkha, see Caplan 1995 and Chisholm 2015.

10. In the mid-1950s they made up a good number of a group of rebels associated with the Nepali Congress agitating against the monarchy for more political rights (Karki and Seddon 2003, 6).

11. Des Chenes 1991, 10.

12. Recruitment of Nepali troops took place almost exclusively in India until 1953 when India banned the practice, forcing the British to recruit within Nepal. I was able to track down one of these first recruiters who came to Nepal in 1953 and still resided outside Pokhara.

13. "Maoist Statements & Documents" 2003, 184.

14. This is in part due to difficulties in the current British Army with retention rates, an issue discussed further in chapter 24.

15. IDG Security n.d.

16. Gurkha Security Services n.d.

17. As will be discussed further below, many companies began supplying one Gurkha with experience in the British Army to oversee ten or more Nepalis who had no military training as a cost-saving device as the war went on.

18. This conflict is discussed in more detail below. For a thorough analysis of the conflict, see Thapa and Sijapati 2004.

19. Office of the United Nations High Commissioner for Human Rights 2012.

20. Colin Powell did visit in 2002 to demonstrate American support for the government, but this was several years into the conflict and America provided only $12 million in actual military aid. For more on the U.S. role, see Thapa and Sijapati 2004, chap. 6.

21. Martin 2012, 223.

22. See Thier 2006/2007.

23. For more on the constitution writing process, see Martin 2012.

24. Another constitutional assembly was elected in 2013, but it too initially failed to agree on a document.

25. Sijapati, Bhattarai, and Pathak 2015. The World Bank estimates that 2 million Nepalis emigrated for work abroad in 2013 and while numerically this means Nepal is thirtieth globally in terms of immigration, the fact that almost 30 percent of their GDP is made up of remittances means makes it the third most dependent country globally on returned labor wages, behind Tajikistan and Kyrgyzstan (World Bank 2016).

26. While the United Nations did set up a special mission in Kathmandu following the peace treaty, it was clear that no international actor was willing to commit much political capital to stabilizing the country politically or economically. For more on the role of the United Nations in Nepal during the conflict, see Martin 2012.

27. See Coburn and Larson 2014.

28. Buckley and Ramzy 2015 and Hindman and Poudel 2015.

29. This figure comes from using the term *Madhesis* in its broadest sense, including Muslims and the numerous indigenous living in the Terai. Political opponents are less likely to include these other groups in the definition of *Madhesis* and the number is closer to 12 percent if only non-Dalit caste Hindus are counted (Lawoti 2012, 131).

30. Specifically, the new constitution drew provincial boundaries to divide the Madhesis and undermined their ability to unite politically. The constitution also had a clause requiring both parents to be Nepali citizens for the child to be granted citizenship. For Madhesis, who often had stronger family ties across the border in India than those living in Nepal's hills, the clause raised the possibility of widespread disenfranchisement.

CHAPTER 3

1. The account here is pulled together by interviews with twelve survivors of the attack along with Afghan and international news sources in an attempt to reconcile some of the discrepancies in the reporting.

2. Reported casualty numbers initially contradicted each other; for example, see Ahmed, 2013 and Stancati 2013. A DynCorp (2013) press release later stated that three of its employees had been killed. The others had worked for other companies despite the fact that the compound was primarily used by DynCorp. The Pinnacle attack, which I spent much time looking into, is a striking example of how difficult it is to confirm casualty rates among international contractors. It is even more difficult to get reliable figures on Afghans killed or wounded.

3. See TV News 2013.

4. These rumors of escaped Taliban following attacks were common. Particularly after the 2016 attack on American University, elaborate stories about how attackers had escaped in an ambulance that used a coded siren to call the attackers circulated widely despite insistences by the authorities that they had all been killed.

5. See Siegel 2014 and Sege 2016. Despite all my best efforts, I have yet to find an explanation for the Star Wars reference in the company's name.

6. All quotes are from "Pinnacle Hotel Services, Afghanistan" 2012.

7. C3PO International, LTD. v. DynCorp International LLC, 2015, 3.

CHAPTER 4

1. No one I interviewed was sure exactly how the Nepali media heard about the attacks so quickly, but the fact that many of the contractors relied on Facebook to communicate with their families at home as well as with each other, makes it seem likely that the report was picked up through social media posting.

2. See Thakuri n.d. ProPublica also did a series of articles, "Disposable Army," focusing on contractors in Iraq and Afghanistan that focused on related issues (available at https://www.propublica.org/series/disposable-army).

3. See Parajuli 2004.

4. Thakuri n.d.

5. It is often these deeper layers of sub- and sub-subcontractors that are most difficult to track. To give a sense of the scope of subcontracting, KBR in 2007 in Iraq had a primary contract from the U.S. Department of Defense and in turn contracted out to over two hundred subcontracting companies. Many of these subcontracts then subcontracted further (Krahmann 2010, 205).

6. For example, documents by the Commission of U.S. Wartime Contracting, probably the most complete U.S. government report on some of the issues produced for Congress, noted the cost of contractor fatalities as one of the main reasons for cutting down on certain types of contracting by companies. See its 2011a report.

7. Referring to the case of injury or death for non-American contractors in Iraq, Isenberg (2009, 56) notes, "If nobody reports the incident to the Department of Labor and the family does not file a claim, it has no way of knowing what happened. In short, the system is totally based on the transparency and honesty of the contracting firms to do what they law says it must do."

8. Axelrod 2014, 223.

9. Hodge 2011, 204–207.

10. For an account of these hearings as well as their place in the evolving role of security contractors, see Hagedorn 2014, chaps. 7–8.

11. CACI employee Eric Flair (2016) also produced one the few memoirs of either the wars in Afghanistan or Iraq by a contractor.

12. Business and Human Rights Resource Centre n.d.; Isenberg 2009, 102–103.

13. U.S. Department of the Army 2004, 726.

14. This labor migration had significant social implications, as well as economic ones, with 53 percent of all households having at least one absent member working abroad. Since the majority

of those abroad were men, this led to villages where the only males were old men or young boys. Some suggested that this had even changed agricultural patterns: women preferred crops like tomatoes and other vegetables over wheat and rice, since vegetables could be grown closer to home and required less heavy labor. Such agricultural shifts only made it more likely that Nepali men would continue to seek work abroad, making the national economy more reliant on contracting work like this. For more, see Sijapati, Bhattarai, and Pathak 2015.

CHAPTER 5

1. I was first given this estimate from the spokesperson for the Maiti Nepali organization, but I verified it with several other sources as well.

2. Vora (2013, 23) describes Indians in Dubai as in a state of "permanent temporariness," generated in part by the granting of some basic rights, particularly to well-educated immigrants, while making citizenship impossible. The Nepalis I interviewed described similar states of temporariness but also emphasized Dubai as a site of transit and a gateway to other opportunities in both conflict zones like Afghanistan and in the West. Vora's ethnography focuses on middle- and upper-class Indians and detailed accounts of South Asian laborers in the United Arab Emirates are rare and tend to be produced by international nongovernmental organizations (e.g., Human Rights Watch November 2006) or by journalists (e.g., Kumar 2016).

3. Due to state restrictions, research in the Gulf on migrant work is limited. For some of the best available studies, see Bruslé 2010, 2012 and Bajracharya and Sijapati 2012.

4. Sijapati, Bhattarai, and Pathak 2015, 14.

5. The Labor Ministry had recently instituted a new policy in an attempt to reform the industry with a new regulation called "free visa, free flight," which essentially meant that the employer, not the employee, was supposed to pay for the visa and flight fees upfront. The idea sounded appealing, but the colonel was concerned it would make potential foreign companies look for workers in places like Bangladesh and Sri Lanka instead. The law was meant to crack down on bribes being extracted by the firms, but since it cut down on the ways the firms could actually make money, this made additional demands for "gifts" even murkier and increased the potential for less experienced workers to be exploited.

6. The legal ambiguity created by these regulations is discussed more thoroughly in chapter 10.

7. Pattisson 2015

8. Referring to the period before the Maoist insurgency but still largely applicable today, Thapa and Sijapati (2004, 191) call U.S. strategic interest in Nepal "always almost non-existent."

9. "UN Chief Calls for Lifting of Blockade on Indo-Nepal Border" 2015.

CHAPTER 6

1. These figures on IDG come from Pingeot 2014, 7 and were confirmed by IDG employees I interviewed.

2. This, I heard in several interviews, was often traumatic for their children who may have lived in Singapore for their entire lives

3. Raj 2007, 54.

4. Changes in the treatment of Nepalis in the British Army, particularly regarding their right to settlement and citizenship, is discussed more fully in chapter 22.

5. Stier 2009. For a fascinating analysis of the gendered and highly sexualized discourse around this scandal, see Hendershot 2015.

6. Isenberg 2009, 24, 68–69.

7. Aegis had also been found not to be providing workers at the proper qualification level on a U.S. government contract in Iraq in 2005 (Isenberg 2009, 71).

8. U.S. Department of State 2014.

9. Those below this status were usually classified as third-country nationals.

CHAPTER 7

1. Despite the rather timeless way that the doko test was described in some of literature on Gurkhas, it has been used for evaluating recruits only since the 1960s.

2. The 236 number is based on the number of new recruits that the Royal Gurkha Brigade needed to maintain its current number of approximately 3,600 Nepali soldiers, given retirements and other losses. The 3 percent success rate takes into account only those who formally enter the selection process. Some reports suggest the number of young men who train and do other forms of preparation but don't actually make it to the selection process is as high as 28,000 (Mazumdaru 2014). Based on my observations, there were certainly more than the 8,000 who actually prepared in Pokhara for the process, though the 28,000 estimate seemed inflated to me.

3. This number is from a briefing I received inside the camp.

4. The recruitment process had been thrown into disarray in 2015 by the two earthquakes that spring. Instead of prepping for recruitment for several months, the British camp had coordinated a hundred British engineers who had come over in the wake of the earthquake to deliver disaster relief. Providing support to these engineers took up much of the small staff's resources. As a result of these efforts, recruiting started later in the year than normal, and instead of holding regional selections in Dharan and Pokhara over three months, the entire process had been condensed to a month and was held exclusively in Pokhara.

5. Testing for the Singapore Police Force, which was still run by the British, was going on simultaneously, and there were 125 more recruits at the camp for that process who were mostly kept separate from the British recruits. The 6,000 applicants in 2015 were fewer than the almost 8,000 in 2014, which the British officers attributed to some of the confusion about the delayed process following the 2015 earthquakes. Those numbers also don't include the recruits who dropped out before this stage in the process, several of whom I had interviewed in the days before. The geographic split was left over from the treaty the king of Nepal had agreed to provide the British with recruits. It also banned the British from taking any recruits from Kathmandu Valley, though several of the young men told me it was not hard to fake these documents or get their birth certificate rewritten to reflect the home of their grandfather or some other relative who did not live in Kathmandu.

6. These numbers were given to us by British officers at the camp, but are also available on the British Army website and published widely as a part of the British government's attempts at transparency around the recruitment process.

7. Kesang Tseten captured this emotional moment powerfully in his film.

CHAPTER 8

1. Foreign and Commonwealth Office n.d. 14. For an account of these campaigns, see Farwell 1984, chaps 9 and 10.

2. See Parker 1999, chap. 8.

3. Farwell 1984, 120.

4. Cross 1986, 85.

5. By Cross's tenure, the British were taking fewer and fewer soldiers, and he rejected two or three recruits for everyone who was accepted. By 2015 it was up to thirty-six rejected for every recruit accepted.

6. When Dawa was standing with me on several occasions, this was prefaced with the speaker turning to him and saying some version of, "I'm sorry, but I'm sure you understand . . ." before insulting Nepalis more generally. Dawa, a committed researcher, would then nod, and the speaker would proceed to make more racially tinged remarks.

7. This also increased the social distance between the Nepali recruits and the white British officers at the camp.

8. The Gurkha Welfare Scheme also gave grants to a training center set up to run a series of skill development programs. The training center's facility was outside the British base, though just around the corner from the compound. Like the Gurkha Welfare Scheme, its orderly offices em-

ployed primarily retired Nepali Gurkhas who had returned to Nepal. Their head, a retired Gurkha who had served in the British Army for twenty-six years, told me that he liked to get in touch with these soldiers before they retired and then have them come straight to work for him when leaving the British Army. At the time of my visit, approximately 160 trainees were in shorter programs for semiskilled laborers doing masonry, carpentry, and electrical repair, as well as a handful of longer-term, skilled training in automobile repair and veterinary medicine.

9. During World War II and later during the Malay Emergency and the war in Borneo, British recruitment surged, with many Nepalis joining to fight in the conflict, but not remaining in the service afterward. These veterans had not served long enough to qualify for pensions. However, if they could prove they were "destitute," they qualified for up to 9,000 Nepali rupees a month ($90), from a foundation set up to support Gurkhas that was independent of British government funds but still distributed by the GWS.

10. Part of this, White (2016) argues, is the way in which contractors are associated with criminality as opposed to victimhood.

11. One of the few places the case of PTSD is considered in civilian aid workers is in O'Donnell 2017.

12. The initial settlement laid out a series of conditions for settlement, but since the Gurkhas had moved their headquarters from Hong Kong to the United Kingdom in 1997, this year marked the key cutoff for eligibility (BBC 2008). The campaign eventually won all Gurkhas with four or more years the right of settlement, though debate around the politics of the settlement and whether it truly helped these retirees has continued in the years since the ruling. See "Gurkhas v Government: Guess Who Won?" 2009; BBC 2009; and chapter 22, this volume).

13. For more on the life of Nepali Gurkhas in Britain, see Choudhary 2015.

14. Cowan 2009. According to the *Himalayan Times* ("THT 10 Years Ago" 2016), remittances from British Gurkhas to the national bank were 5.26 billion rupees in 2004–2005 and 3.22 billion rupees in the first seven months of 2007, a significant drop when inflation is taken into consideration, though less than Cowan cites. At the same time, however, these numbers do not note whether the money transferred actually stays in Nepal.

15. The one exception to this were Afghans eligible for a Special Immigrant Visa. This program is discussed further in chapter 23.

CHAPTER 9

1. Brown 2007, 737; U.S. Department of State 2007.

2. Verité 2015, 122.

3. Isenberg 2009, 94–95.

4. This also created something of an odd social imbalance. Since many of the companies that hired retired British and Indian Gurkhas hired only Nepalis, who worked together only speaking Nepali, their English did not get better, and perhaps it got worse. Many of the Nepali DynCorp employees, however, became very good at English, it seemed, despite the fact they usually came from poorer backgrounds than the British and Indian Gurkhas. During the intervention, where the ability to speak English was perhaps the most essential skill, this allowed many of the DynCorp contractors to advance in ways that their Gurkha counterparts who were not as skilled with English could not.

5. "Supreme Group Holdings Sarl in—Contracting Profile" n.d.

6. U.S. Department of Defense 2013.

7. de Jong 2013.

8. U.S. Department of Justice 2014.

9. Supreme Group 2015.

10. Records are not very reliable, but since most of these firms are significantly smaller than the primary contractors, it is most likely that Supreme still leads in terms of number of deaths (U.S. Department of Labor n.d.c.).

11. While it is not clear whether Supreme owed these back taxes or whether it was an attempt

by the Afghan government to extort funds, it did become obvious that not all Supreme's workers should have been exempted from visa and work permit requirements. One U.S. study of the issue concluded that "the Afghan government requires contractors to receive annual visas and work permits for each non-Afghan employee working in Afghanistan. While some bilateral agreements between various U.S. government agencies and the Afghan government may exempt certain U.S. personnel from requirements to obtain visas, other agreements are silent on the matter." Since Supreme was working on many contracts, some for countries other than the United States, many of their employees were not exempted by these bilateral agreements. Furthermore, Supreme had an incentive not to pay the Afghan government for either visas or work permits, since the study also found that on average, the cost for a worker to have one year of such permits was $1,138 (SIGAR 2013, 2).

12. Multiple interviewees suggested companies looking to keep their workers close also benefited from encouraging fear about what lay outside those compounds. Several laborers told me that their bosses had explained that "all Afghans are Taliban," suggesting that there would be awful consequences if they tried to leave their compounds. For the most part, security contractors seemed somewhat better informed about the reality of the security situation, perhaps because they at least sat looking out at the daily lives of Afghans walking by their compounds, most of whom had little interest in attacking or even interacting with them. Even among these contractors, I was often surprised at how little they knew of the wider security situation.

13. In response to similar accusations made in an Al Jazeera America piece, Supreme responded that "all of its non-managerial employees in Afghanistan are hired directly, without the involvement of recruiters." Al Jazeera reporters countered: "In our reporting, we spoke to five current or former Supreme employees, all of whom worked in non-managerial positions on U.S. bases in Afghanistan between 2009 and 2013. All five told us they had paid recruitment fees to agents, ranging from $1,000 to $3,350" (Black and Kamat 2014). Accounts I collected strongly support the reporters' perceptions of the process, with some I interviewed paying even higher fees to brokers for positions at Supreme.

14. This became an issue in other attacks, such as the attack on Nepali security contractors on their way to guard the Canadian embassy. The security contractors were given weapons only when on duty and thus were unarmed during the attack that killed fifteen. For more, see Coburn 2016b.

15. There were numerous other accounts I found during the year of workers arriving to much lower salaries than they were initially promised with no real recourse. This was not only for lower-level jobs; one security supervisor at ArmorGroup said that he had been promised $2,800, only to receive $900 in his first paycheck. There was no real legal recourse in Afghanistan, where the court system was slow and corrupt and sure to favor a powerful international contractor over a worker with no local connections.

CHAPTER 10

1. One analyst concluded, "The use of unregistered companies and armed support groups operating outside the legal framework, with little oversight or accountability, appears to have been widespread" (Sherman 2015, 105).

2. Government of Nepal Ministry of Labour and Employment n.d.b.

3. Government of Nepal Ministry of Labour and Employment n.d.a.

4. Sijapati, Bandita, Bhattarai, and Pathak 2015, 18.

5. After I left Nepal, a suicide attack on a group of Nepalis working for the Canadian embassy in Kabul killed thirteen and led the government to announce it was banning (or perhaps rebanning) Nepalis working in Afghanistan, Iraq, and Syria ("Govt Imposes Ban on Afghanistan as Work Destination" 2016). A later government study then recommended that the ban be lifted ("Lift the Ban for Foreign Job-Seekers" 2016), but in the meantime, many of the Nepalis I knew already working in Afghanistan seemed to continue with their work unencumbered. These various announcements and reports did nothing to clarify for the average worker what they should and should not be doing regarding work in Afghanistan.

6. Since these compounds were of questionable legality, there was no formal registry of them, but as I surveyed contractors I found fifteen distinct compounds and was assured by contractors interviewed that there were a good number more than this.

7. I was later able to visit this compound in fall 2016, a year after interviewing Pratik for the first time.

8. See Coburn 2016a, chap. 9.

9. Rasmussen 2014.

10. Some of the lessons and reflections from these conversations can be found in Coburn 2017 and Coburn and Sherpa 2018.

CHAPTER 11

1. de Waal 2010, 16.

2. For a good overview of this period, see de Waal 2010, chap. 7.

3. For more on the Georgia Train and Equip Program, see Hodge 2011, chap. 2.

4. Walsh 2004.

5. When Georgian troops later clashed with Russian forces in South Ossetia, Putin accused these trainers of trying to directly meddle in Georgian political affairs (Clover and Sevastopulo 2008).

6. Hodge 2011, 56.

7. "Bush Hails Georgia as "Beacon of Liberty"" 2005.

8. NATO Public Diplomacy Division 2013.

9. Saakashvili December 19, 2009.

10. For more on these ties, see Gvosdev 2000, chaps. 7, 8.

11. Lins de Albuquerque and O'Hanlon 2007; and Campbell and O'Hanlon 2007; Civil Georgia 2008.

12. U.S. Department of State 2011.

13. For a full account, see King 2008.

14. de Waal 2010, 212.

15. de Waal 2010, 216–217.

16. Georgia had previously had sent fifty troops to the country in 2005, but this was a temporary mission providing security for the elections (Civil Georgia 2014).

17. These battalions were both withdrawn by July 2014 (Civil Georgia 2014).

18. The public's backlash against initial troop casualties had led the ministry to withhold details about attacks and generally deny information requests. Through two separate channels, I had sought permission to speak with active officials at the Georgian Ministry of Defense. Both times I was told I would receive permission shortly, but later both requests remained unanswered. However, with such a high turnover rate of governments, finding former officials from earlier governments was fairly easy.

19. This of course ignores the large number of noncitizens serving in the U.S. military, with a reported 70,000, or 4 percent of the U.S. military who were non-Americans in 2011 (McIntosh and Sayala 2011, 14).

20. McCarrel and Jardine 2015.

21. Other countries took other approaches to slow the leak of soldiers into the contracting ranks. For example, both Britain and Sri Lanka offered officers "leaves of absence" for them to do contracting work. In the case of Britain, these "sabbaticals" were yearlong (Isenberg 2009, 61).

22. Civil Georgia 2013.

23. Mullen 2013 and Janashia 2013. Later evidence suggested the video had actually been uploaded in Georgia and there was some debate as to whether the video was actually produced by the Taliban or had come from political opponents to the Saakashvili government.

24. "Mourning Day in Georgia" 2013.

25. Civil Georgia 2015.

26. Civil Georgia 2015.

CHAPTER 12

1. Al Jazeera 2015.

2. Tharoor 2014 and BBC 2014.

3. "Erdoğan Opens 'Beştepe People's Mosque' in Presidential Palace" 2015.

4. A majority of the Turks I interviewed brought up Turkey's historical ties, whereas only one Nepali I spoke with brought up the role of Nepali soldiers in the various Anglo-Afghan wars.

5. See Barfield 2010, chap. 4.

6. Provincial reconstruction teams (PRTs) were one of the central pillars of the international strategy in Afghanistan. Most were essentially large military bases that were expected to oversee and provide support for reconstruction and development in areas. Their attempts to blend military and development resources generated a significant amount of controversy and were much discussed in policy circles. Johnson and Leslie (2005, 19) wrote, "Nowhere is the confusion as to political, military and assistance agendas—each of which has been invoked as an objective of the international engagement—better illustrated than here." (See also Hodge 2011.) Interestingly from my perspective, up until this point, none of my interviews had referenced working on PRTs, despite the fact that I knew some of them had done this. For Nepali workers, PRTs were simply another base where they might be employed, even while they were objects of consternation for many of those above them.

7. Having recently visited the Nepali military museum in Kathmandu and the Gurkha museum in Pokhara, both of which carefully documented the role of Nepali Gurkhas who had been in some of the fiercest fighting during the battle of Gallipoli, I found it interesting, if not surprising, to see that the room dedicated to the battle of Gallipoli made no mention of the involvement of Nepalis.

8. As economic evidence of this, the significance of EU trade with Turkey had peaked in 1999 when it accounted for 56 percent of all of Turkey's foreign trade and by 2008 had dipped to 41 percent, with Russia replacing Germany as Turkey's largest trade partner (Kirişci and Kaptanoğlu 2011, 705)

9. Kirişci and Kaptanoğlu 2011, 711.

10. Kirişci 2012, 320.

11. Kirişci (2011 51–52) labels this shift as a move from a "win-lose" approach to Turkey's foreign policy to "win-win."

12. See Kirişci 2012.

13. Bache Fidan 2016, 199.

14. Bache Fidan 2016, 121, 123.

15. Hodge 2011, chap. 4, includes a history of the evolution of PRTs more generally in Afghanistan.

16. Kaya 2013, 24.

17. Çolakoğlu and Yegin 2014, 32, and Coburn 2016a.

18. Çolakoğlu and Yegin 2014, 33.

19. See Coburn 2016a, chap. 6.

20. Quoted in Kaya J 2013, 23.

21. Shaheed 2016.

CHAPTER 13

1. See NATO Resolute Support Mission 2016.

2. Bache Fidan 2016, 122.

3. On his visits to Afghanistan, the deputy head of the South Asia Desk primarily stayed at the Turkish embassy, meeting with his counterparts at other embassies and occasionally visiting the bases where Turkish troops were stationed. He traveled around the country more than many of his American counterparts did, who tended to be confined to their embassy in Kabul. Still, he acknowledged that his view of the conflict and the role of Turks in it was limited to those bases

where they were stationed. Notably, he did not travel much to large U.S.-dominated bases, which did not have Turkish troops on them, but he did host many Turkish contracting firms.

4. Çolakoğlu and Yegin 2014, 34.

5. Thanks to Zeynep Tuba Sungur for tracking these data down for me.

6. Kirişci 2011, 37.

7. Axlerod 2014, 284. This was the first major rehabilitation of the U.S. embassy in Kabul. Later renovations and expansion were projected to be completed in 2016 at a cost of $770 million (Gibbons-Neff 2014).

8. There was a rich set of tales that Afghans and lower-level international contractors like to tell about corruption among larger, particularly American, contractors about elaborate corruption. These were almost always categorically denied by the companies themselves. In reality, finding examples of international corruption was not difficult despite efforts to disguise it. See Coburn 2016a, chap. 6, and Wissig 2012.

9. Alden 2006.

10. These happened at regular intervals following the U.S.-led invasion but always seemed to be at times when tensions between the government and the international community were highest. See, for example, Friel 2007 and King 2010.

11. This also could have had something to do with the fact that the Turk was Muslim and the Afghan government perceived the drinking of alcohol by a Muslim as a graver offense than the consumption by a non-Muslim.

12. P. W. Singer tracks several examples of security companies contracting out to subcontractors that they had essentially created, particularly the case of Executive Outcomes (2007, chapter 7). Executive Outcomes was officially a subsidiary of Strategic Resource Corporation, which in turn owned twenty other companies that mostly contracted themselves out to Executive Outcomes. Many of these were "stay-behind asset protection," and they would remain in the country after Executive Outcomes claimed to have ceased its military operations (Singer 2007, 104). I found several similar cases during my research. Often, however, when these new subcontractors split off, they would simply take the original contractor's employees, meaning the experience of the workers themselves changed little, with the only major consequence being that the total bill to the U.S. government would increase as the money subcontracted to a smaller, and often less reliable, firm.

13. The acquisition of ArmorGroup by G4S demonstrates some of the conflicts between differing approaches to security by private companies. G4S, the largest private security firm in the world, supplies far more unarmored security guards to guard, for example, banks and parking lots in over a hundred countries around the world. Its subsidiary Wackenhut Cooperation also supplies private guards to eighteen military bases in the continental United States in the place of traditional military police (Axlerod 2014, 343). In 2012, it was the second largest employer in the world behind Walmart (Hagedorn 2014, 228). Many of its guards come from Nepal, and it has a sizable office in Kathmandu. I later interviewed a high-ranking manager at G4S, who had gone to Afghanistan to access the ArmorGroup contracts after their acquisition. He concluded that the paramilitary style approach to security necessary in Afghanistan, which ArmorGroup employed, did not fit G4S's typical model. So while G4S continued to fulfill ArmorGroup's contracts, it did not seek out much additional work in Afghanistan.

14. The rumors were that the Taliban was attempting to poison American troops, and so the U.S. government did not trust food and other supplies from Afghanistan .

15. I could not find anyone in the U.S. government who had worked with Afghan Services Ltd. to confirm whether they believed that Akash's company was Afghan, but considering some of the other cases I had gathered during previous research (Coburn 2016a), I would not have been surprised if they had made this assumption.

CHAPTER 14

1. Interestingly, several of these companies seem to have employed an Afghanization approach in which they would hire a Turkish master carpenter or master rebar worker but would

assign each of these two or three Afghan counterparts to train. After a year or two on the job, the Afghans would be competent enough to do the job themselves, the company would give them a small raise, and it would send the more expensive Turkish workers home to save on costs. These cost-savings measures reshaped the demographics of the international presence in the country over the fifteen years of intervention.

2. Yüksel Holdings 2012.

3. See in particular the case of Executive Outcomes (EO) Diamondworks Ltd, which ran a group of mines in Sierra Leone, while EO was being contracted by Sierra Leone to fight off rebel groups. Essentially this meant that the Sierra Leonean government was paying a company to protect its own assets. For more, see Singer 2003, chap. 7; Axlerod 2014, 207–210; and McFate 2015, 38–39.

4. Avant (2005, 67) argues that much of this is the fact that private security firms tend to "operate with a small full-time contingent and a large database from which to draw teams to carry out contracts. . . . PSCs have a fluid structure that can rapidly dissolve and recreate themselves as need be." All this make security firms more difficult to monitor and regulate.

5. For a more extensive analysis of this process, see Wissing 2012.

6. The fact that 99 of Turkey's 356 generals were arrested and charged in the follow-up from the coup is evidence that Erdoğan understood just how important those ties were. Morris, Gibbons-Neff, and Mekhennet 2016.

7. Several of the officials I interviewed suggested that this was due in part to the fact that military officers were so nervous about being accused of corruption that hiring a Turkish firm would raise suspicions of nepotism.

CHAPTER 15

1. While legally clear, this does not mean that it was not politically contested.

2. Healy 2013.

3. State support for citizens and military, however, was debated. The dispute over the Bilateral Security Agreement between the United States and the Karzai government, which dominated political debates in the lead-up to the 2014 elections, was in part over whether U.S. soldiers could be prosecuted in Afghanistan.

4. U.S. Department of State 2008.

5. Around 2014, the U.S. embassy started providing more regular updates to citizens about security threats. During most of the time that I lived there, however, the information provided by the embassy was almost nonexistent.

6. Relatedly, President Karzai made several attempts to rein in expanding private security companies in 2010 in part by arguing that they were undermining Afghan national sovereignty (Vogt and Faiez 2010 and Sherman 2015, 109–111).

7. U.S. Senate Committee on Homeland Security and Governmental Affairs 2011. Contrack was one of the largest contractors that the Army Corps of Engineers relied on; between 2007 and 2009, it accounted for 11 percent of all the funds allocated (SIGAR 2010, 7).

8. Contrack-Watts n.d.

9. Shalizi 2012.

10. This so-called green-on-blue violence became a central concern for U.S. military officials and in 2012 accounted for 15 percent of coalition deaths (Roggio and Lundquist 2017).

11. For other examples of these conflicts, see Coburn 2016a, chap. 8.

12. Contrack is owned by an Egyptian company, and it's possible this was related to the accusations of spying, but this didn't matter much for Teer, who did not realize Contrack had Egyptian ties and more generally was simply confused by the entire process.

13. An example of Teer Magar's confusion around the process was that at some point, he was told that there was an Afghan law that those who are arrested are not allowed to contact anyone on the outside for thirty-five days. No such law exists, but with no lawyer or translator, it's easy to see how misconceptions could shape his understanding of the situation.

14. Teer asked me to contact the ICRC representative so I could better understand some of the technical aspects of the case, which Teer said he himself still did not fully understand. While the representative responded to my email, he refused to discuss the case based on ICRC policy of confidentially despite Teer's request.

15. Teer Magar was never given a copy of his contract, it seems, even while he showed me some of the other documents the company had given him. He was unsure why it had stopped paying the salary after twenty-two months, and I couldn't find in any of the contracts from other contractors that I did get copies of any clause that suggested a reduced or temporary continuation of payment during imprisonment. It seems like this was temporary benevolence on the part of some Contrack official perhaps feeling guilty for not supplying any other assistance to Teer.

16. This fits into other accounts I have gathered where Nepali workers are more likely to blame the manpower firms that arrange their employment than the Nepali government, which is supposed to be regulating it. In fact, in a video released by their captors, one of the twelve Nepalis killed in Iraq is heard saying shortly before his execution: "Prahlad Gir is the culprit, don't let him go," referring to the broker who had arranged the job from which they had been kidnapped (Parajuli 2004, 23).

17. It was also never clear to Teer Magar whether he had been formally pardoned or whether they had simply allowed him to leave the country.

18. There were other media reports on Nepali women being trafficked to Syria, but these still relied on one or two single cases (Pattisson 2016).

CHAPTER 16

1. See Nissenbaum, Peker, and Marson 2015.

2. This account comes from interviews of Arda and Hakkan both individually and separately, as well as some media coverage that largely confirmed the details of Arda and Hakkan's stories.

3. This, I later discovered, was not strictly true, and a Turkish engineer had been killed after being kidnapped in Kabul in 2004, but it does reflect attitudes of Turkish contractors I interviewed who generally claimed to feel safe in Afghanistan (Gall 2004). Turkish captives, however, were not always released in a timely fashion, and two years before, four Turkish engineers were held by the Taliban for over eight months before being released (Voice of America 2011).

4. The helicopter had landed in the Azra district of Logar, close to the border with Pakistan, though Arda did not know this at the time (BBC 2013a).

5. All of the Turkish officials I asked about Arda's case had either never heard of it or would not comment on the incident.

6. I later found that the U.S. military announced the release of the Kyrgyz flight engineer a month later, but unsurprisingly released no details about ransom (Druzin 2013). I could find no follow-up information on the Russian pilot, and Arda was not sure what had happened to him.

7. In other cases, Turks had paid ransom to the Taliban for their release. See, for example, "Kidnapped Turkish Engineers Freed" 2008.

8. Achakzai 2007. South Korea never confirmed the ransom and South Korean troops returned to Afghanistan at the request of the United States in 2010 (Axe 2010).

9. Sudarshan 2008. Interestingly the account is almost exclusively of the work done by the Indian mission in Baghdad with almost no information about the experience that the hostages themselves had while in captivity.

10. Parajuli 2004, 27. The Nepali press also pointed to the fact that Kenya had, under similar circumstances, sent a delegation to Iraq. The killings of these Nepali workers led to the only significant public demonstrations in Nepal in opposition to the ongoing private security contracting practices.

11. They still faced discrimination by Sunnis, however, since Ismailis are generally considered a Shia subgroup.

12. BBC 2013a.

13. Arda gave most of the credit for their release to the Turkish ambassador, as opposed to

the Foreign Ministry more generally. He thought that his personal connections were probably instrumental in securing his release. Apparently the ambassador prided himself on the fact that before Arda and his colleagues were kidnapped, sixteen other Turks had been kidnapped and each one had been released safely.

CHAPTER 17

1. The $100,000 ransom seems high, but the number was quoted to me by multiple sources. It's possible that it had been embellished with repeated tellings.

CHAPTER 18

1. Farwell 1984, 44.

2. Regulations changed later, forcing the applicant to appear at the embassy when applying for a visa, though when I heard about this later, I could not confirm that this regulation was being regularly enforced. This was the type of restriction that experienced brokers could often work around.

3. In comparison with Hom's case, for example, many of the younger recruits had an advantage in the fact that they were more technically savvy and usually had smart phones and international SIM cards, which they used to communicate with each other and with their brokers. Such technology also kept contractors in India or Afghanistan much more connected with their families at home in 2016 than they were in 2003, for instance.

4. According to a couple of the people we spoke with, this certificate came from the Nepali embassy in Delhi but previously had come from the Nepali embassy in Islamabad.

5. For a while, the officials at the Delhi airport were also preventing Nepalis from flying directly to Kabul if they did not have work permits from the Nepali government. It was not entirely clear to me if either of these practices were legal.

CHAPTER 19

1. See, for example, Price 2013.

2. See Hanifi 2008, chaps. 1 and 3, on the role of the fruit trade in British attempts to transform the Afghan economy and establish Kabul as a colonial capital of the British Empire.

3. This Hindu belief was raised by only a few of the Nepalis I interviewed, perhaps because they had been traveling these routes for so long already. Historically, however, it was an issue, and religious authorities in Kathmandu made special dispensations for Nepalis traveling to Europe during World War I (Farwell 1984, 88).

4. See Haidari 2016 and "India Allows Duty-Free Import of Goods from Afghanistan" 2011.

5. They were not alone, and living on a university campus in Delhi, I found it hard to avoid the near constant student political activism, which was a sharp contrast from anything I had seen in Nepal, or the United States for that matter. Right after my arrival, several student leaders at Jawaharlal Nehru University, more commonly referred to as JNU, Delhi University's neighbor in the south of the city, were arrested on sedition charges. Many students I spoke with interpreted the scrutiny of the JNU students as further evidence of BJP's pushing its right-wing nationalist agenda internally while promoting a business-led agenda abroad that refused to engage Pakistan. Several tense days followed with police searching for the students, several of whom were said to still be hiding out on the sprawling campus, while other students organized new protests in their support. Counterprotests were organized, claiming that historically left-leaning JNU was a "hotbed of traitors," while just outside my guesthouse at Delhi University, others demonstrated in support of the students.

. BBC 2016.

6. For a particularly nuanced account of various understandings of "corruption" in Afghanistan, see Arbabzadah 2013, 104–106.

7. This task force was led by H. R. McMaster, later Donald Trump's national security advisor.

8. See Coburn 2016a, chap. 8.

CHAPTER 20

1. For an ethnographic analysis of the relationship between ethnicity, religion, and migration from a variety of communities in Kerala, see Kurien 2002.

2. Special Inspector General for Afghan Reconstruction 2016.

3. See Guha 2007, chap. 5.

4. Almost everyone I interviewed in India said that they far preferred working for international firms than for Indian companies, which were more likely to underpay and underresource their Kabul offices. Dinesh, who had worked over the past decade for four different companies in Kuwait and Afghanistan but was currently back in India looking for work, told me he would be hesitant to work for an Indian firm again. The most obvious reason was that international firms tended to pay better, but it was more than this, he said. Indian firms were too disorganized and amateurish. They were always running out of cash and paying their workers late. Ironically, he said, he felt far more respected by management at large international companies than at Indian ones. They also were much more helpful in setting him up for his next job

5. The Nepali government, like other countries in the region, had a program that allowed government officials to take extended leaves of absence to work for the United Nations or related organizations in other countries where there was an international intervention, while preserving their positions at home. These officials could also get permission to work for all sorts of contractors, like Chemonics and IRD, working on development, state-building, or counterinsurgency programs in Afghanistan. I heard from a couple of sources that the cap was supposed to be three years at a time, but in actuality, most could find ways around these regulations, with some I interviewed having spent more than the last ten years abroad, while maintaining their positions in the Nepali government. It's also worthwhile to point out that the Nepali military had no such program despite the fact that Nepali officers stood to gain at least as much from international experience as, say, an expert on local governance or elections.

6. Carol Upadhya and A. R. Vasavi 2008 raise some comparative questions about the IT industry in India, which often produces unsubstantiated narratives about social mobility.

7. Relatively little on research on the privatization of security focuses on gender. The one notable exception is a Eichler 2015.

8. For more on the deep connection between U.S. military bases abroad and the trafficking of sex workers, see Gillem 2007, chap. 3; Vine 2015, chap. 9; and Hughs, Chon, and Ellerman 2013.

9. Scamwarners 2014. I also found the break times a little unbelievably high from an American perspective. It listed: "WORK PERIODS: Monday to Fridays, Time: 9.30am to 5.30pm with break period between 12.00 noon to 2.00pm," which might have been a normal Indian lunch but didn't seem reasonable in Manhattan.

CHAPTER 21

1. This population numbers here are disputed. Official 2011 census data reported that 7.6 percent of the Rushmoor District, which includes both Aldershot and Farnborough (also heavily Nepali), are Nepali (Rushmoor Borough Council 2016). This number is likely to have missed many of the more recent arrivals or family members on extended visits, and often the press reports that 10 percent of Aldershot's population is Nepali (e.g., Hollingshead 2011), though this too may be an underestimate of the actual number of Nepalis residing in the area at any one time.

2. Rushmoor Borough Council 2016 and Chapple 2015.

3. FSI Worldwide n.d.

4. This was a form of semiretirement that was particularly appealing since it allowed soldiers to continue living in military housing. Such arrangements made it expensive to try to lure them abroad to work in conflict zones.

5. See U.K. National Audit Office 2006 and Shute and Oliver 2014.

6. Choudhary (2015, 24–28) found some similar results while interviewing the wives of Gurkha soldiers. All of those in the youngest generation had married their husbands by choice, were struggling to develop their own careers, typically in the health care industry, and were dealing with

a certain amount of disillusionment around the challenges that living in the United Kingdom brought with it. Most telling, she concludes: "This is the group which sees itself as least likely to go back to Nepal" (28).

7. Smith 2015.

8. Carroll 2012 provides an insider's account of this process, while Choudary 2015 accesses some of the repercussions of the policy changes.

9. Choudhary 2015 provides a detailed look at many of the complex implications of these various immigration issues.

10. According to the U.K. Office of National Statistics, 31,000 out of 67,000 Nepalis residing in the United Kingdom, or 46.27 percent, are the dependents of active or retired Gurkhas (Choudhary 2015, 4). A decade before, the number would have been much closer to 100 percent.

11. Of course Nepalis in the military were accessing social services as well, but within the context of the British military bureaucracy and therefore much less visible to outsiders.

12. Rushmoor Borough Council 2016.

13. "Gurkha Heroes Are 'Overwhelming' Our Town" 2011.

14. "Gurkha Heroes Are 'Overwhelming' Our Town" 2011.

15. "Gerald Howarth Attacks Joanna Lumley's `Disgraceful' Gurkhas Campaign" 2012. Howarth is also known for his opposition to same-sex marriage and his views that "the aggressive homosexual community . . . see this as but a stepping stone to something even further (BBC 2013b)."

16. Pride 2015.

17. Choudhary 2015, 14–18.

CHAPTER 22

1. For more on the background on the program see Coburn and Sharan 2016.

2. The program document that states that Afghans who "were employed by or on behalf of the U.S. government" does not make clear whether this includes subcontractors as well (U.S. Department of State n.d.). In practice, as I interviewed some thirty Afghan contractors who had made it to the United States, I found many had worked for companies I had heard about in Nepal and elsewhere, such as DynCorp and Chemonics. Almost all were working as for a contracting company, not as direct employees of the U.S. government (there were relatively few Afghans who were U.S. government employees—even translators were usually contractor). I did not find any contractors working on secondary, tertiary, or lower levels of contracts, who had made it through the application process. This confusion is further evidence of some of the problems of defining who is and who is not a "contractor" for the U.S. government.

3. Bruno 2015.

4. U.S. Department of State and Department of Homeland Security 2016.

5. U.S. Department of State and Department of Homeland Security 2016.

6. U.S. Department of State and Department of Homeland Security 2016.

7. For more, see Coburn and Sharan 2016.

8. Coburn and Sharan 2016.

9. Miller 2010.

10. All of the research for this book was conducted before the Trump administration's travel ban that first was rolled out in early 2017. The Afghans I spoke with after it went into effect were unsurprisingly opposed to it despite the fact that Afghanistan was not one of the countries affected.

11. Peters, Schwartz, and Kapp 2015, 3.

12. Malikyar 2015.

13. For an account of Afghan refugees in Turkey trying to travel to Europe, see Arjomand 2016.

CHAPTER 23

1. Many of the studies on whether contracting saves the government money are contentious

a

and contradictory. For example, the Defense Science Board Task Force estimated in 1996 that by 2002, contracting would cost the Department of Defense $7 to $12 billion dollars a year. A follow up report by the Government Accountability Office, however, concluded that these estimates were greatly overstated (Dunigan 2011, 11). As several of the cases in this book suggest, the real costs of contracting are difficult to calculate, particularly in the complex landscape of a counterinsurgency campaign. The few studies of actual cost savings of defense contracting are decidedly mixed. One particularly alarming study by the RAND Corporation estimated that after the Army outsourced its Reserve Officer Training Corps program to the private firm MPRI, the actual cost per instructor increased by $10,000 per year. Avant 2005, 117–118.

2. This almost certainly would have required a reinstitution of the draft. Perhaps not coincidentally, former military officers from the conflict have notably called for a draft or some sort of equivalent that Stanley McChrystal (2016), former commander of ISAF, has referred to as a national "service year" for all young people.

3. Prince 2017. The fact that Prince is brother of Secretary of Education Betsy Devos has contributed to charges of nepotism in the private security business.

4. For more, see Wissig 2012.

5. For a compelling articulation of this argument, see Stanger 2009.

6. "Lift the Ban for Foreign Job-Seekers" 2016.

7. See International Committee of the Red Cross 2009.

8. For a complete account of the issues surrounding the signing of the Montreux Document, see Hagedorn 2014, chaps. 9 and 11.

9. U.N. Office on Drugs and Crime 2013.

10. There are several programs doing work trying to track and publicize some of the impacts and true boundaries of the American empire. This include the Costs of War Project at Brown University and the American Empire Project published by Metropolitan Books, which include several books cited here, such as Gopal 2015, Vine 2015, and Flair 2016.

11. Hager and Mazzetti 2015. For more on the market for Latin American contractors, see Mani 2015.

12. Duffield 2014, chap. 7.

13. Alibaba.com n.d. I assume that each "pack" represented one worker, since that would be logical if the price range were $300 to $600 per month, though it was unlikely, for that rate, that the contractors had any significant military experience. After sending a request for more information through Alibaba, the seller assured me that the government ban on Nepalis in Afghanistan was not an issue. In a follow-up email from a Chinese company matched based on my inquiry, I was asked if I was interested in ordering any "smart robots" in addition to the Nepali guards I had inquired about.

SOURCES

Achakzai, Saeed Ali S. 2007. "South Korea Paid $20 Million Ransom: Taliban Leader." *Reuters*, September 1.

Ahmed, Azam. 2013. "Suicide Attack at Afghan Base Kills at Least 9." *New York Times*, July 2.

Alden, Edward. 2006. "Bottled Up: Why Coke Stands Accused of Being Too Cosy with the Karimovs." *Financial Times*, June 14.

Alibaba.com. N.d. "Nepali Workers and Ex-Gurkha Security Guard from Nepal." Labour & Employment. https://www.alibaba.com/product-detail/Nepali-Workers-and-Ex-Gurkha-Security_139362828.html, accessed September 13, 2016.

Al Jazeera. 2015. "Peace Rally Bombing Stokes Tensions in Turkey." October 12.

Arbabzadah, Nushin. 2013. *Afghan Rumour Bazaar: Secret Sub-Cultures, Hidden Worlds and the Everyday Life of the Absurd*. London: Hurst.

Arjomand, Noah. S. 2016. "Afghan Exodus: Smuggling Networks, Migration and Settlement Patterns in Turkey." Afghan Analysts Network, September 10.

Axe, David. 2010. "South Korea's Secret War." *Diplomat*, April 27.

Axlerod, Alan. 2014. *Mercenaries: A Guide to Private Armies and Private Military Companies*. Los Angeles: Sage.

Avant, Deborah. 2005. *The Market for Force: The Consequences of Privatizing Security*. New York: Cambridge University Press.

Bache Fidan, Christina. 2016. "Turkish Business in the Kurdistan Region of Iraq." *Turkish Policy Quarterly* 14 (4): 117–126.

Bajracharya, Rooja, and Bandita Sijapati. 2012. "The Kafala System and Its Implications for Nepali Domestic Workers." Policy Brief 1, March. Kathmandu, Nepal: CESLAM.

Barfield, Thomas. 2010. *Afghanistan: A Cultural and Political History*. Princeton, NJ: Princeton University Press.

BBC. 2008. "Gurkhas Win Right to Stay in UK." September 30.

———. 2011. "Was Lumley Campaign Good for Gurkhas?" July 31.

———. 2013a. "Four Turks Held Hostage in Afghanistan by Taliban Freed." May 12.

———. 2014. "Turkey President Erdogan's Palace Costs to Soar." November 14.

———. 2016. "Kanhaiya Kumar: India `Sedition' Student May Be Expelled." March 15.

Belasco, Amy. 2014. "The Cost of Iraq, Afghanistan and Other Global War on Terror Operations Since 9/11." Washington, DC: Congressional Research Service, December 8.

Black, Samuel, and Anjali Kamat. 2014. "After 12 Years of War, Labor Abuses Rampant on U.S. Bases in Afghanistan." *Al Jazeera America*, March 7.

Brown, Amy Kathryn. 2007. "Baghdad Bound: Forced Labor of Third-Country Nationals in Iraq." *Rutgers Law Review* 60:737.

Bruno, Andorra. 2015. "Iraqi and Afghan Special Immigrant Visa Programs." Washington, DC: Congressional Research Service, January 20.

Bruslé, Tristan. 2010. "Who's in a Labour Camp? A Socio-Economic Analysis of Nepalese Migrants in Qatar." *European Bulletin of Himalayan Research* 35–36:154–170.

———. 2012. "What Kind of Place Is This? Daily Life, Privacy and the Inmate Metaphor in a Nepalese Workers' Labour Camp Qatar." *South Asian Multidisciplinary Academic Journal* 6.

Buckley, Chris, and Austin Ramzy. 2015. "Frustration Grows in Nepal as Earthquake Relief Trickles In." *New York Times*, April 28.

"Bush Hails Georgia as 'Beacon of Liberty.'" 2005. *Guardian*, May 10.

Business and Human Rights Resource Centre. N.d. "Abu Ghraib Lawsuits Against CACI, Titan. Now L-3."

CACI. N.d. "What We Do." CACI. http://www.caci.com/whatwedo.shtml.

Campbell, Jason, and Michael O'Hanlon. 2007. "Iraq Index: Tracking Variables of Reconstruction and Security in Post-Saddam Iraq." Washington, DC: Brookings Institute, December 21.

Caplan, Lionel. 1995. *Warrior Gentlemen: "Gurkhas" in the Western Imagination*. Providence, RI: Berghahn Books.

Carroll, Peter. 2012. *Gurkha: The True Story of a Campaign for Justice*. London: Biteback Publishing.

Chandrasekaran, Rajiv. 2012. *Little America: The War Within the War for Afghanistan*. New York: Knopf.

Chapple, James. 2015. "Aldershot Prepares for Dalai Lama Return to Open Buddhist Centre." *GetHampshire*, June 25.

Chisholm, Amanda. 2015. "From Warriors of Empire to Martial Contractors: Reimagining Gurkhas in Private Security." In *Gender and Private Security in Global Politics*, edited by Maya Eichler. New York: Oxford University Press.

Choudhary, Neha. 2015. "The Gurkha Wives of the United Kingdom: Challenges to Social Integration." Master's thesis, London School of Economics.

Civil Georgia. 2008. "Georgia Extends Troop Deployment in Iraq." March 21.

———.2013. "Seven Georgian Soldiers Die in Afghan Truck Bomb Attack." June 6.

———.2014. "Georgian Troops End Mission in Helmand." July 17.

———.2015. "Georgian Soldier Killed in Afghanistan." September 23.

Clover, Charles, and Demetri Sevastopulo.2008. "U.S. Military Trained Georgian Commandos." *Financial Times*, September 5.

Coburn, Noah. 2011. *Bazaar Politics: Pottery and Power in an Afghan Market Town*. Stanford, CA: Stanford University Press.

———. 2016a. *Losing Afghanistan: An Obituary for the Intervention*. Stanford, CA: Stanford University Press.

———. 2016b. "Caught in Kabul: Both Nepal and Govts of Donor Countries Have Failed to Provide Any Protection for Migrant Workers." *Kathmandu Post*, July 4.

———. 2017. "The Guards, Cooks, and Cleaners of the Afghan War: Migrant Contractors and the Cost of War." Providence, RI: The Watson Institute.

Coburn, Noah, and Anna Larson. 2014. *Derailing Democracy in Afghanistan: Elections in an Unstable Political Landscape*. New York: Columbia University Press.

Coburn, Noah, and Timor Sharan. 2016. "Out of Harm's Way? Perspectives of the Special Immigrant Visa Program for Afghanistan." Washington, DC: Hollings Center.

Coburn, Noah, and Dawa Sherpa. 2018. "The Cost of British Recruitment on Nepali Youth," with Dawa Sherpa. *The Record*, January 31.

Çolakoğlu, Selçuk, and Mehmet Yegin. 2014. "The Future of Afghanistan and Turkey's Contributions." Ankara: International Strategic Research Organization, September.

Coll, Steve. 2005. *Ghost Wars: The Secret History of the CIA, Afghanistan and Bin Laden from the Soviet Invasion to September 10, 2001*. London: Penguin Books.

Commission on Wartime Contracting in Iraq and Afghanistan. 2011a. "Transforming Wartime Contracting: Controlling Costs, Reducing Risks." Final report to Congress. Arlington, VA: Commission on Wartime Contracting, August.

———. 2011b. "At What Risk? Correcting Over-Reliance on Contractors in Contingency Operations." Second interim report to Congress. Arlington, VA: Commission on Wartime Contracting, February 24.

"Contrack International Buys Watts Construction for an Undisclosed Price." 2013. *Pacifica Business News*, December 24.

Contrack-Watts. N.d. "Projects: ANA Camp Scorpion SOF Training Facility and Strategic Airlift Apron, Runway, and Rotary Wing." http://contrackwatts.com/projects/, accessed September 3, 2016.

Cowan, Sam. 2009. "Ochterlony's Men." *Himal Southasian* (May).

Cross, J. P. 1986. *In Gurkha Company: The British Army Gurkhas, 1948 to the Present.* London: Arms and Armour Press.

C3PO International, LTD. v. DynCorp International LLC. 2015. 14–564. February 24, 5th Circuit.

Crawford, Neta S. 2016. "US Budgetary Costs of Wars Through 2016: $4.79 Trillion and Counting Summary of Costs of the US Wars in Iraq, Syria, Afghanistan and Pakistan and Homeland Security." Costs of War Project. Providence, RI: Watson Institute, September.

Des Chene, Mary Katherine. 1991. "Relics of Empire: A Cultural History of the Gurkhas: 1815–1987." PhD dissertation, Stanford University.

Druzin, Heath. 2013. "Taliban Release Kyrgyz Chopper Crewman." *Stars and Stripes*, June 11.

Duffield, Mark. 2014. *Global Governance and the New Wars: The Merging of Development and Security*, rev. ed. London: Zed Books.

Dunigan, Molly. 2011. *Victory for Hire: Private Security Companies' Impact on Military Effectiveness.* Stanford, CA: Stanford University Press.

Dunigan, Molly, and Ulrich Petersohn, eds. 2015. *The Markets for Force: Privitization of Security across World Regions.* Philadelphia: University of Pennsylvania Press.

DynCorp. 2013. "Three Killed in Attack in Kabul Afghanistan." Press release, July 3. http://www.dyn-intl.com/news-events/press-release/three-killed-in-attack-in-kabul-afghanistan/.

"DynCorp International—Contracting Profile." N.d. Inside Gov. http://government-contractors.insidegov.com/l/312967/Dyncorp-International-in-Fort-Worth-TX, accessed April 8, 2016.

Eichler, Maya, ed. 2015. *Gender and Private Security in Global Politics.* New York: Oxford University Press.

"Erdoğan Opens `Beştepe People's Mosque in Presidential Palace." 2015. *Daily Sabah*, July 3.

Farwell, Byron. 1984. *The Gurkha.* New York: Norton.

Fergusson, Niall. 2004. *Colossus: The Price of America's Empire.* New York: Penguin.

Finkel, David. 2013. *Thank You for Your Service.* New York: Picador.

Flair, Eric. 2016. *Consequence.* New York: Holt.

Food and Agriculture Organization of the United Nations. 2015. "Nepal Earthquakes: Situation Report." Kathmandu, Nepal: FAO, June 16.

Foreign and Commonwealth Office. N.d. "First World War Centenary: WW1 Victoria Cross Recipients from Overseas."

Friel, Terry. 2007. "Afghan Police Hunt Booze, Smokes, Chips in Raids." *Reuters*, February 28. http://www.reuters.com/article/us-afghan-lawless-idUSSP15307120070228, accessed September 6, 2016.

FSI Worldwide. N.d. "About Us: History." http://www.fsi-worldwide.com/index.php/about-us/history#.V8q_ED4rK2x, accessed September 3, 2016.

Gall, Carlotta. 2004. "Body of Kidnapped Turkish Worker Is Found on an Afghan Hilltop." *New York Times*, December 16.

"Gerald Howarth Attacks Joanna Lumley's `Disgraceful' Gurkhas Campaign." 2012. *Huffington Post*, March 27.

Gibbons-Neff, Thomas. 2014. "The Proposed Construction at Kabul Embassy Is Behind Schedule and over Budget." *Washington Post*, July 8.

Gillem, Mark L. 2007. *America Town: Building the Outposts of Empire.* Minneapolis: University of Minnesota Press.

Giustozzi, Antonio, ed. 2012. *Decoding the New Taliban: Insights from the Afghan Field*. New York: Columbia University Press.

Gopal, Anand. 2015. *No Good Men Among the Living: America, the Taliban, and the War Through Afghan Eyes*. New York: Picador.

Government of India Ministry of Home Affairs. 2011. "C-1 Population by Religious Community." Office of the Registrar General and Census Commission. http://www.censusindia.gov.in/2011census/C-01.html, accessed September 14, 2016.

Government of Nepal Ministry of Labour and Employment. N.d.a. "Year Report FY 2070–71." Department of Foreign Employment, Kathmandu. http://dofe.gov.np/new/uploads/article/year2070–71.pdf, accessed June 2, 2016.

———. Government of Nepal Ministry of Labour and Employment. N.d.b. "Recognized Countries." Department of Foreign Employment, Kathmandu. http://www.dofe.gov.np/new/pages/details/28, accessed June 2, 2016.

"Govt Imposes Ban on Afghanistan as Work Destination." 2016. *Kathmandu Post*, June 23. http://kathmandupost.ekantipur.com/news/2016–06–23/govt-imposes-ban-on-afghanistan-as-work-destination.html, accessed August 24, 2016.

Gregson, Jonathan. 2002. *Massacre at the Palace: The Doomed Royal Dynasty of Nepal*. New York: Hyperion.

Guha, Ramachandra. 2007. *India After Gandhi: The History of the World's Largest Democracy*. London: Picador.

"Gurkha Heroes Are 'Overwhelming' Our Town, Say Aldershot Councillors and MP." 2011. *Daily Mail*, February 11.

Gurkha Security Services. N.d. "History of Gurkhas." http://gurkhasecurityservices.co.uk/history-of-gurkhas/, accessed September 18, 2016

"Gurkhas v Government: Guess Who Won?" 2009. *The Economist*, April 30.

Gvosdev, Nikolas K. 2000. *Imperial Policies and Perspectives towards Georgia, 1760–1819*. London: Macmillan.

Hagedorn, Ann. 2014. *The Invisible Soldiers: How America Outsourced our Security*. New York: Simon and Schuster.

Hager, Emily, and Mark Mazzetti. 2015. "Emirates Secretly Sends Colombian Mercenaries to Yemen Fight." *New York Times*, November 25.

Haidari, M. Ashraf. 2016. "Afghanistan Celebrates India's Post-Independence Achievements." *Diplomat*, August 18.

Hanifi, Shah Mahmoud. 2008. *Connecting Histories in Afghanistan: Market Relations and State Formation on a Colonial Frontier*. Stanford, CA: Stanford University Press.

Healy, Jack. 2013. "Soldier Sentenced to Life Without Parole for Killing 16 Afghans." *New York Time*, August 23.

Hedgpeth, Dana. 2007. "Army Splits Award Among 3 Firms." *Washington Post*, June 28.

Hendershot, Chris. 2015. "Heteronormative and Penile Frustrations: The Uneasy Discourse of the ArmorGroup Hazing Scandal." In *Gender and Private Security in Global Politics*, edited by Maya Eichler. New York: Oxford University Press.

Hindman, Heather, and Bijaya Raj Poudel. 2015. "Can Nepal's Youth Build Back Better and Differently?" *Cultural Anthropology*, October 14.

Hodge, Nathan. 2011. *Armed Humanitarians: The Rise of the Nation Builder*. New York: Bloomsbury.

Hollingshead, Ian. 2011. "The Gurkhas in Aldershot: Little Nepal." *Telegraph*, February 21.

Hughs, D. M., K. Y. Chon, and D. P. Ellerman. 2013. "Modern-Day Comfort Women: The U.S. Military, Transnational Crime, and the Trafficking of Women." *Violence against Women* 13 (9): 901–922.

Human Rights Watch. 2006. "Building Towers, Cheating Workers: Exploitation of Migrant Construction Workers in the United Arab Emirates." *Human Rights Watch* 18(8E). https://www.hrw.org/sites/default/files/reports/uae1106webwcover.pdf, accessed September 12, 2016.

IDG Security. N.d. Homepage. http://idg-security.com/, accessed September 18, 2016.

Ignatieff, Michael. 2003. *Empire Lite: Nation-Building in Bosnia, Kosovo and Afghanistan.* London: Vintage.

"India Allows Duty-Free Import of Goods from Afghanistan." 2011. *Economic Times,* June 3,

International Committee of the Red Cross. 2009. "The Montreux Document." Geneva, Switzerland: ICRC, August. https://www.icrc.org/eng/assets/files/other/icrc_002_0996.pdf, accessed September 14, 2016.

Isenberg, David. 2009. *Shadow Force: Private Security Contractors in Iraq.* Westport, CT: Praeger Security International.

———. 2011. "The Unknown Contractor." *Huffington Post,* May 25.

Janashia, Eka. 2013. "Georgian Soldiers Killed in Afghanistan" .*The Central Asia-Caucasus Analyst,* June 12.

Johnson, Chris, and Jolyon Leslie. 2005. *Afghanistan: The Mirage of Peace.* London: Zed Books.

de Jong, David. 2013. "Supreme Owner Made a Billionaire Feeding the U.S. War Machine." *Bloomberg Business,* October 7.

Joya, Omar, Claudia Nassif, Aman Farahi, and Silvia Redaelli. 2015. "Afghanistan Development Update." Washington, DC: World Bank, October.

Karki, Arjun, and David Seddon. 2003. "The People's War in Historical Context." In *The People's War: Left Perspectives,* edited by Arjun Karki and David Seddon. New Delhi: Adroit Publishers.

Kaya, Karen. 2013. "Turkey's Role in Afghanistan and Afghan Stabilization." *Military Review* (July–August).

"KBR Inc. in—Contracting Profile," N.d. Inside Gov. http://government-contractors.insidegov.com/l/610902/KBR-Inc-in, accessed August 26, 2016.

King, Charles. 2008. "The Five-Day War: Managing Moscow After the Georgia Crisis." *Foreign Affairs* 87:6.

King, Laura. 2010. "Afghan Enforcement of Liquor Ban Rankles Foreigners." *Los Angeles Times,* April 26.

Kirişci, Kemal. 2011. "Turkey's `Demonstrative Effect' and the Transformation of the Middle East." *Insight Turkey* 13 (2): 33–55.

———. 2012. "Turkey's Engagement with Its Neighborhood: A `Synthetic' and Multidimensional Look at Turkey's Foreign Policy Transformation." *Turkish Studies* 13 (3): 319–341.

Kirişci, Kemal, and Neslihan Kaptanoğlu. 2011. "The Politics of Trade and Turkish Foreign Policy." *Middle Eastern Studies* 47 (5): 705–724.

Koenig, Bryan. 2015. "DynCorp Says It Owes Nothing in C3PO's $6.6M Contracting Suit." *Law360* (2015).

Krahmann, Elke. 2010. *States, Citizens and the Privatization of Security.* New York: Cambridge University Press.

Kumar, Hari. 2016. "Thousands of Indian Workers Are Stuck in Saudi Arabia as Kingdom's Economy Sags." *New York Times,* August 1.

Kurien, Prema A. 2002. *Kaleidoscopic Ethnicity: International Migration and the Reconstruction of Community Identities in India.* New Brunswick, NJ: Rutgers University Press.

Lawoti, Mahendra. 2012. "Ethnic Politics and the Building of an Inclusive State." In *Nepal in Transition: From People's War to Fragile Peace,* edited by Sebastian von Einsiedel, David M. Malone, and Suman Pradhan. New York: Cambridge University Press.

"Lift the Ban for Foreign Job-Seekers." 2016. *Himalayan Times,* August 24.

Lins de Albuquerque, Adriana, and O'Hanlon, Michael. 2007. "Iraq Index: Tracking Variables of Reconstruction and Security in Post-Saddam Iraq." Washington, DC: Brookings Institute, December 21.

Lutz, Catherine. 2001. *Homefront: A Military City and the American 20th Century.* Boston: Beacon Press.

———, ed. 2009. *The Bases of Empire: The Global Struggle Against U.S. Military Posts.* New York: New York University Press.

MacFarlane, Alan. 1976. *Resources and Population: A Study of the Gurungs of Nepal.* Cambridge: Cambridge University Press.

Macleish, Kenneth. 2013. *Making War at Fort Hood: Life and Uncertainty in a Military Community*. Princeton, NJ: Princeton University Press.

Maclellan, Nic. 2006. "Fiji, the War in Iraq, and the Privatisation of Pacific Island Security." APSNet Policy Forum, Nautilus Institute, April 6.

Malikyar, Helena S. 2015. "Afghanistan: The Other Refugee Crisis." *Al Jazeera*, September 16.

Mani, Kristina. 2015. "Diverse Markets for Force in Latin America: From Argentina to Guatemala." In *The Markets for Force: Privitization of Security Across World Regions*, edited by Molly Dunigan and Ulrich Petersohn. Philadelphia: University of Pennsylvania Press.

Mann, Michael. 2003. *Incoherent Empire*. New York: Verso.

"Maoist Statements and Documents,". 2003. In *The People's War: Left Perspectives*, edited by Arjun Karki and David Seddon. New Delhi: Adroit Publishers.

Martin, Ian. 2012. "The United Nations and Support to Nepal's Peace Process: The Role of the UN Mission in Nepal." In *Nepal in Transition: From People's War to Fragile Peace*, edited by Sebastian von Einsiedel, David M. Malone, and Suman Pradhan.. New York: Cambridge University Press.

Mathema, Kalyan Bhakta. 2011. *Madhesi Uprising: The Resurgence of Ethnicity*. Kathmandu: Mandala Book Point.

Mazumdaru, Srinivas. 2014. "Gurkhas: Nepalese Warriors in World War I." *Deutsche Welle*, May 13.

McCarrel, Ryan, and Bradley Jardine. 2015. "Can Georgia Care for Its Afghan War Veterans?" *Diplomat*, October 22.

McChrystal, Stanley. 2016. "You Don't Have to Wear a Military Uniform to Serve Your Country." *Atlantic*, July 20.

McFate, Sean. 2015. *The Modern Mercenary: Private Armies and What They Mean for World Order*. New York: Oxford University Press.

McIntosh, Molly, and Seema Sayalam. 2011. "Non-Citizens in the Enlisted U.S. Military." CAB D0026449.A1, Center for Naval Analyses, December.

Miller, Judith. 2010. "The Afghans of Fremont: Anxious, Uprooted—and Under Surveillance." *City Journal* (Autumn).

Morris, Loveday, Thomas Gibbons-Neff, and Souad Mekhennet. 2016. "Turkey Is Expected to Curb Military Power as Purge Expands." BBC, July 19.

"Mourning Day in Georgia: Bodies of Soldiers Fallen in Afghanistan, Brought to Georgia." 2013. *Georgia Journal*, May 16.

Mukhopadhyay, Dipali. 2014. *Warlords, Strongmen Governors and the State in Afghanistan*. New York: Cambridge University Press.

Mullen, Mark. 2013. "Georgian Mission in Afghanistan and Threatening Video to Georgian Soldiers." *Georgian Journal*, June 21.

National Reconstruction Authority. 2016. "Nepal Earthquake 2015: Post Disaster Recovery Framework." Kathmandu, Nepal: Government of Nepal, May.

NATO Public Diplomacy Division. 2013. "Backgrounder: Deepening Relations with Georgia." May 30. Brussels.

NATO Resolute Support Mission. 2016. "Troop Contributing Nations." June.

Neumann, Ronald E. 2009. *The Other War: Winning and Losing in Afghanistan*. Washington, DC: Potomac Books.

———. 2015. "Failed Relations Between Hamid Karzai and the United States: What Can We Learn?" Special Report. Washington, DC: USIP, May.

Nissenbaum, Dion, Emre Peker, and James Marson. 2015. "Turkey Shoots Down Russian Military Jet." *Wall Street Journal*, November 24.

Nordstrom, Carolyn. 2004. *Shadows of War: Violence, Power and International Profiteering the Twenty-First Century*. Berkeley: University of California Press.

Odom, William, and Robert Dujarric. 2004. *America's Inadvertent Empire*. New Haven, CT: Yale University Press.

O'Donnell, Kelly. 2017. "Unbreakable? Recognizing Humanitarian Stress and Trauma." Global Genevam. October 10.

Office of the United Nations High Commissioner for Human Rights. 2012. "Nepal Conflict Report 2012: Executive Summary." Geneva: United Nations, October.

Parajuli, John Narayan. 2004. "The Sound of Fury." *Nation Weekly*, September 12.

Parker, John. 1999. *The Gurkhas: The Inside Story of the World's Most Feared Soldiers*. London: Headline Publishing Group.

Partlow, Joshua. 2016. *A Kingdom of Their Own: The Family Karzai and the Afghan Disaster*. New York: Knopf.

Pattisson, Pete. 2015. "In Nepal, $1Bn Impact of Strikes over Constitution `Worse Than Earthquakes.'" *Guardian*, October 5.

———. 2016. "Nepalese Women Trafficked to Syria and Forced to Work as Maids." *Guardian*, January 1.

Peters, Heidi, Moshe Schwartz, and Lawrence Kapp. 2015. "Department of Defense Contractor and Troop Levels in Iraq and Afghanistan: 2007–2015." Washington DC: Congressional Research Service, December 1.

———. 2017. "Department of Defense Contractor and Troop Levels in Iraq and Afghanistan: 2007–2017." Washington DC: Congressional Research Service, April 28.

Pingeot, Lou. 2014. "Contracting Insecurity: Private Military and Security Companies and the Future of the United Nations." Bonn and New York: Rosa Luxemburg Stiftung and Global Policy Forum, February.

"Pinnacle Hotel Services, Afghanistan." 2012. YouTube video. June 28. https://youtu.be/ZtWgR6CboW4, accessed December 27, 2015.

Price, Gareth. 2013. "India's Policy Towards Afghanistan." Asia ASP 2013/04. London: Chatham House, August.

Pride, Tom. 2015. "UKIP Candidate Calls Gurkhas "Parasites" and "Mercenaries." *Pride's Purge*. Wordpress blog, April 28.

Prince, Erik. 2017. "The MacArthur Model for Afghanistan: Consolidate Authority into One Person." *Wall Street Journal*, May 31.

Raj, Prakash A. 2007. *Crisis of Identity in Nepal*. Varanasi, India: Pilgrims Publishing.

Rashid, Ahmed. 2008. *Descent into Chaos: The U.S. and the Disaster in Pakistan, Afghanistan and Central Asia*. New York: Viking.

Rasmussen, Sune Engel. 2014. "Four Taliban Militants Dead in Attack on Foreign Workers in Kabul." *Guardian*, November 19.

Rathaur, Kamal Raj, 2001. "British Gurkha Recruitment: A Historical Perspective." *Voice of History* 16:2.

Reuters. 2008. "Kidnapped Turkish Engineers Freed: Afghan Official." *World News*, July 21. http://www.reuters.com/article/us-afghan-abduction-idUSISL26548320080721, accessed September 11, 2016.

Roggio, Bill, and Lisa Lundquist. 2017. "Green-on-Blue Attacks in Afghanistan: The Data." *Long War Journal*, March 21. http://www.realcleardefense.com/articles/2017/03/21/green-on-blue_attacks_in_afghanistan_the_data_111015.html, accessed August 1, 2017.

Rosenberg, Matthew, and Michael D. Shearoct. 2015. "In Reversal, Obama Says U.S. Soldiers Will Stay in Afghanistan to 2017." *New York Times*, October 15.

Rushmoor Borough Council. 2016 "Ethnic Diversity and Migration Data Sheet." Strategy Engagement and Organisational Development Team, August. http://www.rushmoor.gov.uk/CHttpHandler.ashx?id=11199&p=0, accessed September 2, 2016.

Saakashvili, Mikheil. 2009. "Georgia and the War in Afghanistan: Why the Young Democracy Is Sending Nearly 1,000 Troops to the War Effort." *Washington Post*, December 19.

Scamwarners. 2014. "Contrack International Construction/Fake Job Scam Fraud." *Forum Post*, February 18. https://www.scamwarners.com/forum/viewtopic.php?f=34&t=82377, accessed August 25, 2016

Schneirderman, Sara. 2015 . *Rituals of Ethnicity: Thangmi Identities Between Nepal and India*. Philadelphia: University of Pennsylvania Press.

Schwartz, Moshe. 2014. "Reform of the Defense Acquisition System: Statement of Moshe Schwartz, Specialist in Defense Acquisition Before the United States Senate Committee on Armed Services." Washington, DC: Congressional Research Service, April 30.

Sege, Adam. 2016. "DynCorp says C3PO"s Trimmed $6.6M Contract Void." *Law360*, January 5.

Shaheed, Anisa. 2016. "No Plan to Close Afghan-Turk Schools: Education Minister." *Tolo News*, August 4

Shalizi, Hamid. 2012. "Taliban Say Suicide Bomber Targeted U.S. Company in Kabul." *Reuters*, December 17

Sherman, Jake. 2015. "The Markets for Force in Afghanistan." In *The Markets for Force: Privitization of Security across World Regions*, edited by Molly Dunigan and Ulrich Petersohn. Philadelphia: University of Pennsylvania Press.

Shute, Joe, and Mark Oliver. 2014. "Why Britain's Armed Forces Are Shrinking by the Day and Does It Really Matter?" *Telegraph*, June 25.

Siegel, David. 2014. "DynCorp Hit With $9M Suit Over Afghan Base Subcontract." *Law360*,

Sijapati, Bandita, Ashim Bhattarai, and Dinesh Pathak. 2015. "Analysis of Labour Market and Migration Trends in Nepal." Kathmandu: GIZ and ILO.

Singer, P. W. 2007. *Cooperate Warriors: The Rise of the Privatized Military Industry*, updated ed. Ithaca, NY: Cornell University Press.

Smith, Joshua., 2015. "The Palace in Aldershot Welcomes Guests for Official Opening Event." *GetHampshire*, December 20.

Special Inspector General for Afghan Reconstruction. 2010. "DOD, State and USAID Obligated Over $17.7 Billion to About 7,000 Contractors and Other Entities for Afghanistan Reconstruction During Fiscal Years 2007–2009." SIGAR Audit 11–4. May 27.

———. 2013. "Alert 13–3." SIGAR letter to Congressional Committees. June 28.

———. 2015. "$36 Million Compound and Control Facility at Camp Leatherneck, Afghanistan: Unwanted, Unneeded and Unused." Office of Special Projects, SIGAR-15–57-SP. May. Washington, DC.

———. 2016. "Review Letter: USAID-Supported Health Facilities in Kabul." SIGAR-16–09-SP. January 5. Washington, DC.

———. July 2016. "Afghanistan's Information and Communications Technology Sector: U.S. Agencies Obligated over $2.6 Billion to the Sector, But the Full Scope of U.S. Efforts Is Unknown." SIGAR 16–46AR/Afghanistan's ICT Sector, Washington, DC.

Stancati, Margherita. 2013. "Death Toll Grows From Taliban Attack on Kabul Compound." *Wall Street Journal*, July 3.

Stanger, Allison. 2009. *One Nation Under Contract: The Outsourcing of American Power and the Future of Foreign Policy*. New Haven, CT: Yale University Press.

Stier, Ken. S. 2009. "Afghan Embassy Scandal's Link to Cost-Cutting Security." *Time*, September 11.

Stillman, Sarah. 2011. "The Invisible Army: For Foreign Workers on U.S. Bases in Iraq and Afghanistan, War Can Be Hell." *New Yorker*, June 6.

Sudarshan, V.. 2008. *Anatomy of an Abduction: How the Indian Hostages in Iraq Were Freed*. Delhi: Penguin Books.

Stiglitz, Joseph, and Linda J. Bilmes. 2012. "Estimating the Costs of War: Methodological Issues, with Applications to Iraq and Afghanistan." In *The Oxford Handbook of the Economics of Peace and Conflict*, edited by Michelle Garfinkel and Stergios Skaperdas. New York: Oxford University Press.

Strick van Linschoten, Alex, and Felix Kuehn. 2012. *An Enemy We Created: The Myth of the Taliban–Al Qaeda Merger in Afghanistan*. Oxford: Oxford University Press.

Suhrke, Astri. 2011. *When More Is Less: The International Project in Afghanistan*. New York: Columbia University Press.

Supreme Group. 2015. "Supreme Group's Chief Ethnics and Compliance Officer Speaks at 12th European Business Ethics Forum. March 2. http://www.supreme-group.net/Supreme-Chief-Ethics-and-Compliance-Officer-Speaks-at-12th-European-Business-Ethics-Forum, accessed February 16, 2016.

"Supreme Group Holdings Sarl in—Contracting Profile." N.d. Inside Gov. http://government-contractors.insidegov.com/l/602603/Supreme-Group-Holding-Sarl-in, accessed February 16, 2016.

Thakuri, Rajendra. N.d. "Finally, Justice Served to Families of 12 Nepalis Killed in Iraq." Nepal America Legal Information Center. https://anlus.wordpress.com/2008/05/22/finally-justice-served-to-families-of-12-nepalis-killed-in-iraq/, accessed May 31, 2016.

Thapa, Deepak, and Bandita Sijapati. 2004. *A Kingdom Under Siege: Nepal's Maoist Insurgency, 1996 to 2004*, updated edition, New York: Zed Books.

Tharoor, Ishaan. 2014. "The White House Would Be a Tiny Wing of Turkey's New Presidential Palace." *Washington Post*, October 30.

Thier, J. Alexander. 2006/2007. "The Making of a Constitution in Afghanistan." *New York Law School Review* 51:558–579.

"THT 10 Years Ago: Leaving Ex-British Gurkhas to Cost Nepal Dearly." 2016. *Himalayan Times*, October 6.

Towell, Pat, and Daniel Else. 2012. "Defense: FY2013 Authorization and Appropriations." Washington DC: Congressional Research Service, September 5.

TV News 1. 2013. "World at 6." YouTube. July 2. https://youtu.be/A3k4QZTeCfI, accessed, December 27, 2015.

de Waal, Thomas. 2010. *The Caucasus*. New York: Oxford University Press.

"UN Chief Calls for Lifting of Blockade on Indo-Nepal Border." 2015. *Times of India*, November 11

U.K. National Audit Office. 2006. "Ministry of Defence: Recruitment and Retention in the Armed Forces." Report by the Comptroller and Auditor General, HC 1633-I Session 2005–2006. London: TSO, November 3.

U.N. Assistance Mission in Afghanistan, and United Nations Office of the High Commissioner for Human Rights. 2016. "Annual Report 2015: Protection of Civilians in Armed Conflict." Kabul, Afghanistan: UNAMA and UNHCR. February.

U.N. Office on Drugs and Crime. 2013. "Honouring Corporate Creativity and Courage in the Fight against Human Trafficking." UNODC. January.

U.S. Army Sustainment Command Public Affairs.2007. "ASC Selects LOGCAP IV Contractors." June 28. https://www.army.mil/article/3836/ASC_selects_LOGCAP_IV_contractors/, accessed September 17, 2016.

U.S. Department of the Army. 2004. "The Mikolashek Report: The Inspector General, Detainee Operations Inspection." In *The Torture Papers: The Road to Abu Gharib*, edited by Karen Greenberg and Joshua Drate. New York: Cambridge University Press.

U.S. Department of Defense. 2010. "Contractor Support of U.S. Operations in the USCENTCOM Area of Responsibility, Iraq and Afghanistan." Deputy Assistant Secretary of Defense, Program Support. December 15.

———. 2013. "Mr. Daniel R. Blair, Deputy Inspector General for Auditing, Department of Defense Inspector General before the Subcommittee on National Security, Committee on Oversight and Reform. "Contracting to Feed U.S. Troops in Afghanistan: How Did the Defense Department End Up in a Multi-Billion Dollar Billing Dispute?" April 17.

———. 2014. "Report of the Defense Science Board: Task Force on Contractor Logistics in Support of Contingency Operations," Washington D.C.: Office of the Under Secretary of Defense for Acquisition, Technology, and Logistics, June.

———. N.d. "U.S. Casualty Status," http://www.defense.gov/casualty.pdf, accessed June 29, 2016.

U.S. Department of Justice. 2014. "Defense Contractor Pleads Guilty to Major Fraud in Provision of Supplies to U.S. Troops in Afghanistan: Agrees to Pay $434 Million in Criminal Penalties and to Settle False Claims Act Allegations." Office of Public Affairs, Justice News. December 8.

U.S. Department of Labor. N.d.a. "About the Defense Base Act Case Summary Reports," Office of Workers" Compensation Programs, Division of Longshore and Harbor Workers Compensation." https://www.dol.gov/owcp/dlhwc/lsaboutdbareports.htm, accessed May 5, 2016.

———. N.d.b. "Defense Base Act Case Summary by Nation (09/01/2001—3/31/2016." Office of Workers" Compensation Programs, Division of Longshore and Harbor Workers Compensation/" https://www.dol.gov/owcp/dlhwc/dbaallnation.htm.

———. N.d.c. "Defense Base Act Case Summary by Employer (09/01/2001—12/31/2015." Office of Workers" Compensation Programs, Division of Longshore and Harbor Workers Compensation/" http://www.dol.gov/owcp/dlhwc/dbaallemployer.htm, accessed Feb 17, 2016.

U.S. Department of State. 2007. "Iraqi PM Chief of Staff Discusses Ministerial Conference, DynCorp Ban." Cable 07BAGHDAD927_a to U.S. Secretary of State. March 16, Released by the WikiLeaks Public Library of U.S. Diplomacy.

———. 2008. "Travel Warning—Afghanistan February 06, 2008." Washington, DC: Bureau of Consular Affairs, February 6.

———. 2011. "Foreign Operations and Assistance Fact Sheet." Office of the Coordinator of U.S. Assistance to Europe and Eurasia. April.

———. 2014a. "Operationalizing Counter/Anti-Corruption Study." Joint Chiefs of Staff, Suffolk, Virginia: Joint and Coalition Operational Analysis, February 28.

———. 2014d. "Audit of Bureau of Diplomatic Security Worldwide Protective Services Contract," Office of Inspector General, Task Order 10, AUD-MERO-15–03. October.

———. N.d. "Special Immigrant Visas for Afghans—Who were Employed by/on Behalf of the U.S. Government." Bureau of Consular Affairs. https://travel.state.gov/content/visas/en/immigrate/afghans-work-for-us.html, access August 24, 2016.

U.S. Department of State and Department of Homeland Security. 2016. "Status of the Afghan Special Immigrant Visa Program," Joint Report. July. https://travel.state.gov/content/dam/visas/SIVs/Report%20of%20the%20Afghan%20SIV%20Program%20-%20July%202016.pdf, accessed August 24, 2016.

U.S. Senate Committee on Homeland Security and Governmental Affairs. 2011. "Afghanistan Reconstruction Contracts: Lessons Learned and Ongoing Problems." Contractors Panel, Ad hoc Subcommittee on Contracting Oversight. June 30.

Upadhya, Carol and A.R. Vasavi, eds. 2008. In an Outpost of the global economy: Work and workers in India's information technology industry. New Delhi: Routledge.

Verité. 2015. "Strengthening Protections Against Trafficking in Persons in Federal and Corporate Supply Chains." Amherst, MA. January

Vine, David. 2015. *Base Nation: How U.S. Military Bases Abroad Harm America and the World*. New York: Holt.

Vine, David. 2009. *Islands of Shame: The Secret History of the U.S. Military Base on Diego Garcia*. Princeton, NJ: Princeton University Press.

Vogt, Heidi, and Rahim Faiez. 2010. "Afghan Starts to Close Private Security Firms." Associated Press. October 3.

Voice of America. 2011. "Afghan Taliban Releases 4 Kidnapped Turkish Engineers." September 3.

Vora, Neha. 2013. *Impossible Citizens: Dubai's Indian Diaspora*. Durham, NC: Duke University Press.

Walsh, Nick Paton. 2004. "U.S. Privatises its Military Aid to Georgia." *Guardian*, January 6.

White, Adam. 2016. "Private Military Contractors as Criminals/Victims." In *Palgrave Handbook of Criminology and War*, edited by Ross McGarry and Sandra Walklate. London: Macmillan.

White House. 2014. "Statement by the President on the End of the Combat Mission in Afghanistan." Office of the Press Secretary. December 28.

Wissing, Douglas. 2012. *Funding the Enemy: How U.S. Taxpayers Bankrolled the Taliban*. Amherst, NY: Prometheus Books.

World Bank. 2016. *Migration and Remittances Factbook 2016*, 3rd ed. Washington, DC: KNOMAD.

Woodward, Bob. 2010. *Obama's Wars*. New York: Simon and Schuster.

Yüksel Holdings. September 2012. "Riseability II: Yüksel"s Sustainability Report."

INDEX

Abdullah Abdullah, 281

Abu Ghraib prison, 76–77

Afghan contractors, 30, 31, 222, 344, 355n5, 360n21, 375n2; as immigrants to United States, 20, 34, 317–22, 324–31, 337; and Special Immigrant Visa (SIV) program, 317–22, 324–31, 337

Afghanistan: Bilateral Security Agreement, 371n3; bribery in, 100, 141, 148, 159, 201, 206, 215, 216, 226, 285, 288–89; British invasion, 42–43, 336, 358n2, 373n2; Commander Emergency Response Programs, 191; constitution of, 48–49; contractors killed in, 12, 15–16, 27, 29, 57–58, 141; corruption in, 38, 100, 141, 142–43, 148, 159, 200–201, 205–6, 215, 216, 221, 222–24, 284–89, 341, 360n35, 367n15, 370n8; economic conditions, 4–5, 25, 36, 37, 198, 205, 208, 215–16, 278–80; green-on-blue violence in, 225, 371n10; Helmand province, 174, 177, 226, 227, 305, 309; Herat, 282; vs. Iraq, 37, 152, 154, 160, 235, 292, 293, 321, 332; Islam in, 24, 188, 357n1, 370n11; Jowzjan, 184, 190; Karzai government, 26, 30, 36, 37, 78, 133, 201, 219, 222, 229, 281, 282, 371nn3,6; Khost, 236; length of war in, 39; vs. Malaya, 116–17; media coverage of, 2, 5, 9, 11, 12, 14, 16, 17, 29, 57–58, 75, 76, 77–78, 88, 89, 140–41, 174, 250, 355n15, 363n1; Ministry of Defense, 120; Ministry of Justice, 326; Ministry of the Interior, 139, 163; Ministry of Urban Development, 201; National Directorate of Security (NDS), 226, 227; vs. Nepal, 24–25, 88, 89, 114, 159, 202, 322; Northern Alliance, 36; number of Afghan civilians killed, 9–10; Panjshir Valley, 240; Pashtuns in, 36, 37, 43, 188; patronage in, 36, 221; political conditions, 48–49, 50, 182–83, 213, 219–20, 281, 341; provincial reconstruction teams (PRTs), 190–91, 369n6; refugees from, 330, 331; relations with Georgia, 276, 281; relations with India, 13, 276–78, 279–80, 282–83; relations with Pakistan, 25; relations with United Kingdom, 12, 25; relations with United States, 20, 86, 198, 218–19, 253, 279, 281, 358n2, 371n3; Soviet occupation, 276, 280, 336; vs. Turkey, 182–83; U.S. invasion of 2001, 20, 86, 198, 279, 281, 358n2; U.S. soldiers killed in, 9, 12, 29, 355n14, 371n10; U.S. surge in, 1–2, 12, 37–38, 87, 160, 173, 336, 355n2; U.S. troop levels in, 1, 11, 15, 16, 28, 37, 38, 355n2, 356n16, 376n2; visas for, 9, 19, 77, 80, 129, 131, 133, 141, 148, 149, 152, 159, 161, 163, 168, 206, 210, 251–52, 253, 257, 258, 264, 267, 268–69, 341, 366n11, 373n2; Wardak, 184, 190, 191; warlords in, 36; women in Afghanistan, 24. See also Bagram Airbase; Kabul; Taliban

Afghan National Army (ANA), 225, 236, 243, 245, 247, 287–88

Afghan National Police (ANP), 26, 57, 58, 241; detainment by, 100, 202, 204, 217, 246; raids by, 8, 202, 203, 204, 253, 296, 370n10; relations with contractors, 8, 100, 138, 143, 148, 202, 204–5, 219–20, 253, 254

Afghan Services Ltd., 205, 370n15